ESSENTIALS OF SEMICONDUCTOR DEVICE PHYSICS

Jasmina Novakovic

Toronto Academic Press

ESSENTIALS OF SEMICONDUCTOR DEVICE PHYSICS

Jasmina Novakovic

Toronto Academic Press
4164 Lakeshore Road
Burlington ON L7L 1A4
Canada
www.tap-books.com

Email: orders@arclereducation.com

ISBN: 978-1-77956-341-5

ABOUT THE AUTHOR

Jasmina Novaković is an accomplished physicist and educator with a rich background in applied and computer physics. Holding a bachelor's degree in Applied and Computer Physics, Jasmina has dedicated her career to making complex scientific concepts accessible and engaging. With years of experience as a physics teacher, she has a proven track record of inspiring curiosity and fostering a deep understanding of the subject among her students. Her passion for physics extends beyond the classroom. Jasmina is a committed advocate for science education and has actively participated in popularizing physics through hands-on experiments and educational initiatives. Her work includes conducting interactive sessions and workshops aimed at demystifying physics for learners of all ages. In addition to her scientific expertise, Jasmina is a creative writer and translator fluent in English and Spanish, with native proficiency in Serbian. This linguistic skill set enhances her ability to communicate complex ideas and stories across cultures. This unique combination of skills positions her as a versatile professional capable of addressing a wide range of projects with precision and cultural sensitivity. Jasmina is also committed to community service, having volunteered in education, disaster relief, and legal aid. Her diverse background in physics, writing, and community support provides a unique and enriched perspective in all her endeavors.

Table of Contents

List of Figures

List of Tables

PREFACE

Semiconductor materials, devices, and systems have become indispensable pillars supporting the modern world, deeply ingrained in various facets of our daily lives. These advanced technologies are foundational to numerous applications across different industries, profoundly influencing how we live, work, and communicate.

In computing, semiconductors are the backbone of microprocessors and memory units. They enable the rapid calculations and data storage capabilities that power everything from personal computers to large-scale data centers. Modern computing relies heavily on semiconductor technology to perform complex tasks at incredible speeds, facilitating everything from everyday internet browsing to advanced scientific research. The ever-increasing demand for faster and more efficient computing solutions drives continual advancements in semiconductor fabrication and design, ensuring that processors become more powerful and memory units more capacious over time.

When it comes to power conversion, semiconductor-based components such as transistors and diodes play a critical role in optimizing energy efficiency. These components are vital in applications ranging from renewable energy systems to electric vehicles. In renewable energy systems, semiconductors help convert solar energy into electrical power with high efficiency, thereby making solar panels and other renewable technologies more viable and cost-effective. In electric vehicles, semiconductor devices manage power conversion and distribution, ensuring that the vehicles operate efficiently and reliably. The advancements in semiconductor technology contribute significantly to reducing energy losses and enhancing the overall performance of power systems.

In the field of optoelectronics, semiconductor technologies have revolutionized lighting and detection solutions. The development of energy-efficient and long-lasting light-emitting diode (LED) technology is one of the most notable achievements in this area. LEDs have transformed the way we illuminate our surroundings, providing brighter, more reliable, and energy-saving lighting options for homes, businesses, and public spaces. Beyond lighting, semiconductor-based optoelectronic devices have enhanced the precision and sensitivity of various detection and sensing applications. For example, semiconductor sensors are now integral to medical devices, environmental monitoring systems, and industrial automation, offering unparalleled accuracy and reliability.

In the ever-expanding domain of information processing, semiconductors are at the heart of information storage, communication devices, and complex integrated circuits. These technologies facilitate the seamless exchange and long-duration storage of data, underpinning the functionality of smartphones, network infrastructure, and data servers. Semiconductor memory devices, such as flash drives and solid-state drives (SSDs), provide fast and reliable data storage solutions that are crucial for both personal and enterprise-level computing needs. Meanwhile, integrated circuits enable the miniaturization and enhancement of electronic devices, making them more powerful and efficient.

Contents/Structure of the Book

This book is structured into ten comprehensive chapters, each aimed at providing a thorough understanding of semiconductor physics and its applications in electronic devices.

Chapter 1 introduces the basic concepts of statistical physics relevant to semiconductor physics. It covers topics such as statistical distributions, energy levels, and the behavior of particles in a semiconductor material. These concepts form the foundation for understanding the behavior of electrons and holes in semiconductors.

Chapter 2 focuses on the properties and characteristics of semiconductors. It covers topics such as band theory, carrier concentration, and mobility, providing a detailed explanation of how these properties influence the behavior of semiconductor devices.

Chapter 3 introduces the p-n junction, which is the basic building block of many semiconductor devices. It explains the formation of the p-n junction, its electrical characteristics, and its role in diode and transistor operation.

Chapter 4 explores photovoltaic devices, with a focus on solar cells. It covers the principles of solar energy conversion, the operation of different types of solar cells, and the factors affecting their efficiency.

Chapter 5 discusses the properties and applications of metal-semiconductor and semiconductor heterojunctions. It explains how these junctions are formed and how they are used in devices such as diodes and transistors.

Chapter 6 provides an in-depth look at the metal-oxide-semiconductor field-effect transistor (MOSFET). It covers the operation of the MOSFET, its characteristics, and its applications in integrated circuits.

Chapter 7 explores the bipolar transistor, another important semiconductor device. It covers the operation of the bipolar transistor, its characteristics, and its applications in amplifiers and other electronic circuits.

Chapter 8 focuses on the junction field-effect transistor (JFET). It explains the operation of the JFET, its characteristics, and its applications in amplifiers and switching circuits.

Chapter 9 discusses semiconductor devices used in optical applications. It covers topics such as light-emitting diodes (LEDs), laser diodes, and photodetectors, explaining their principles of operation and applications.

Chapter 10 explores semiconductor devices used in microwave and power applications. It covers topics such as microwave transistors, power diodes, and thyristors, explaining their operation and applications in high-frequency and high-power circuits.

This book provides a comprehensive overview of semiconductor physics and its applications in electronic devices. It is suitable for students, researchers, and professionals seeking to understand the principles underlying semiconductor devices and their practical applications.

Audience

This book is particularly suitable for students, offering invaluable insights for undergraduate and graduate students in electrical engineering, physics, materials science, and related fields. The book serves as a comprehensive textbook for courses in semiconductor devices, providing both theoretical foundations and practical applications.

Academics and researchers in semiconductor physics, device engineering, and related disciplines will also find this book beneficial, as it presents detailed explanations and the latest research findings. It offers a solid basis for teaching and conducting advanced research in these areas.

Furthermore, professional electrical engineers working in the semiconductor industry or related fields will find this book useful for understanding the principles underlying semiconductor devices. It provides insights into device design, fabrication, and characterization, which are essential for developing new technologies and products.

Professionals in the electronics industry, including manufacturers of semiconductor devices and electronic products, will also benefit from the insights provided in this book. It offers a comprehensive overview of semiconductor physics and its applications in electronic devices.

Learning Features

- Start-of-chapter learning objectives
- Key terms in each chapter
- Summary Points at the end of each chapter
- Review Questions at the end of each chapter
- Case Studies and Activities in each chapter for solving real problems
- References at the end of each chapter

CHAPTER

1 Concepts of Statistical Physics

LEARNING OBJECTIVES

After studying this chapter, you will be able to:

- Determine the essentials of statistical physics
- Explain the fundamentals of statistical mechanics
- Access the method of MONTE CARLO in statistical physics.

KEY TERMS FROM THIS CHAPTER

Approximations
Bose-einstein statistics
Equilibrium
Fermi-dirac statistics
Macroscopic
Microscopic interpretation
Neuroscience
Atomic motion
Entropy
Fermi-dirac distribution
Free energy
Macroscopic parameters
Molecules
Pauli-exclusion

1.1. INTRODUCTION

A subfield of physics known as statistical physics deals with large populations and approximations by using statistical techniques. It connects the behavior of individual atoms and molecules to the observable properties of materials, building a bridge between the macroscopic and microscopic worlds. The fundamental tenet of statistical physics is that, although it is impossible to pinpoint a system's exact state when it consists of numerous particles, statistical laws allow for the highly accurate prediction of the system's collective behavior. The ensemble, a vast collection of virtual replicas of a system, each representing a potential state the system might be in, is one of the core ideas of statistical physics. With ensembles, physicists can compute the average properties of the system using statistical techniques. Different physical conditions and constraints, such as constant energy, temperature, or chemical potential, are suited for different types of ensembles, such as the microcanonical, canonical, and grand canonical ensembles. A fundamental aspect of statistical physics is the laws of thermodynamics. An isolated system's energy is constant, according to the first law of energy conservation. The notion of entropy, a gauge of disorder or randomness, is introduced in the second law. It states that the total entropy in an isolated system can never drop with time. The idea of equilibrium is thus introduced, whereby the macroscopic characteristics of a system become time-invariant. Through the use of the Boltzmann formula, $S = k_B \ln \Omega$, where Ω is the number of microscopic configurations that correspond to a macroscopic state, k_B stands for Boltzmann's constant, and S is the entropy, statistical mechanics offers a microscopic interpretation of entropy.

The distribution functions, which reveal the distribution of particles across different energy states, are an additional important consideration. For classical particles, the Maxwell-Boltzmann distribution, for fermions (particles that obey the Pauli Exclusion Principle) and bosons (particles that do not obey the Pauli Exclusion Principle), the Fermi-Dirac distribution are the most notable distributions. Various assumptions regarding the indistinguishability of the particles and the quantum states they can occupy are made in order to derive these distributions. Additionally, statistical physics makes extensive use of fluctuations. However, statistical physics predicts tiny deviations from these averages, called fluctuations, even though thermodynamic quantities are average properties. Due to their potential to cause phenomena like critical opalescence, these are particularly important in small systems or close to critical points. Phase transitions, in which a system changes state (e.g., phase transitions in a system), are best understood using the theory of fluctuations. Changes in the external environment, such as temperature or pressure, from liquid to gas. Non-equilibrium phenomena, or situations in which a system is not in thermodynamic equilibrium, are also included in statistical physics literature. Diffusion, thermal conduction, and viscosity are examples of transport processes that involve this. Often characterized by the Langevin and Fokker-Planck equations or the Boltzmann transport equation, non-equilibrium statistical mechanics examines the rates at which processes take place and how systems approach equilibrium.

1.2. ESSENTIALS OF STATISTICAL PHYSICS

The goal of statistical physics is to use the laws of mechanics to study the macroscopic parameters of an equilibrium system based on its microscopic characteristics. It is the area of physics where a formula for calculating free energy is developed. The fact that matter is made up of atoms is used in statistical physics. It provides a general expression for the free energy based on an understanding of the microscopic laws that control atomic motion and, most importantly, an additional statistical physics law. A conceptual bridge connecting the macroscopic and microscopic views is made possible by statistics in physics. By looking at the statistical distribution of particle velocities, for instance, we can learn more about gases and how macroscopically observable quantities like pressure, volume, and temperature relate to one another.

1.2.1. Application of Statistical Physics

Thermal equilibrium and non-equilibrium states are both subjects to study in statistical physics. Compared to thermodynamics, which examines the macroscopic system in equilibrium from a macroscopic perspective without taking the microscopic parameters into account, this method is distinct.

Following are the applications of statistical physics:

- The first application of statistical physics was concentrated on the distribution of molecules in an assembly. It was applied in Maxwell's distribution of molecular velocity.
- Gibbs explained thermodynamics with the help of statistical physics.

The foundation of statistical physics is the notion that the number of ways a given macroscopic state can be constructed from its microscopic constituents determines its statistical weight. Whether or not the constituents can be distinguished determines how many different ways a macroscopic state can form. The constituents in Maxwell-Boltzmann's statistics are identifiable, and the statistical weight of a given particle distribution among the single-particle states is equal to the product of the number of ways to select the particles for each energy range and the number of ways to assign these particles to the microscopic states in that range. By maximizing the statistical weight, one can obtain the Maxwell-Boltzmann distribution law. Two other types of statistics can be used to describe macroscopic systems made up of indistinguishable components. When the constituents of a macroscopic system do not obey the Pauli-exclusion principle, Bose-Einstein statistics are applicable, while Fermi-Dirac statistics apply in the opposite case.By maximizing the statistical weight, the distribution laws for each of these two types of particles are once more obtained. (Brian & Anderson 1969) The intrinsic spin or angular momentum of particles and the type of statistics they adhere to have been found to be significantly correlated. Half-integer spin particles follow Fermi-

Dirac statistics, whereas zero or integral spin particles follow Bose-Einstein statistics. The condensation of a macroscopic system into its lowest quantum state and the radiation field inside a black body can both be described by Bose-Einstein statistics. Understanding the relationship between the dynamical properties of charge carriers and the electronic properties of metals and semiconductors is made possible by Fermi-Dirac statistics.

1.2.2. Need for Statistical Description in Physics

Statistics used in physics give a conceptual link between the macroscopic and the microscopic view.

- Particle distribution at various energies as a function of temperature is described by Maxwell-Boltzmann statistics. Diffusion is one of the many processes that can be understood using this method.
- Different concepts can be understood more deeply when thermodynamics is approached statistically. For example, it is simple to understand temperature statistically as the average kinetic energy of atoms in a bulk material.
- To derive the path-integral formulation of quantum physics, statistics have been applied to describe processes like Brownian motion, and this has proven to be useful.
- Testing theories and estimating intervals on aggregate data can be done with a practical set of tools that the study of statistics offers. It serves as the foundation for carefully planning experiments, interpreting data, and correlating information, all of which support the advancement of contemporary science.

A subfield of physics known as statistical physics develops from the principles of statistical mechanics. It makes use of statistical and probability theory techniques. In order to solve physical problems, it typically makes use of mathematical techniques for handling large populations and approximations. It can be used to characterize a wide range of domains where randomness is inherent. Numerous issues in the disciplines of physics, biology, chemistry, and neuroscience are among the many applications of statistical physics. The primary task is to clarify, in terms of the physical principles governing atomic motion, the aggregate properties of matter. The results of thermodynamics are developed phenomenologically in statistical mechanics. These findings originate from an analysis of the underlying microscopic systems using probability. Classical mechanics was one of the first areas of physics to use statistical methods. Its focus is on how particles or objects move in response to external forces.

1.2.3. The Aims of Statistical Mechanics (SM)

The field of statistical mechanics is a subfield of physics that integrates the concepts and methods of statistics with the laws of classical and quantum mechanics. It is particularly relevant to the study of thermodynamics. By utilizing the characteristics and actions of the microscopic components of those systems, it seeks to forecast and elucidate the measurable attributes of macroscopic systems. For instance, temperature is a quantitative indicator of the energy distribution among atomic particles, and thermal energy is the energy of these particles in disordered states according to statistical mechanics. In order to focus on the average behavior of a large number of particles of the same kind rather than the behavior of each individual particle within a macroscopic substance, statistical mechanics heavily relies on the laws of probability. After relativity theory and quantum theory, statistical mechanics (SM) is the third pillar of contemporary physics. Its goal is to explain how the dynamical laws governing the microscopic components of physical systems and the probabilistic assumptions made about them account for the macroscopic behavior of these systems (Batterman & Robert, 1998). Equilibrium, the central idea of SM, is one facet of that behavior.

Let's use a common example to clarify the main concerns about equilibrium. Imagine a gas that is only allowed to exist on the left side of a container that has a dividing wall (Figure 1.1a). The gas is in equilibrium, and its macro-physical characteristics—such as pressure, temperature, and volume—have not changed noticeably. Now, as you abruptly remove the dividing wall (see Figure 1.1b), the gas begins to spread throughout the whole volume that is available. As seen in Figure 1.1c, the gas is no longer in equilibrium. When all of the available space is uniformly filled, the gas stops spreading (see Figure 1.1d). The gas has now achieved a new equilibrium. This process is an approach to equilibrium because the spreading process results in a new equilibrium. A salient feature of the equilibrium approach is its apparent irreversibility: systems transition from a state of non-equilibrium to equilibrium, but not the other way around; gases diffuse to fill the container uniformly, but they do not gather naturally in the left half of the container. This is known as thermodynamic behavior since an irreversible approach to equilibrium is frequently connected to thermodynamics. The main objective of SM is to characterize the equilibrium state and explain how and why a system approaches equilibrium. These two issues are sometimes attributed to distinct theories (or distinct sections of a larger theory), in which case they are denoted as equilibrium SM and non-equilibrium SM, respectively.

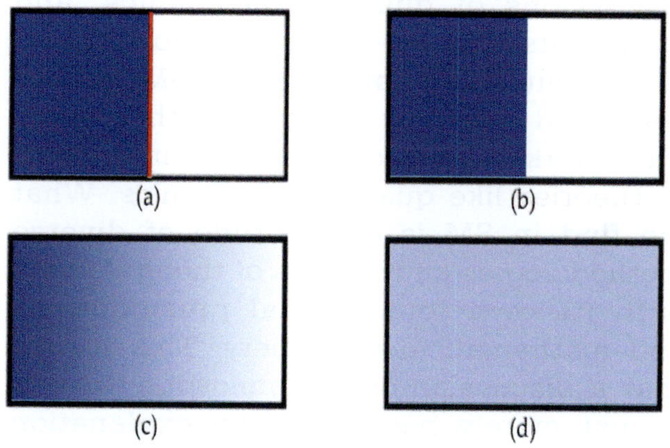

Figure 1.1. *Concerning equilibrium with a standard example.*

Source: https://plato.stanford.edu/entries/statphys-statmech/figure1c.svg

Even though equilibrium takes center stage, SM naturally addresses other topics as well, like phase transitions, the entropy costs of computation, and the mixing of substances. In philosophical contexts, SM has also been used to clarify the nature of time's direction, how probabilities are interpreted in deterministic theories, the state of the universe just after the Big Bang, and the possibility of knowing the past. All of these topics will be briefly discussed below, but because equilibrium is so important to SM, the majority of this post will analyze the theoretical foundations of both equilibrium and non-equilibrium SM.

1.2.4. The Theoretical Landscape of SM

There is an immediate problem with philosophical discussions in SM. A widely accepted theory and its formalism can serve as the foundation for philosophical initiatives in many branches of physics. For example, the Hilbert space formulation of the theory can be the starting point of philosophical discussions of quantum mechanics, and arguments can be developed from there. In SM, things are not the same. SM has not yet found a widely recognized theoretical framework or canonical formalism, in contrast to theories like quantum mechanics. What we find in SM is a multitude of diverse methodologies and schools of thought, each with its own foundational presumptions and mathematical machinery. This means that a discussion of the philosophy of SM cannot merely begin with an explanation of the fundamental ideas of the theory before moving on to various theories of interpretation. Our assignment is to categorize various methods first, then talk about how each operates; the relationship between them is then a further question.

Different approaches can be taken, and classifying and labeling approaches present their own set of challenges. Despite the theoretical diversity of SM, the majority of its approaches can be categorized under one of three major theoretical headings. These are referred to as the "Boltzmann Equation" (BE), "Gibbsian SM" (GSM), and "Boltzmannian SM" (BSM) (David Hoffman & John Dahler, 1969). The term "BSM" is a bit misleading since it could imply that Boltzmann exclusively (or mostly) supported this specific method, but in reality, he helped to develop a wide range of theoretical stances (for a summary of his contributions to SM, see the entry on Boltzmann's work in statistical physics). But these names have become commonplace, so we continue to use "BSM" despite its unfortunate past.

1.2.5. Statistical Physics and Complexity

The goal of statistical physics is to elucidate how the interactions between the constituent parts of the patterns and structures we observe in the macroscopic world come about. In the twenty-first century, applying statistical physics to systems that are out of equilibrium will be a significant task. Statistical physics examines the quantity of microscopic objects, such as molecules, atoms, bacteria, and vehicles, that could result in macroscopic occurrences. In the twenty-first century, the main task for statistical physics is to characterize nonequilibrium systems in which the conventional understandings of thermal equilibrium and the thermodynamic limit might no longer be valid. One still doesn't know a great deal about nonequilibrium systems in comparison to equilibrium systems. However, nonequilibrium systems are present in all aspects of nature: both the states of open quantum systems and complex or biophysical systems necessitate the application of nonequilibrium statistical mechanics. We carry out a broad program of basic research on model systems here in Edinburgh, addressing the numerous novel principles of non-equilibrium physics. Our research also directly relates to many other fields, including biology, ecology, and linguistics.

One important theoretical toolkit that we employ to comprehend complex systems is statistical mechanics. Although we also work on more traditional condensed matter, the systems in question include many non-traditional areas like ecosystem dynamics and traffic jamming. Large deviation functions are extensions of classic thermodynamic concepts like free energy in modern nonequilibrium statistical mechanics. In fact, models of nonequilibrium systems are frequently defined only in terms of stochastic dynamics rather than any thermodynamic concepts; as a result, new theoretical and dynamical simulation techniques must be developed in order to study these models.

1.2.5.1. Fundamental Nonequilibrium Systems and Phase Transitions

Systems that are not in equilibrium can be broadly characterized by the random movements of their constituent particles. In a unique scenario where the dynamics defy detailed balance, the system will eventually settle into an equilibrium state that is characterized by the Boltzmann distribution. Stochastic dynamics, on the other hand, typically do not adhere to detailed balance, and the system will settle into a non-equilibrium stationary state with currents. The nature and potential of these states are still unknown to us (Figure 1.2).

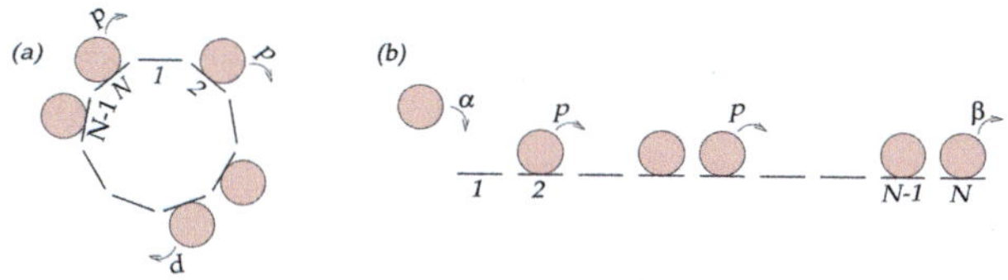

Figure 1.2. *Dynamics of the asymmetric exclusion process.*

Source: https://www.ph.ed.ac.uk/sites/default/files/images/icmcs/asep.jpg

We can clarify the characteristics of nonequilibrium states by looking at and solving basic mathematical models. For instance, the figure above depicts the "asymmetric exclusion process," a straightforward lattice gas. Particles cannot pass one another in a lattice; they move along it stochastically. A nonequilibrium stationary state is reached when particles are injected at one end and removed at the other. This system serves as the foundation for simulating a variety of general transport issues, such as traffic from vehicles and biophysical phenomena like ribosomes—large, complex molecules—moving along and translating RNA. It has been astonishingly demonstrated that nonequilibrium phase transitions take place in this system. That is, significant modifications to the bulk behavior of the system can be achieved by adjusting the boundary injection and extraction rates.

Here in Edinburgh, we have also found another nonequilibrium phase transition: the real-space condensation transition, in which a large number of microscopic constituents come together in one place. Condensation has been found to occur in a surprising variety of contexts, including wealth condensation in macroeconomics, where a single agent amasses a finite portion of the system's wealth, hub-formation in networks, jamming in vehicular flows, and coalescence in granular gases. This phenomenon is related to the quantum Bose-Einstein condensation. Clarifying unexpected aspects of this phenomenon, like the peculiar dynamics of moving condensates, has been a recurring theme in our work.

1.2.5.2. Statistical Physics of Evolutionary Dynamics

In biology, where DNA molecules and more complex organisms are replicated, and in culture, where agents learn new behaviors from each other, evolution—a theory of change by replication—occurs (as illustrated in figure 1.3). The manner and timing of replication events are erratic, which makes this process intrinsically random. Furthermore, the dynamics lack the equilibrium time-reversal symmetry that characterizes physical systems. From a statistical physics perspective, evolving populations represent a class of nonequilibrium dynamical systems that are highly intriguing.

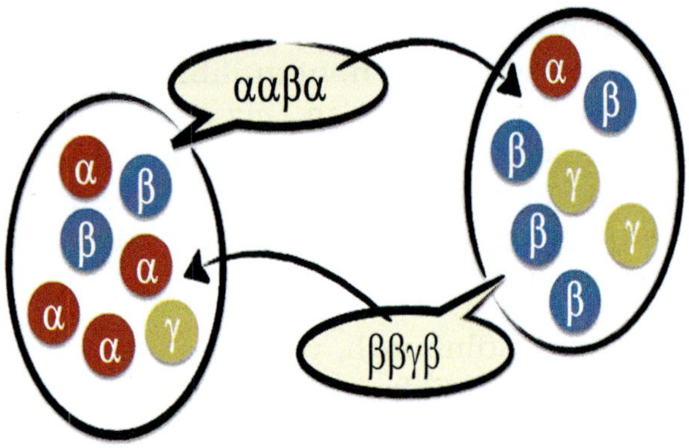

Figure 1.3. *Simple cultural evolutionary model where linguistic behavior is replicated and transmitted between agents.*

Source: https://www.ph.ed.ac.uk/sites/default/files/images/icmcs/usm.jpeg

Specifically, the fluctuations in replication dynamics are not in equilibrium, which gives rise to a range of intriguing macroscopic effects that we have not yet completely figured out. For instance, we have found that endemic fluctuations in evolving populations can introduce cycles into a system whose deterministic limit only has stable fixed points, whereas weak fluctuations can eliminate cyclic behavior that exists in a deterministic system. Simultaneously, these models have applications to a wide range of real-world phenomena, especially in language dynamics, where we have looked at how children learn word meanings and how the English dialect in New Zealand developed (Figure 1.4).

1.2.5.3. Statistical-Physics Models of Biological Populations

Numerous applications of statistical physics are found in the modeling of biological population dynamics and evolution. Phase transitions can happen even in zero- or one-dimensional systems because these systems are out of equilibrium due to the replication and growth of organisms.

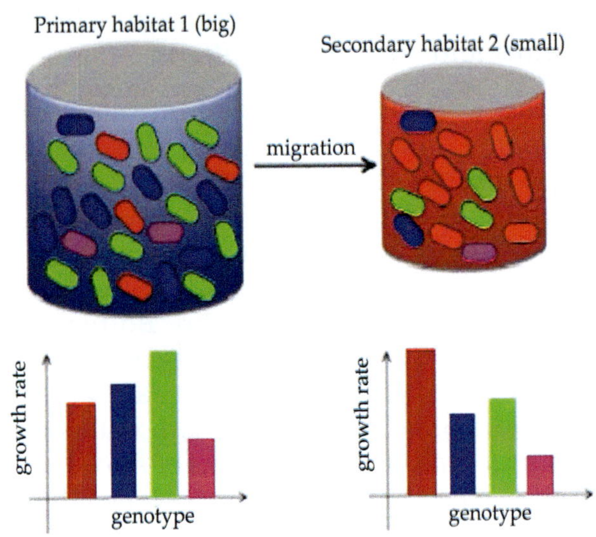

Figure 1.4. *Source-sink model in which a dynamical phase transition occurs at a critical migration rate.*

Source: https://www.ph.ed.ac.uk/sites/default/files/images/icmcs/source-sink.jpg.

Examples include models of migrating organisms, bacterial colloids sedimenting due to gravity, and the extraction of membrane tubes by molecular motors. Condensation, localization, jamming, and many other issues that arise in modeling chemical reactions, like traveling waves, are all addressed by the same mathematical framework that is used to model these systems. This framework is also used to address other phenomena in statistical and condensed matter physics.

1.2.5.4. Statistical Physics Models of Biological Cells

Many biochemical reactions take place inside biological cells. These enable the cell to produce new cell components, absorb energy from food, and respond to stressors like temperature changes or drug inflow. We are able to construct models for these processes by applying methods from statistical physics. For instance, we forecast the configuration of DNA within biological cells using sophisticated computer simulation methods. We also create basic dynamical models to study how bacteria respond to antibiotics and create network-based algorithms to study the best biochemical pathways for energy metabolism (Figure 1.5).

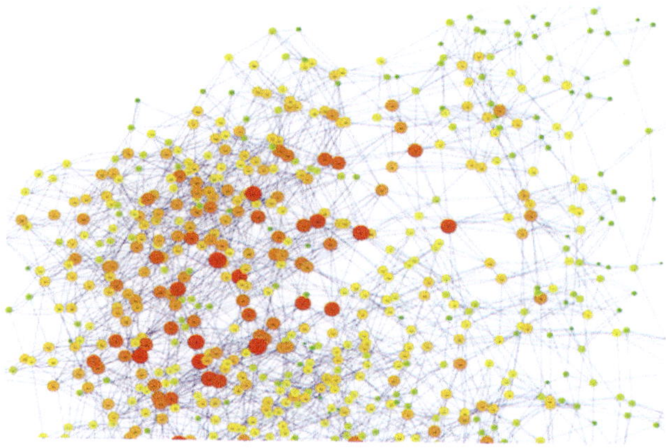

Figure 1.5. *A diagrammatic representation of the network of all possible chemical reaction pathways for 4-carbon molecules such as those that the cell uses for energy metabolism.*

Source: https://www.ph.ed.ac.uk/sites/default/files/images/icmcs/Steve_network.png

Networks and Agents in Physics and Ecology

The evolution of interacting objects in nonequilibrium settings is the focus of this work. Frequently, the objects are not just passive entities, but rather dynamic "agents" that function based on a learnable strategy. Furthermore, the interactions among agents dictate how each one reacts to and influences its surroundings. According to evolutionary game theory, which defines evolution, an object's ability to survive and reproduce depends on its ability to accomplish a specific goal. This could be selling a product, avoiding a predator, or assimilating into a specific culture. From a computational perspective, this means that instead of using data fitting, the evolutionary process can determine the numerical **values** of many parameters with complex interpretations. In the large number long-term limit, these individual-based models can depict behavior that is far richer than the corresponding continuum theories to which they reduce. This methodology could be applied to a variety of projects, such as studying the dynamics behind flock and city development, understanding the rise and fall of civilizations, and comprehending boom and bust economics. Based on the methods previously discussed, the simulation of the spread of Neolithic farming is depicted in the figure. The lower panel depicts the spread of the farming technology adopters, while the top panel depicts the demic spread from the genetic offspring of the original Middle Eastern farmers.

1.3. FUNDAMENTALS OF STATISTICAL MECHANICS

The theory of statistical mechanics explains the relationship between the description of a microscopic state that varies around an average state and the ideas from macroscopic observations (such as temperature and pressure). It links thermodynamic quantities (like heat capacity) to behavior at the microscopic level whereas, in classical thermodynamics, one can only measure and tabulate these quantities for different materials. Systems that are out of balance can also be studied using statistical mechanics. Modeling the speed of irreversible processes driven by imbalances at the microscale is a significant subfield of non-equilibrium statistical mechanics. Such processes include heat-producing particle flows and chemical reactions. The fundamental understanding gained from using non-equilibrium statistical mechanics to examine the most straightforward non-equilibrium scenario of a steady state current flow in a system of many particles is known as the fluctuation-dissipation theorem. Statistical mechanics is a subfield of physics that combines the laws of classical and quantum mechanics with the concepts and practices of statistics, especially as they relate to thermodynamics. It seeks to use the characteristics and actions of the microscopic components of macroscopic systems to predict and explain the measurable properties of those systems. For instance, thermal energy is understood by statistical mechanics to be the energy of atomic particles in disordered states, and temperature is a quantitative indicator of the energy distribution among these particles. Statistical mechanics heavily relies on the laws of probability, focusing instead on the average behavior of a large number of particles of the same kind rather than the behavior of each individual particle within a macroscopic substance.

One tactic that would most certainly fail is to write down and solve the Schrödinger equation for 10^{23} particles. That is generally not achievable with even 23 particles, much less 10^{23}. What would you do with the system's wave function, even if you could find it? Nobody really cares about the positions of individual particles. We're looking for answers to far simpler, almost childlike queries concerning what's in the box. How can we start answering these kinds of questions starting from the fundamental laws of physics? For centuries, from the 1600s to the 1900s, scientists were discovering "laws of physics" that govern different substances. Is it wet? Is it hot? What color is it? Is the box in danger of exploding? What happens if we squeeze it, pull it, or heat it up? These laws number in the hundreds and are primarily named after their discoverers. Boyle's and Charles's laws, which are typically combined to form the ideal gas law, relate pressure, volume, and temperature of gases; the Stefan-Boltzmann law determines the amount of energy emitted by a hot object; Wien's displacement law determines the color of the hot object; the Dulong-Petit law determines the energy required to heat a lump of material;

Curie's law describes how a magnet loses its magnetic properties when it comes into contact with a flame; and so on. However, it is now clear that these laws are not essential. In certain instances, they can be deduced from a combination of statistical reasoning and Newtonian mechanics. In other situations, quantum mechanics must also be considered. However, we will see how to derive them from first principles in every situation.

Determining the intriguing properties of a mixture of 10^{23} particles will be a major focus of this course. One of the themes that will come up again is $10^{23} \neq 1$. More is different in that certain fundamental ideas only become apparent when we examine a vast array of particles; they are not apparent in the fundamental laws of physics. Temperature is one extremely basic example. This is not a fundamental idea: discussing the temperature of a single electron is illogical. However, discussing the physics of the world we live in today would be incomplete without discussing temperature. This demonstrates how terminology used to explain physics on one scale differs greatly from terminology used on other scales. You'll observe a number of related emergent quantities, such as phase transitions, which are instances in which abrupt, discontinuous changes in the structure of matter result from the combined action of the smooth, continuous laws of physics. In the past, statistical mechanics methods have shown to be an invaluable resource for comprehending the fundamental principles of physics. Not only is the field's history closely linked to the first indications of atom existence, but quantum mechanics was also discovered through the use of statistical techniques to interpret the light spectrum released by heated objects. Physics is not a finished subject, though. Many significant systems in nature, such as black holes and high-temperature superconductors, are still poorly understood. Our objective is to use the limited information we have about these systems—their macroscopic properties—to dissect the underlying mechanisms at play. This task will require the tools that we will learn in this course.

1.3.1. The Microcanonical Ensemble

The statistical ensemble used to represent the potential states of a mechanical system with a precisely defined total energy is called a microcanonical ensemble. It is assumed that the system is isolated in the sense that it is unable to exchange particles or energy with its surroundings. As a result, the system's energy will always remain precisely known due to energy conservation. All possible states of the system maintain the same energy, composition, volume, and shape. The first step in analyzing the properties of matter in a real-world problem would be to write down the basic equations and attempt to solve them mathematically. Even though some people attempt to employ this strategy, they are the industry's failures. We'll begin by thinking about an isolated system, E, that has a fixed energy. Although almost everything applies equally well to classical systems, we will use the language of quantum mechanics to describe our system for the discussion. Only systems with a few degrees of freedom were covered in the first two quantum mechanics courses you took.

These are defined by a Hamiltonian, H^, and the goal is usually to solve the time-independent Schrodinger equation

$$\hat{H}|\psi\rangle = E|\psi\rangle$$

In this course, we will still look at systems that are defined by a Hamiltonian, but now with a very large number of degrees of freedom, say N~10^{23}. The energy eigenstates $|\psi\rangle$ are very complicated objects since they contain information about what each of these particles is doing. They are called microstates.

In actuality, it is frequently very challenging to record the microstate that describes each of these particles. More importantly, though, is that it is typically completely boring. Since real macroscopic systems are not described by a single pure quantum state, the wave function for a macroscopic system rarely captures the relevant physics. They interact with their surroundings, which are continuously agitating and buffeting them. The system experiences a tiny perturbation each time it is jogged, and there is a chance that it will change states. The transitions will only occur to states of equal (or very nearly equal) energy if the perturbation is very small. However, there can be a large number of microstates with the same energy E when there are 10^{23} particles. We don't need to be familiar with the specifics of any one state in order to comprehend the physics of these systems. We must be aware of every state's minute details. Maintaining track of the dynamics that cause the various states to transition would be incredibly laborious. Rather, we will utilize statistical techniques. The system will be explained in terms of a probability distribution across the possible quantum states. Stated differently, the system exists in a mixed state as opposed to a pure state.

Since we have fixed the energy, there will only be a non-zero probability for states which have the specified energy E. We will denote a basis of these states as $|n\rangle$ and the probability that the systems sit in a given state as p(n). Within this probability distribution, the expectation value of any operator $\hat{O}$ is

$$\langle\hat{O}\rangle = \sum_n p(n)\langle n|\hat{O}|n\rangle$$

Determining the appropriate probability distribution p(n) for large systems is our immediate objective. First of all, the kinds of situations we can discuss will be severely limited. We will only talk about systems that have been inactive for a while. This guarantees that all of the system's energy and momentum have been dispersed among its numerous particles and that any remnants of the unique initial conditions the system was initially subjected to have long since vanished. From an operational perspective, this implies that the probability distribution remains independent of time, guaranteeing that the macroscopic observables' expectation values remain constant over time. When a system is in equilibrium, we say so. A glass of water left alone will eventually reach equilibrium, but the atoms inside are still moving. It is important to keep in mind that just because a system is in equilibrium, it does not mean that all of its parts have stopped moving. It is now possible for us to declare the central tenet of statistical mechanics. It is the notion that we ought to approach every state equally and with the simplest mindset imaginable. Or, more precisely:

All possible microstates are equally likely for an isolated system in equilibrium. Given the current state of our knowledge, a democratic approach appears to be the most rational choice. Observe how we have given ourselves some leeway by including the word "accessible." This describes any possible state that the system may experience as a result of minor disruptions.

We will interpret it to mean all states with the same energy E for the time being. We'll see situations later on where we define an accessible state with even more limitations.

Let us introduce some notation. We define

$\Omega(E)$ = Number of states with energy E

The probability that the system with fixed energy E is in a given state $|n\rangle$ is simply

$$p(n) = \frac{1}{\Omega(E)} \quad (1)$$

The probability that the system is in a state with some different energy $E' \neq E$ is zero. This probability distribution, relevant for systems with fixed energy, is known as the microcanonical ensemble. Some comments:

- $\Omega(E)$ is a usually ridiculously large number. For example, suppose that we have $N \sim 10^{23}$ particles, each of which can only be in one of two quantum states – say "spin up" and "spin down." Then the total number of microstates of the system is $2^{10^{23}}$. This is a silly number. In some sense, numbers this large can never have any physical meaning! They only appear in combinatorial problems, counting possible eventualities. They are never answers to problems which require you to count actual existing physical objects. One slightly facetious way of saying this is that numbers this large can't have physical meaning because they are the same no matter what units they have (If you don't believe me, think of $2^{10^{23}}$ as a distance scale: it is effectively the same distance regardless of whether it is measured in microns or lightyears.
- In quantum systems, the energy levels will be discrete. However, with many particles, the energy levels will be finely spaced and can be effectively treated as a continuum. When we say that $\Omega(E)$ counts the number of states with energy E, we implicitly mean that it counts the number of states with energy between E and E + δE, where δE is small compared to the accuracy of our measuring apparatus but large compared to the spacing of the levels.

We phrased our discussion in terms of quantum systems, but everything described above readily carries over to the classical case. In particular, the probabilities p(n) have nothing to do with quantum indeterminacy. They are due entirely to our ignorance

We recall the definition of this ensemble – it is that set of microstates which for given *N, V* have an energy in the interval $[E, \Delta E]$.

The number of such microstates is proportional to the phase space volume they inhabit. And we found some reason to suspect that this volume, its logarithm, may be identified as that property which the thermodynamicists have dubbed entropy and denoted by *S (N, V, E)*. This can hold only if S has the two essential properties of entropy:

- If the system is divided into two subsystems that may freely exchange energy, then the equilibrium state is the one in which the available energy $E=E_1 + E_2$ is distributed such that

$$\left.\frac{dS(N_1,V_1,E_1)}{dE_1}\right|_{E_1=E_1^*} \overset{?}{=} \left.\frac{dS(N_2,V_2,E_2)}{dE_2}\right|_{N_2=E=-E_1^*}$$

The quantity $T=[dS(N,V,E)/dE]^{-1}$ is commonly called temperature.

- In equilibrium the entropy S is additive:

$$S(N_1,V_1,E_1)+S(N_2,V_2,E_2)\overset{?}{=}S(N_1+N_2,V_1+V_2,E_1+E_2)$$

1.3.1.1. Applicability

The microcanonical ensemble, which describes the possible states of an isolated mechanical system when the energy is precisely known but provides no additional information about the internal state, is sometimes regarded as the fundamental distribution of statistical thermodynamics because of its justification in elementary principles like the principle of indifference. Additionally, in some special systems, the evolution is ergodic; in these cases, starting from a single state of energy E, the microcanonical ensemble equals the time-ensemble (a time-ensemble is the ensemble formed of all future states evolved from a single initial state). The microcanonical ensemble is not representative of an experimentally feasible scenario in real life. A real physical system has some degree of energy uncertainty because of uncontrollable elements in the system's setup. In addition to the challenge of locating an experimental analog, the requirement of fixed energy makes it hard to perform calculations that allow logically independent components of the system to be examined independently. Furthermore, there are disagreements over how to define terms like temperature and entropy in the microcanonical ensemble. Systems in thermal equilibrium with their surroundings are characterized by uncertainty in their energy, which is represented by the canonical ensemble or the grand canonical ensemble, depending on whether the system is also in equilibrium with particle exchange with its surroundings.

1.3.2. The Canonical Ensemble

Systems with a fixed energy E are described by the microcanonical ensemble. We can infer the equilibrium temperature T from this. Nevertheless, this is frequently not the optimal way to approach a system. A glass of water resting on a table, for instance, has a distinct average energy. However, due to its interactions with the surroundings, the energy is always changing. It is usually more appropriate to consider such systems as being at a fixed temperature T, from which the average energy can be calculated. In order to model this, we will take a large heat reservoir, denoted as R, in contact with another system S. It is assumed that the reservoir is at some equilibrium temperature, T. The word "reservoir" refers to the fact that S energy is negligible in comparison to R. In particular, S can cheerfully donate or absorb energy from the reservoir without causing a change in the surrounding temperature, T.

How are the energy levels of S populated in such a situation? We label the states of S as $|n\rangle$, each of which has energy E_n. The number of microstates of the combined systems S and R is given by the sum over all states of S,

$$\Omega(E_{\text{total}}) = \sum_n \Omega_R(E_{\text{total}} - E_n) \equiv \sum_n \exp\left(\frac{S_R(E_{\text{total}} - E_n)}{k_B}\right)$$

I stress again that the sum above is over all the states of S, rather than over the energy levels of S. (If we'd written the latter, we would have to include a factor of $\Omega_S(E_n)$ in the sum to take into account the degeneracy of states with energy E_n). The fact that

R is a reservoir means that $E_n \otimes E_{total}$. This allows us to Taylor expand the entropy, keeping just the first two terms,

$$\Omega(E_{\text{total}}) \approx \sum_n \exp\left(\frac{S_R(E_{\text{total}})}{k_B} - \frac{\partial S_R}{\partial E_{\text{total}}}\frac{E_n}{k_B}\right)$$

But we know that $\partial S_R/\partial_E total = 1/T$, so we have

$$\Omega(E_{\text{total}}) = e^{S_R(E_{\text{total}})/k_B} \sum_n e^{-E_n/k_BT}$$

We now apply the fundamental assumption of statistical mechanics — that all accessible energy states are equally likely — to the combined system + reservoir. This means that each of the $\Omega(E_{total})$ states above is equally likely. The number of these states for which the system sits in $|n\rangle$ is $e^{S_R/k_B}e^{-E_n/k_BT}$. So the probability that the system sits in a state $|n\rangle$ is just the ratio of this number of states to the total number of states,

$$p(n) = \frac{e^{-E_n/k_BT}}{\sum_m e^{-E_m/k_BT}}$$

This is the Boltzmann distribution, also known as the canonical ensemble. Notice that the details of the reservoir have dropped out. We don't need to know $S_R(E)$ for the reservoir; all that remains of its influence is the temperature T.

The exponential suppression in the Boltzmann distribution means that it is very unlikely that any of the states with $E_n \leq k_BT$ are populated. However, all states with energy $E_n \leq k_BT$ have a decent chance of being occupied. Note that as $T \to 0$, the Boltzmann distribution forces the system into its ground state (i.e., the state with the lowest energy); all higher energy states have vanishing probability at zero temperature.

1.3.2.1. The Partition Function

Since we will be using various quantities a lot, it is standard practice to introduce new notation. Firstly, the inverse factor of the temperature is universally denoted,

$$\beta \equiv \frac{1}{k_BT}$$

And the normalization factor that sits in the denominator of the probability is written,

$$Z = \sum_n e^{-\beta E_n} \tag{2}$$

In this notation, the probability for the system to be found in state $|n\rangle$ is

$$p(n) = \frac{e^{-\beta E_n}}{Z} \tag{3}$$

Rather remarkably, it turns out that the most important quantity in statistical mechanics is Z. Although this was introduced as a fairly innocuous normalization factor, it actually contains all the information we need about the system. We should think of Z, as defined in (2), as a function of the (inverse) temperature β. When viewed in this way, Z is called the partition function.

We will see lots of properties of Z soon. But we'll start with a fairly basic, yet important, point: for independent systems, Z's multiply. This is easy to prove. Suppose that we have two systems which don't interact with each other. The energy of the combined system is then just the sum of the individual energies. The partition function for the combined system is (in, hopefully, obvious notation).

$$Z = \sum_{n,m} e^{-\beta(E_n^{(1)}+E_m^{(2)})}$$
$$= \sum_{n,m} e^{-\beta E_n^{(1)}} e^{-\beta E_m^{(2)}}$$
$$= \sum_{n} e^{-\beta E_n^{(1)}} \sum_{m} e^{-\beta E_m^{(2)}} = Z_1 Z_2$$

1.3.2.1.1. A Density Matrix for the Canonical Ensemble

In statistical mechanics, the inherent probabilities of the quantum world are joined with probabilities that arise from our ignorance of the underlying state. The correct way to describe this is in terms of a density matrix, ˆρ. The canonical ensemble is really a choice of density matrix,

$$\hat{\rho} = \frac{e^{-\beta \hat{H}}}{Z}$$

If we make a measurement described by an operator $\hat{O}$, then the probability that we find ourselves in the eigenstate $|\phi\rangle$ is given by

$$p(\phi) = \langle \phi|\hat{\rho}|\phi\rangle$$

For energy eigenstates, this coincides with our earlier result (3). We would not use the language of density matrices in this course, but it is an elegant and conceptually clear framework to describe more formal results.

1.3.2.1.2. Energy and Fluctuations

Let's see what information is contained in the partition function. We will start by thinking about the energy. In the microcanonical ensemble, the energy was fixed. In the canonical ensemble, that is no longer true. However, we can happily compute the average energy,

$$\langle E\rangle = \sum_n p(n)E_n = \sum_n \frac{E_n e^{-\beta E_n}}{Z}$$

But this can be very nicely expressed in terms of the partition function by

$$\langle E\rangle = -\frac{\partial}{\partial\beta}\log Z$$

We can also look at the spread of energies about the mean — in other words, about fluctuations in the probability distribution. As usual, this spread is captured by the variance,

$$\Delta E^2 = \langle (E-\langle E\rangle)^2\rangle = \langle E^2\rangle - \langle E\rangle^2$$

This too can be written neatly in terms of the partition function,

$$\Delta E^2 = \frac{\partial^2}{\partial\beta^2}\log Z = -\frac{\partial\langle E\rangle}{\partial\beta} \quad (4)$$

There is another expression for the fluctuations that provides some insight. Recall our definition of the heat capacity (9) in the microcanonical ensemble. In the canonical ensemble, where the energy is not fixed, the corresponding definition is

$$C_V = \left.\frac{\partial\langle E\rangle}{\partial T}\right|_V$$

Then, since $\beta = 1/k_BT$, the spread of energies in (4) can be expressed in terms of the heat capacity as

$$\Delta E^2 = k_B T^2 C_V \tag{5}$$

This little equation actually hides two important points. It connects two quite different quantities at the beginning. The variations in the system's energy are shown by ΔE on the left side. On the right side, the heat capacity C_V tells us how much energy the system can take in. When C_V is high, the system can absorb a large amount of energy without a significant rise in temperature. Equation (5) shows that the system's ability to release or take in energy is linked to its fluctuations. This is the first step in understanding the fluctuation-dissipation theorem, which leads to a broader conclusion. Another key point to remember from (5) is how the size of fluctuations changes as the number of particles N in the system increases.

Typically, $E \sim N$ and $C_V \sim N$. Which means that the relative size of the fluctuations scales as

$$\frac{\Delta E}{E} \sim \frac{1}{\sqrt{N}}$$

The limit $N \rightarrow \infty$ is known as the thermodynamic limit. The energy becomes peaked closer and closer to the mean value $\langle E \rangle$ and can be treated as essentially fixed. But this was our starting point for the microcanonical ensemble. In the thermodynamic limit, the microcanonical and canonical ensembles coincide.

All the examples that we will discuss in the course will have a very large number of particles, N, and we can consider ourselves safely in the thermodynamic limit. For that reason, even in the canonical ensemble, we will often write E for the average energy rather than $\langle E \rangle$.

1.4. MONTE CARLO IN STATISTICAL PHYSICS

In statistical physics, or statistical mechanics, Monte Carlo refers to the application of the Monte Carlo method to problems. Evaluating a multivariable integral is the main reason statistical physicists utilize the Monte Carlo method. The typical problem starts with a system that obeys Boltzmann statistics, is at a certain temperature, and has a known Hamiltonian. We try to track the "time dependence" of a model in a Monte Carlo simulation (Santra & Ray, 2011) where growth or change does not occur in a strictly predetermined manner (e.g., based on Newton's equations of motion) but rather randomly, relying on a series of random numbers that are produced throughout the simulation. The simulation will not produce the same results with a second, different sequence of random numbers, but it will produce values that are within a "statistical error" of those obtained with the first sequence. This category includes a very wide range of problems: percolation is the process of gradually adding particles to an empty lattice by randomly adding a particle with each "tick of the clock." Subsequently, numerous inquiries could be made concerning the ensuing "clusters," which consist of adjacent inhabited locations. A lot of focus has been on figuring out the "percolation threshold," i.e., the threshold number of occupied sites at which an "infinite percolating cluster" initially manifests itself. A percolating cluster is one that extends from one of a (macroscopic) system's boundaries to the other. These objects' characteristics are relevant to a variety of physical issues, including the conductivity of random mixtures, flow through porous rocks, and the behavior of diluted magnets. Another illustration is diffusion-limited aggregation (DLA), in which a particle moves randomly through space, one step at a time, until it comes into contact with a "seed" mass and adheres to it. A study of this mass's growth can then be conducted when numerous random walkers are released.

In statistical mechanics, we might be trying to sample a phase space region to estimate some of the model's properties, but we might not be traveling in phase space in the same direction as an exact solution to the model's time dependence would. To compute the thermal averages of (interacting) many-particle systems, keep in mind that equilibrium statistical mechanics relies on Monte Carlo simulations, which accurately account for statistical fluctuations and their effects in such systems. We won't go into additional detail about many of these models here because they will be covered in greater detail in later chapters. One way to improve a Monte Carlo (Newman & Barkema 2001) estimate's accuracy is to probe phase space more thoroughly. This can be achieved by extending the calculation's duration to obtain a greater number of samples. In contrast to the use of numerous analytical methods (e.g., perturbation theory, for which it might be unfeasible to extend to a higher order), increasing the precision of Monte Carlo findings is feasible both in theory and in actual use.

We want to calculate thermal averages of the form

$$\langle A \rangle = \frac{1}{Z}\mathrm{Tr} A e^{-H/T} = \sum_x A(x)P(x)$$

Here Tr = $\sum_x$ denotes a sum over all states x, where x denotes a state, i.e., a multidimensional vector in phase space containing all microscopic coordinates of all particles (we will usually suppress vector notation for states). The Hamiltonian H(x) is the energy of the system in the state x. T is the temperature. The Boltzmann distribution P(x) = (1/Z) exp(−H(x)/T) is the probability of being in state x, Z = exp(−F/T) = Tr exp(−H/T) is the partition function, and F = −T lnZ is the free energy. We will also use the notation P(x) = $e^{-S(x)}$, where S is the" action."

One may immediately think of two possible ways to simulate $\langle A \rangle$:

- *Method 1:* Try to sum over all states numerically by a deterministic method. If Tr is the integral over particle positions, the number of mesh points in a deterministic integration method explodes with increasing particle number. This just does not work.
- *Method 2:* Generate configurations at random using direct sampling. In most cases, this will not work because:

1. The factor exp(−H/T) varies exponentially so almost all generated configurations will give negligible contributions to averages.
2. The normalization constant Z is unknown and has to be calculated in the same way, which is another source of errors.
3. In complicated cases, methods to directly generate configurations are sometimes not available. (This is the case, for example, in many-fermion systems, where it would be of great interest to be able to directly sample configurations.)

The way out of these difficulties is to use importance sampling and generate configurations distributed according to the Boltzmann distribution. MC estimates of thermal averages have the form

$$\langle A \rangle = \frac{1}{N}\sum_{x_i \in P, i=1,\dots,N} A(x_i) \pm \frac{\sigma}{\sqrt{N}}$$

$$\sigma^2 = \frac{1}{N}\sum_i A(x_i)^2 - \left[\frac{1}{N}\sum_i A(x_i)\right]^2$$

Note that we can always use any distribution P' here, instead of the Boltzmann distribution P. The averages then become

$$\langle A \rangle = \frac{\frac{1}{N}\sum_{x_i \in P', i=1,\dots,N} \frac{A(x_i)e^{-H(x_i)/T}}{P'(x_i)}}{\frac{1}{N}\sum_{x_i \in P', i=1,\dots,N} \frac{e^{-H(x_i)/T}}{P'(x_i)}}$$

and similarly for the error. This is more complicated to use, since the normalization factor in the denominator has to be calculated. However, sometimes the Boltzmann distribution can be unsuitable for sampling of states, and then this form becomes useful. The remaining problem is to sample states that are distributed according to the Boltzmann distribution. This is solved by the method of Metropolis et al., But before we describe this, we will again consider throwing stones.

1.4.1. Stones, Markov Process Sampling, and pi

Let's return to our stone-throwing exercise for the computation of π. Since every throw in that method is independent of every other throw, it is equivalent to direct sampling. We're going to employ an additional sampling technique that works like this: Markov process sampling. Envision a large square, perhaps 100 by 100 meters. Toss one stone at a time. After that, proceed to the spot where it landed and toss the following stone at random. Moreover, this should converge to π/4. What should we do, though, if a stone misses the square? Should we keep throwing and hope to enter the square again, or should we discard the missed throw and try again until a stone hits? Both of these approaches are incorrect and provide the wrong answers; the proper course of action is to add a stone on top of the one we were throwing from and then try again. Although this is the right approach, it will result in more stones being placed close to the square's edges.

1.4.2. The Metropolis Method

The series of states x_1, x_2, x_3....., constitutes a Markov process. such that a transition probability w ($x_i \rightarrow x_{i+1}$) specifies the probability distribution for x_{i+1} from the previous state x_i. Since the index i can be thought of as a time step t, it will be very helpful to think of the Markov process as a dynamic process in phase space. The Master equation then provides the probability distribution's time evolution.

$$P(x,t+1) - P(x,t) = \sum_{y \neq x} [P(y,t)w(y \rightarrow x) - P(x,t)w(x \rightarrow y)]$$

This equation can be expressed as follows: P (x, t + 1) – P (x, t) represents the change in the probability P(x,t) of being in state x from time t to time t+1. This change must be equal to the sum of all transitions from any state y into x, which occur with probability P (y, t) w (y → x), less the sum of transitions from any state y into x, which occur with probability P (x, t) w (x → y). In general, the Master equation holds true, but our goal is to construct w so that the Master equation yields the desired P.

In other words, we want to construct the Markov process, i.e., choose w, such that after sufficiently many time steps the probability distribution for the states approaches the desired distribution P(x) = $e^{-S(x)}$.

The following two conditions turn out to be sufficient to make the Markov process approach the desired distribution P(x) = $e^{-S(x)}$ after a larg number of steps:

1. Ergodicity: Any state in the system is reachable by the transition probability in the Markov process in a finite number of steps.
2. Detailed Balance (or Micro-reversibility): The transition probability w (x → y) to go from state x to y obeys

$$e^{-S(x)}w(x \rightarrow y) = e^{-S(y)}w(y \rightarrow x)$$

The proof is straightforward and has two steps:

STEP 1: Show that $e^{-S(x)}$ is an equilibrium solution. This means that if P(x) = $e^{-S(x)}$ gives the distribution for state x, then in the next step of the Markov process, the distribution is the same:

$$P(y) = \int dx\, e^{-S(x)} w(x \rightarrow y) = e^{-S(y)}$$

This follows from detailed balance and the normalization condition $\int dx w(x \rightarrow y) = 1$.

$$P(y) = \int dx\, e^{-S(x)} w(x \to y) = e^{-S(y)} \int dx\, w(y \to x) = e^{-S(y)}$$

This proves that the desired distribution P(x) is a stationary distribution. Note that this corresponds to setting all terms in the Master equation to zero! This is clearly a sufficient but not necessary condition for a stationary distribution.

STEP 2: Show that the Markov process approaches e $^{-S(x)}$. As a measure of the distance between two distributions $P_A(x)$ and $P_B(x)$ we use $D = \int dx\, |P_A(x) - P_B(x)|$. If the distribution at one step is M(x), the derivation from equilibrium is

$$D_{\text{old}} = \int dx\, |M(x) - e^{-S(x)}|$$

The distribution at the next step, $P(y) = \int dx\, M(x) w(x \to y)$, has deviation from equilibrium:

$$\begin{aligned} D_{\text{new}} &= \int dy | \int dx\, M(x) w(x \to y) - e^{-S(y)} | \\ &= \int dy | \int dx\, [M(x) - e^{-S(x)}] w(x \to y) \\ &\leq \int dy \int dx\, |M(x) - e^{-S(x)}| w(x \to y) \\ &= \int dx\, |M(x) - e^{-S(x)}| = D_{\text{old}} \end{aligned}$$

where we used the normalization condition $\int w = 1$ and detailed balance. This shows that D → 0 when N → ∞. Strict equality above holds in equilibrium: M(x) = e $^{-S(x)}$ or if some states are not accessible to w (x → y), which is excluded by hypothesis.

The derivation of the Markov process's convergence towards the Boltzmann distribution is now complete. It should be noted that no claims are made regarding the rate at which the generated distribution approaches the intended equilibrium distribution. Building the transition probability w (x → y) is the next task. W (x → y) = t (x → y) a (x → y) is what we set, with t (x → y) representing the trial probability, i.e., the likelihood that a Markov process attempts to go from x to y, and a (x → y) is the acceptance probability, i.e., the likelihood that, if attempted from state x, state y will be accepted.

Detailed balance requires:

$$\frac{w(x \to y)}{w(y \to x)} = \frac{t(x \to y)\, a(x \to y)}{t(y \to x)\, a(y \to x)} = \frac{e^{-S(y)}}{e^{-S(x)}}$$

Here there is apparently an infinite freedom in choosing t and a such that the desired distribution e $^{-S(x)}$ is obtained. One usually (but not always) takes the trial probability to be symmetric: t (x → y) = t(y → x). Common choices for the acceptance probability are:

1. The original Metropolis choice is

$$a(x \to y) = \min\left\{1, \frac{e^{-S(y)}}{e^{-S(x)}}\right\}$$

2. In practice the following acceptance probability is often to be preferred (since this is a smooth function)

$$a(x \to y) = \frac{1}{1 + \frac{e^{S(y)}}{e^{S(x)}}}$$

The detailed balance is satisfied by the following algorithm. Using x and a probability of t (x → x_T), create a trial state x_T. Probability-wise, this state is accepted with a probability of (x → x_T). In the Markov process, set the new state to y = x_T if the state is accepted; if not, set the new state to y = x, which is the previous state. If the square is missed, this translates to setting a second stone on top of the first. The subsequent states of the Markov process are correlated, which presents a major challenge for sampling. This implies that determining the simulation's convergence rate and error estimate can be extremely difficult.

MC Algorithm using the Metropolis Method

1. Specify some initial state x_0.
2. Using a trial probability distribution, create a trial state y from x_t, and then accept the trial state using an acceptance probability distribution based on the detailed balance. If approved, set $x_{t+1} = y$ as the new state in the Markov process; if not, set the new state to $x_{t+1} = x_t$, which is the previous state.
3. Repeat from step 2 until enough states have been generated in equilibrium to form accurate thermal averages. Discard initial states to approach the equilibrium distribution.

MC Algorithm for the Ising Model

The Ising model is defined by the Hamiltonian

$$H = -J \sum_{\langle i,j \rangle} S_i S_j$$

where J is a coupling constant, $S_i = \pm 1$ is a" spin" degree of freedom on site of a simple cubic lattice in d dimensions, and $\sum_{\langle i,j \rangle}$ denotes summation over all nearest-neighbor pairs of sites. Thermal averages have the form

$$M = \frac{\mathrm{Tr} \sum_i S_i e^{-H/T}}{\mathrm{Tr} e^{-H/T}}$$

where M is the magnetization, i.e., the average spin, and Tr denotes summation over all states, i.e., sum over all possible combinations of values of spins, ($S_1 = \pm 1$, $S_2 = \pm 1$..). (K. P. N. Murthy 2004)

Monte Carlo simulation is easily implemented by the following algorithm:

1. Start with all spins up: Si = 1 for all i.
2. Select a spin Si at random. The trial move is to flip this spin $S_i \rightarrow -S_i$. Compute w by e.g., w = exp(−ΔH/T) where $-\Delta H = J \Delta S_i \sum_j S_j$, where j are the nearest neighbors to i. If w > r, where r ∈ [0, 1] is a random number, then accept the new state, otherwise, keep the old state. Advance MC" time" to t + Δt. If the trial move was rejected, let the old state be the new state at t + Δt.
3. A "suitable" number of initial states must be skipped in order to approach the equilibrium distribution before averages can be formed. Measure variables of interest on the generated states and add to averages at "suitable" MC time intervals. Continue from step 2 until sufficient information is gathered.

1.4.3. How to Get it to Work

Periodic boundary conditions, such as x+L → x, are frequently used to simulate bulk properties in large systems with a macroscopic number of degrees of freedom and a linear dimension of L. Thus, surface effects are eliminated. A Monte Carlo sweep through the system is a series of Monte Carlo steps where each spin is, on average, attempted to be updated once. One Monte Carlo sweep (Δt = 1) is the unit of time in a Monte Carlo simulation. Later on, more on the idea of Monte Carlo time. While it might seem more natural, choosing spins in a sequential manner as opposed to at random is not natural. Each move is exactly equivalent when using random selection; this is not the case when using sequential updates. There is a temperature limit for an Ising model at which all moves are accepted, at which point sequential selection completely breaks down. A sequential sweep just flips the configuration, starting from a uniform initial state. The three remarks that follow is about downtime.

Before data collection begins, discard t_{eq} approximately 10^3 sweeps to allow the system to approach equilibrium. Should this not be the case, systematic errors from the initial nonequilibrium states will introduce bias into the computed averages. For the averages, skip a certain number of sweeps (approximately 2) in between data measurements. The reason is that subsequent configurations don't provide independent information because they are correlated. More sweeps can be taken in between if data collection is costly (takes up a lot of computer time); if data collection is inexpensive, data can be taken after each sweep. Skipping more than one relaxation MC time is pointless as it would result in the discarding of uncorrelated data. Gather information over t_{sample} approximately 10^5 sweeps. At times, less is sufficient, and at other times, much more is required. A single run shouldn't take very long because we're impatient, the computer might crash, the results might not look good, and the program might need to be changed, etc. Therefore, it is not advisable to take a lot of steps.

- Repeat the simulation N_{runs} times, using different random numbers. This ensures Nruns independent samples, that is, the averages from the independent runs. The error estimate is then

$$\Delta = \pm \frac{\sigma}{\sqrt{N_{\text{runs}}}}$$

where σ is the variance of the averages. R_{erun} until Δ becomes small enough! Usually $N_{runs} \sim 10 - 1000$.

Use lookup tables for time consuming calculations that are performed in the MC loop. For example, define a table $w(\Delta H) = \exp(-\Delta H/T), \Delta H = -J\Delta s_i \sum_j s_j = 0, \pm 4J, \ldots$. For models with continuous spin (XY, Heisenberg), discretize the spin in e.g., ~ 1000 steps, and the same method works.

It is crucial to understand that in the simulation, rejected MC moves are also counted as new configurations. It is incorrect to keep attempting different moves until something is accepted; additionally, the previous configuration adds to the averages when a trial move is turned down. Physically speaking, it should come as no surprise that very few movements are permitted at low temperatures, which translates to few deviations from the ground state. The majority of movements should be tolerated at high temperatures because of the rapid fluctuations. For the first t_{eq} sweeps, no measurements are made, so how are averages like the average energy per spin actually computed? The number of accepted and rejected moves has significant physical meaning. Regarding the subsequent t_{sample} sweeps, the energy is measured, let's say every two sweeps, and the average $E_{sum} \to E_{sum} + E(t)$ is multiplied by the current energy value, E(t). Remember to set the sum to zero (E_{sum}). It is not really necessary to compute the current energy E(t); instead, it should be noted by adding ΔH to E each time a move is approved.

The desired expectation values including the final normalization become:

$$\text{energy density: } e = \frac{E_{\text{sum}}}{N_{\text{samples}} N_{\text{spins}}}$$

$$\text{specific heat: } c = \frac{E_{\text{sum}}^2/N_{\text{samples}} - (E_{\text{sum}}/N_{\text{samples}})^2}{T^2 N_{\text{spins}}} \quad (6)$$

$$\text{magnetization: } m = \frac{M_{\text{sum}}}{N_{\text{samples}} N_{\text{spins}}}$$

$$\text{susceptibility: } \chi = \frac{M_{\text{sum}}^2/N_{\text{samples}} - (M_{\text{sum}}/N_{\text{samples}})^2}{T N_{\text{spins}}}$$

1.4.4. Random Numbers

Consider writing an application that outputs a random value. In fact, this is an unresolved issue! The bad news is that a large number of random numbers are needed by the MC program in order to determine which spin to update and to confirm acceptance. To calculate approximate random numbers, there are various algorithms available, resulting in a sequence that looks like this: r_1, r_2, r_3, etc., through a formula. These numbers are known as pseudo-random numbers because they are not truly random; if you know the formula and the first number, you also know all subsequent generated numbers.

How good are pseudo-random numbers? To mimic "real" random numbers, the pseudo-random numbers should be independent and perfectly uniformly distributed. This turns out to be very difficult to accomplish and is an active area of research! Pseudo-random numbers have two types of defects:

1. Correlations between x_i and x_{i+j}.
2. Finite period: $x_i = x_{i+\text{period}}$.

It appears that the quality of random numbers affects MC. Unbiased random numbers are necessary for the process to converge to the proper equilibrium distribution; otherwise, a biased distribution with biased averages is produced. Simulations with high precision may be able to detect this. Routines for updating collectively are particularly delicate. Recent studies have used the exact solution for finite systems and the MC data for the 2D Ising model to test the quality of random number generators.

1.4.5. Correlation Functions

Usually in a simulation, it is desirable to calculate correlation functions. Spatial and temporal correlations give important information about ordering, the approach to equilibrium, and relaxation of excitations.

1.4.5.1. Spatial Correlations

The spin correlation function is defined by

$$G(r) = \langle S(r)S(0)\rangle$$

The correlation function measures the ordering properties of the system. In an ordered phase the correlation function approaches a constant at large distance:

$$G(r) = \langle S(r)S(0)\rangle \to m^2, \quad r \to \infty$$

where m = $\langle S\rangle$.

In the disordered phase, at temperatures above T_c, relaxation is usually approximately exponential:

$$G(r) \sim e^{-r/\xi}$$

which defines the correlation length $\xi(T)$. At the transition temperature T_c, the correlation length diverges, and the correlation function typically becomes a power law:

$$G(r) \sim \frac{1}{r^{d-2-\eta}}$$

Calculating the correlation function in both the x and y directions and averaging the results may be sufficient in this simulation. The exponents, and so forth, can be calculated in appropriate logarithmic plots by finding the slope through fitting a straight line. The correlation function holds significant information about the system's characteristics, which have an impact on the simulation's convergence properties. The correlation length is small away from a phase transition. Below T^c, in the Ising model, there is little correlation between fluctuations coming from the ordered state, and an efficient averaging over these fluctuations can be achieved in a restricted number of sweeps. Likewise, the spins above T_c exhibit fluctuations and have short correlation lengths, meaning that each sample essentially averages over a large number of fluctuations. Near $T = T_c$, where the correlations become large, this is not the case. The convergence slows down as a result of each sample providing a limited average over various fluctuations.

The specific heat, Eq. (6), measures fluctuations in the energy, and the order parameter susceptibility, Eq. (7), measures fluctuations in the order parameter. These can be directly measured in the simulation and typically diverge at T_c.

1.4.5.2. Temporal Correlations

The correlation time τ (T) is a very useful quantity to calculate in a MC simulation. It is defined via the autocorrelation function

$$C_{AA}(t) = \langle A(t)A(0)\rangle - \langle A\rangle^2$$

The correlation time τ is the time it takes for the autocorrelation function to drop to a fraction of its value $C_{AA}(0)$ at $t = 0$. A convenient definition of τ is given by the integral of the normalized autocorrelation function,

$$\tau = 1 + 2\sum_{t=1}^{\infty} \frac{C_{AA}(t)}{C_{AA}(0)} \tag{8}$$

The sum can in practice be terminated at the first negative value: $C_{AA}(t) \leq 0 \Rightarrow t_{max} = t$.

1.4.5.3. Statistical Error of Correlated Samples

We can now estimate the expected statistical error in the simulation using correlated Markov chain sampling. The number of independent samples obtained in the simulation is approximately N/τ, where N is the total number of samples and τ is the relaxation time. The statistical error of the average can be estimated by

$$\sigma \approx \sqrt{\frac{C_{AA}(0)\tau}{N}}$$

(note that $C_{AA}(0) = \langle A^2\rangle - \langle A\rangle^2$ is the variance of each term in the average).

This relation is instructive to derive:

$$\begin{aligned}\sigma^2 &= \left\langle \left(\frac{1}{N}\sum_{t=1}^{N}(A(t) - \langle A\rangle)\right)^2 \right\rangle \\ &= \frac{1}{N^2}\sum_{t=1}^{N}\underbrace{\left\langle (A(t) - \langle A\rangle)^2\right\rangle}_{=C_{AA}(0)} + \frac{2}{N^2}\sum_{t'=1}^{N}\sum_{t=1+t'}^{N}\langle (A(t') - \langle A\rangle)(A(t) - \langle A\rangle)\rangle\end{aligned}$$

Now let us approximate by changing the upper limit of the last sum from N to N + t,' which is a good approximation if $N \gg \tau$. Using τ from Eq. (8) in this expression leads to

$$\sigma^2 = \frac{C_{AA}(0)}{N} + \frac{2}{N^2}\sum_{t'=1}^{N}\sum_{t=1+t'}^{N+t'}\underbrace{\langle (A(t') - \langle A\rangle)(A(t) - \langle A\rangle)\rangle}_{C_{AA}(t-t')} = C_{AA}(0)\tau/N$$

1.4.5.4. Calculation of the Equilibration Time

The equilibration time (t_{eq}), or the amount of time it takes to approach equilibrium from a non-equilibrium initial state, is related to the relaxation time. The process that was previously outlined for the relaxation time can be used to obtain t_{eq}, where t = 0 represents the simulation's beginning time in this instance. This is the method used to determine how many sweeps in the simulation need to be initially discarded before data collection begins.

1.4.5.5. Calculation of the Relaxation Time

There are two important points to note when the correlation time is computed:

1. The term $\langle A\rangle^2$ is a potential source of systematic error. Suppose that we get in the simulation $\langle A\rangle = X \pm \Delta a$. Squaring gives $\langle A\rangle^2 = X^2 + 2X\Delta X + (\Delta X)^2$. The last term is not uniformly distributed around zero, and gives a bias if it is evaluated in this way. A fix-up is to evaluate $\langle f\rangle$ independently twice and set $\langle A\rangle^2 = X_1X_2 + X_1\Delta X_2 + X_2\Delta X_1 + \Delta X_1\Delta X_2$, which is unbiased. If this point is not noticed, the procedure of repeating the simulation many times and averaging over the results will have systematic errors.
2. The calculation of the sum of the autocorrelation function over time is best organized as follows. Rewrite the sum involved in calculating $\sum_t\langle A(t)A(0)\rangle$ in Eq. (8) in the following way:

$$\sum_t \langle A(t)A(0)\rangle = \sum_{t=0}^{t_{\max}} \frac{1}{N} \sum_{t'=1}^{N} A(t+t')A(t')$$

Change summation order and define $F(t') = \sum_{t=0}^{t_{\max}} A(t+t')$, which gives

$$\sum_t \langle A(t)A(0)\rangle = \frac{1}{N} \sum_{t'=1}^{N} F(t')A(t')$$

The advantage is that this contains no double time summation, but instead only involves updating the sum $F(t') = \sum_{t=0}^{t_{\max}} A(t+t')$ at every time step, which requires storing all the terms A(t + t'),t = 0..,t_{max}. For this calculation to be reliable, $t_{\max} \gg \tau$ must be fulfilled.

ROLE MODEL

ARIANNA WRIGHT ROSENBLUTH

An American physicist who contributed to the development of the Metropolis–Hasting's algorithm. She wrote the first full implementation of the Markov chain Monte Carlo method.

Biography

Arianna Rosenbluth was born in Houston, Texas, on September 15, 1927. She attended university at the Rice Institute, now Rice University, where she received a Bachelor of Science in 1946. During her college days, she fenced competitively and won both the Texas women's championship in foil as well as the Houston men's championship. She qualified for the Olympics, but was unable to compete because the 1944 Summer Olympics were canceled due to World War II and she could not afford to travel to the 1948 games in London.

Rosenbluth obtained her Master of Arts from Radcliffe College in 1947 before beginning her PhD in physics at Harvard University under the supervision of Nobel Laureate John Hasbrouck Van Vleck. At the time, Van Vleck also supervised the future Nobel Laureate P.W. Anderson and the philosopher of science Thomas Kuhn. She completed her thesis, entitled Some Aspects of Paramagnetic Relaxation, in 1949 at the age of 22.

Career

After completing her thesis, Rosenbluth won an Atomic Energy Commission postdoctoral fellowship to Stanford University, which she attended before moving to a staff position at Los Alamos National Laboratory where her research focused on atomic bomb development and statistical mechanics.

Along with Marshall Rosenbluth, she verified analytic calculations for the Ivy Mike test using the SEAC at the National Bureau of Standards. Once the MANIAC I had been completed at Los Alamos, she collaborated with Nicholas Metropolis, Marshall N. Rosenbluth, Augusta H. Teller, and Edward Teller to develop the first Markov chain Monte Carlo algorithm, in particular the prototypical Metropolis–Hasting's algorithm, in the seminal paper "Equation of State Calculations by Fast Computing Machines." In close collaboration with her husband Marshall, she developed the implementation of the algorithm for the MANIAC I hardware, making her the first person to ever implement the Markov chain Monte Carlo method.

Over the next few years, Rosenbluth and Marshall applied the method to novel studies of statistical mechanical systems, including three-dimensional hard spheres and two-dimensional Lennard-Jones molecules and two and three-dimensional molecular chains. After the birth of her first child, Rosenbluth left research to focus on raising her family.

Personal life

While at Stanford University, she met Marshall Rosenbluth and the two married on January 26, 1951. They had four children before divorcing in 1978. In 1956, she moved from Los Alamos to San Diego, California, and then Princeton, New Jersey, before finally settling in the greater Los Angeles area. She kept her married name after the divorce.

Rosenbluth died from complications of COVID-19 in the greater Los Angeles area on December 28, 2020, during the COVID-19 pandemic in California. She was 93.

Researchers had also begun using digital computers to bolster weapons development. Los Alamos hosted one such machine, the Mathematical Analyzer Numerical Integrator and Automatic Computer, or MANIAC. The MANIAC consisted of 1,024 vacuum tubes, and its memory consisted of a thousand 40-bit words. Beyond weapons development, Los Alamos scientists were simply curious how to make their new toy sing.

Arianna and Marshall proposed using the MANIAC to study how solids melt. They framed the problem as a collection of up to 224 rigid, two-dimensional disks, representing simplified molecules, in contact with a heat bath at a fixed temperature. The computer would predict the disks' equilibrium thermodynamic properties such as pressure and density.

They considered simulating the motion of each disk individually, but it was too computationally expensive. Edward Teller recommended they instead use statistical mechanics to calculate the disks' average values. To do this, Arianna and Marshall would generate random configurations of disks allowed for a given energy and temperature. (The element of randomness reminded Los Alamos physicists of a game of chance—hence "Monte Carlo," after the casino.) With enough configurations, they could estimate an average for the thermodynamic properties.

But some molecular arrangements are more probable than others. Their key innovation: Instead of generating completely random configurations, their algorithm forced the computer to sample configurations weighted by their probability.

The two ran the MANIAC during the midnight shift. They had the rare authority to call engineers in the middle of the night to reboot the computer if it crashed, Arianna told Gubernatis in 2003. Arianna did all the programming, as she and Marshall both recounted later. She had learned to program the MANIAC when she verified calculations for the first full-scale test of a hydrogen bomb in 1952.

Interacting with the computer required a detailed understanding of both the machine and the physics. Arianna meticulously coded in assembly language, just one level of abstraction above machine language. She would have, for example, needed to track the physical location of numbers stored in memory to use in calculations.

The Metropolis algorithm was one of the first examples of a "numerical experiment," said Adam Iaizzi, a physicist who dedicated his PhD dissertation to Arianna Rosenbluth. They had devised a new way to use computers beyond simply performing accelerated calculations that humans could already do.

The paper also furthered the understanding of solid-liquid phase transitions, says Gubernatis, a physicist now retired from Los Alamos. The simulation provided early evidence that molecules in a liquid exhibit some structure rather than being entirely disordered, as previously thought.

Arianna published a few more papers about the Metropolis algorithm, but she left physics to raise her four children in support of Marshall's career. Los Alamos was her last professional experience. Their marriage ended in divorce in 1978.

Over the subsequent decades, Arianna lost touch with her former collaborators. "She was simply surprised when I told her how famous this particular paper became," recalled Gubernatis of a 2003 phone call. Arianna never expressed regret about leaving her career, her daughter Jean said. But "she was not the happiest person while we were growing up," said Jean. "I think part of it was that she missed her work because it meant a lot to her." Alan once asked her if she had ever experienced any gender discrimination in physics. "She said that she didn't always feel comfortable being the recipient of so much attention from all the men," he said. Felix Bloch also declined to take her as a graduate student at Harvard because he categorically didn't accept female students. "She shrugged it off matter-of-factly, although I think it annoyed her," he said.

An apocryphal tale about the algorithm's origins may illustrate the era's sexism best. The rumor, passed down among physicists over the years, said that Metropolis, Marshall Rosenbluth, and Edward Teller had devised the algorithm at a cocktail party and that they added their wives' names to the publication as a thank-you for enduring the technical conversation. There is no evidence this is the case.

CLASS ACTIVITY

EXPLORING PHASE TRANSITIONS THROUGH SIMULATION

Objective: The objective of this activity is to introduce students to the concept of phase transitions in statistical physics and to allow them to explore these transitions through computer simulations.

Materials Needed

- Computers with simulation software installed (e.g., MATLAB, Python with NumPy)
- Projector or screen for demonstration
- Whiteboard and markers
- Handouts with instructions and guidelines

Assessment

- Evaluate students based on their participation in the group discussions, the quality of their observations and analyzes, and their understanding of the theoretical concepts demonstrated in the activity.
- Optionally, assign a follow-up assignment where students write a brief report summarising their findings and reflections on the activity.

SUMMARY

- Statistical physics aims to study the macroscopic parameters of a system in equilibrium based on the knowledge of microscopic properties using the laws of mechanics. It is the branch of physics where a method to calculate free energy is formulated.
- Statistical physics can analyze both thermal equilibrium states and non-equilibrium states. This approach differs from thermodynamics, which examines the macroscopic system in equilibrium from a macroscopic perspective without considering the microscopic parameters.
- Statistical mechanics is a branch of physics that combines the principles and techniques of statistics with the laws of both classical and quantum mechanics, especially in the field of thermodynamics.
- Statistical physics seeks to elucidate how the patterns and structures in the macroscopic world around us emerge from the interactions among their constituent parts. A significant challenge in the 21st century is to expand statistical physics to systems that are significantly far from equilibrium.
- Statistical mechanics demonstrates how concepts derived from macroscopic observations (such as temperature and pressure) are linked to the description of a fluctuating microscopic state around an average state.
- A microcanonical ensemble is the statistical ensemble that is used to represent the possible states of a mechanical system which has an exactly specified total energy.
- The microcanonical ensemble is sometimes considered to be the fundamental distribution of statistical thermodynamics, as its form can be justified on elementary grounds such as the principle of indifference: the microcanonical ensemble describes the possible states of an isolated mechanical system when the energy is known exactly, but without any more information about the internal state.
- In statistical mechanics, the inherent probabilities of the quantum world are joined with probabilities that arise from our ignorance of the underlying state.
- Monte Carlo in statistical physics refers to the application of the Monte Carlo method to problems in statistical physics, or statistical mechanics. The general motivation to use the Monte Carlo method in statistical physics is to evaluate a multivariable integral.
- A Markov process is a sequence of states x_1, x_2, x_3.. such that the probability distribution for xi_{+1} is specified from the previous state xi by a transition probability w $(x_i \rightarrow x_{i+1})$. It will be very useful to think of the index i as a time step t, so the Markov process will represent a dynamic process in phase space.

REVIEW QUESTIONS

1. What are the applications of statistical physics?
2. Why is there a need for statistical description in physics?
3. What is the microcanonical ensemble?
4. How are stones, Markov process sampling, and π related?
5. What are random numbers and how are they used in statistical physics?

REFERENCES

1. Anderson, B. D. O. (1969). The inverse problem of stationary covariance generation. *Journal of Statistical Physics, 1*(1), 133–147. https://doi.org/10.1007/BF01007246.
2. Batterman, R. W. (1998). Why equilibrium statistical mechanics works: Universality and the renormalization group. *Philosophy of Science, 65*(2), 183–208. https://doi.org/10.1086/392634.
3. Bennett, C. H. (2003). Notes on Landauer's principle, reversible computation, and Maxwell's demon. *Studies in History and Philosophy of Science Part B: Studies in History and Philosophy of Modern Physics, 34*(3), 501–510. https://doi.org/10.1016/S1355-2198(03)00039-X.
4. Bricmont, J., Ghirardi, G., Dürr, D., Petruccione, F., Galavotti, M. C., & Zanghì, N. (Eds.). (2001). *Chance in Physics: Foundations and Perspectives* (Lecture Notes in Physics 574). Springer Berlin Heidelberg. https://doi.org/10.1007/3-540-44966-3.
5. Hoffman, D. K., & Dahler, J. S. (1969). The Boltzmann equation for a polyatomic gas. *Journal of Statistical Physics, 1*(4), 521–558. https://doi.org/10.1007/BF01024129.
6. Landau, D. P., & Binder, K. (2005). *A guide to Monte Carlo Simulations in Statistical Physics*. Cambridge University Press.
7. Murthy, K. P. N. (2004). *Monte Carlo Methods in Statistical Physics*. University Press.
8. Newman, M. E. J., & Barkema, G. T. (2001). *Monte Carlo Methods in Statistical Physics*. Clarendon Press.

CHAPTER 2

Semiconductors

LEARNING OBJECTIVES

After studying this chapter, you will be able to:

- Give an overview of semiconductors
- Explain the types of semiconductors
- Discuss the mechanical properties of semiconductor

KEY TERMS FROM THIS CHAPTER

Alternating current
Diodes
Electrical conductivity
Insulator
Resistivity
Artificial intelligence
Direct current
Electronic devices
Integrated circuits

2.1. INTRODUCTION

Semiconductors are the essential components of modern electronic devices driving the digital era we currently reside in. composed of materials such as silicon, germanium, and gallium arsenide, these substances possess conductivity levels between those of conductors and insulators, giving rise to their name "semiconductors." An important trait of semiconductors is their capability to modify their conductivity when exposed to external factors like temperature, voltage, or light. This particular feature serves as the foundation for a multitude of electronic gadgets, including transistors, diodes, and integrated circuits (ICs).

Transistors are tiny switches that can be used to control or increase the flow of electrical current in a circuit. They are arguably the most important application of semiconductors. They function as the fundamental components of electronic devices, making it possible to design intricate logic gates and circuits. The digital era began with the miniaturization of electronic devices, which was made possible by the invention of the transistor.

Another essential semiconductor device, diodes, only permit one direction of current flow. Because of this characteristic, they are crucial for photovoltaic cells' conversion of light energy into electrical energy and power supplies' rectifying of alternating current (AC) into direct current (DC). Integrated circuits (ICs) are full electronic circuits made from a single semiconductor material. They are sometimes referred to as microchips or chips.

These tiny marvels allow the construction of powerful yet small electronic devices like computers, digital cameras, and smartphones by packing millions or even billions of transistors, diodes, and other components onto a single chip. Over the years, the semiconductor industry has experienced exponential growth due to ongoing advancements in manufacturing processes, materials science, and device design. Gordon Moore, a co-founder of Intel, developed Moore's Law, which states that as a microchip's transistor count doubles roughly every two years, computing power and efficiency will also rise.

Furthermore, the constant drive towards creating smaller, faster, and more energy-efficient semiconductor devices has sparked creativity in industries such as telecommunications, healthcare, automotive, and aerospace. Cutting-edge technologies like artificial intelligence (AI), Internet of Things (IoT), and 5G networks heavily rely on semiconductor devices to power their functions and pave the way for advancements in the digital realm.

The search for new materials to maintain Moore's Law, addressing environmental issues related to semiconductor manufacturing processes, and maintaining supply chain resilience in the face of geopolitical tensions are just a few of the many difficulties facing the semiconductor industry. To overcome these obstacles and maintain the pace of technological innovation and the continuous spread of semiconductor-enabled solutions in the digital era, industry players, legislators, and researchers must work together.

2.2. OVERVIEW OF SEMICONDUCTORS

Any crystalline solid that falls between an insulator and a conductor in terms of electrical conductivity is referred to as a semiconductor. Semiconductors are used in the production of integrated circuits, diodes, and transistors among other types of electronic devices. These devices' affordability, portability, dependability, and power efficiency have led to their widespread use. They have been employed as discrete parts in solid-state lasers, optical sensors, power devices, and light emitters. They can handle a wide range of current and voltage, and more importantly, they can be easily integrated into intricate yet easily manufactured microelectronic circuits. For the foreseeable future, they will continue to be the essential components of most electronic systems that support computing, communications, signal processing, and control applications in the consumer and industrial markets.

Semiconductors are materials with conductivity properties that fall between conductors like metals and insulators like rubber or glass. They are essential components in electronic devices like transistors, diodes, integrated circuits, and solar cells. One unique trait of semiconductors is their ability to have their conductivity easily altered by adding impurities, a process known as doping. By doping, semiconductors can have an excess of negatively charged electrons or positively charged "holes" created within the material. These excess charges can be used to build various electronic components. Silicon is the most commonly used semiconductor material due to its abundance and ease of manufacturing, et al., like germanium, gallium arsenide, and indium phosphide. Semiconductors have transformed the electronics industry by enabling the development of smaller and more efficient devices, with applications ranging from small-scale components to large-scale systems like solar panels and microprocessors.

2.2.1. History of Semiconductor

The exploration of semiconductors began with experiments on the electrical properties of materials in the 19th century. Thomas Johann Seebeck first noticed the effects of semiconductors in 1821. Michael Faraday discovered that the resistance of silver sulfide decreased when heated in 1833. Alexandre Edmond Becquerel observed the photovoltaic effect in 1839. Willoughby Smith noticed decreasing resistance in selenium resistors when exposed to light in 1873. Karl Ferdinand Braun observed conduction and rectification in metallic sulfides in 1874, a phenomenon previously described in 1835 by Peter Munck af Rosenschold and Arthur Schuster. William Grylls Adams and Richard Evans Day observed the photovoltaic effect in selenium in 1876.

During the first half of the 20th century, significant progress was made in the field of solid-state physics, leading to a unified explanation of various phenomena. Edwin Herbert Hall's demonstration of the Hall effect in 1878, where charge carriers were deflected by a magnetic field, sparked advancements in understanding electron-based conduction in solids following J.J. Thomson's discovery of the electron in 1897.

Karl Baedeker's observation of a reversed Hall effect in copper iodide led to the identification of positive charge carriers in the material. In 1914, Johan Koenigsberger categorized solid materials into metals, insulators, and variable conductors, laying the groundwork for further developments. Josef Weiss introduced the term "Halbleiter" (semiconductor) in his 1910 Ph.D. thesis, while Felix Bloch proposed a theory on electron movement through atomic lattices in 1928.

By the 1930s, theories of conductivity in semiconductors due to impurities and the band theory of conduction were well-established thanks to the work of researchers like B. Gudden, Alan Herries Wilson, and others. Walter H. Schottky and Nevill Francis Mott advanced understanding of metal-semiconductor junctions, while Boris Davydov's theory on the copper-oxide rectifier in 1938 highlighted the role of p-n junctions, minority carriers, and surface states in semiconductors.

There was occasionally a lack of agreement between theoretical predictions (based on evolving quantum mechanics) and experimental findings. John Bardeen subsequently explained this by pointing out that semiconductors exhibit extremely structure-sensitive behavior, meaning that even minute amounts of impurities can cause dramatic changes in their properties. Different experimental results were obtained with commercially pure materials from the 1920s that contained different amounts of trace contaminants. This sparked the advancement of better material refining methods, which led to the production of materials with parts-per-trillion purity in contemporary semiconductor refineries.

Before semiconductor theory offered a roadmap for building increasingly capable and dependable devices, semiconductor-using devices were initially built using empirical knowledge. In 1880, Alexander Graham Bell discovered a way to transmit sound through a light beam using selenium's light-sensitive characteristic. Charles Fritts created a low-efficiency working solar cell in 1883 by coating a metal plate with selenium and a thin layer of gold. In the 1930s, the device was used commercially in photographic light meters. Jagadish Chandra Bose used point-contact microwave detector rectifiers made of lead sulfide in 1904; the cat's-whisker detector, which uses natural galena or other materials, became a common device in the radio development industry. But in use, it was a little erratic and needed to be adjusted by hand for optimal functionality. In 1906, H. J. Round observed light emission when electric current flowed through silicon carbide crystals, demonstrating the working principle of the light-emitting diode. Similar light emission was noticed by Oleg Losev in 1922, but at that time the effect was not useful. As an alternative to vacuum tube rectifiers, power rectifiers—which use copper oxide and selenium—were developed in the 1920s and gained commercial significance.

Galena was used in the first semiconductor devices, such as the crystal detector created in 1874 by German physicist Ferdinand Braun and the radio crystal detector created in 1901 by Bengali physicist Jagadish Chandra Bose. Prior to World War II, lead-sulfide and lead-selenide material research was spurred by infrared detection and communications devices. These gadgets served as voice communication systems, infrared rangefinders, and ship and aircraft detection. Since existing vacuum tube devices could not function as detectors above roughly 4000 MHz, the point-contact crystal detector became essential for microwave radio systems; advanced radar systems relied on the quick response of crystal detectors. During the war, a great deal of research and development was done on silicon materials to create detectors that were consistently of high quality.

2.2.1.1. Early Transistors

The signal could not be amplified by a detector or power rectifier. Many attempts were made to create a solid-state amplifier, and they were successful in creating the point contact transistor, a device that had an amplified output of at least 20 dB. Oleg Losev created successful two-terminal negative resistance radio amplifiers in 1922, but he was killed in the Leningrad Siege. Although it was impractical, Julius Edgar Lilienfeld patented a device in 1926 that looked similar to a field-effect transistor. R. W and R. Hilsch. Pohl used a structure that resembled a vacuum tube's control grid to demonstrate a solid-state amplifier in 1938. While the device showed power gain, its cut-off frequency was only one cycle per second, which was too low for any real-world uses but a useful application of the theory at the time. A. Bell and William Shockley worked at Bell Labs. In 1938, Holden began researching solid-state amplifiers. Russell Ohl discovered the first p–n junction in silicon in or around 1941 when he noticed a specimen that was sensitive to light and had a distinct border separating p–type impurity at one end from n–type impurity at the other. When exposed to light, a slice taken from the specimen at the p–n boundary developed a voltage.

The first working transistor was created by John Bardeen, Walter Houser Brattain, and William Shockley at Bell Labs in 1947. Shockley had initially planned to build a field-effect amplifier using germanium and silicon, but instead invented the point-contact transistor using germanium. In France during the war, Herbert Mataré had also observed amplification between adjacent point contacts on a germanium base. After the war, Mataré's group introduced their Transistron amplifier shortly after Bell Labs announced the transistor. In 1954, Morris Tanenbaum produced the first silicon junction transistor at Bell Labs. However, these early junction transistors were large and difficult to mass produce, limiting their applications to specialized uses.

2.2.2. Properties of Semiconductors

Certain electrical characteristics are present in semiconductors. An insulator is a material that does not conduct electricity, whereas a conductor is a substance that conducts electricity. Substances with properties in between are known as semiconductors. Resistivity is one way to indicate electrical properties. Conductors with low resistance and easy electrical conductivity include copper, silver, and gold. Rubber, glass, and ceramics are examples of insulators that have a high resistance to electrical current flow. Semiconductors' characteristics fall in between these two ranges. For instance, their resistivity may vary with temperature. Almost no electricity flows through them at low temperatures. However, when it gets hotter, electricity can flow through them with ease.

Nearly pure semiconductors don't conduct any electricity at all. However, when certain components are incorporated into semiconductors, electricity flows through them with ease. Elements that make up a semiconductor are referred to as elemental semiconductors, and silicon is a well-known example of one. Contrarily, semiconductors composed of two or more compounds are referred to as compound semiconductors and are utilized in light-emitting diodes, semiconductor lasers, and other applications.

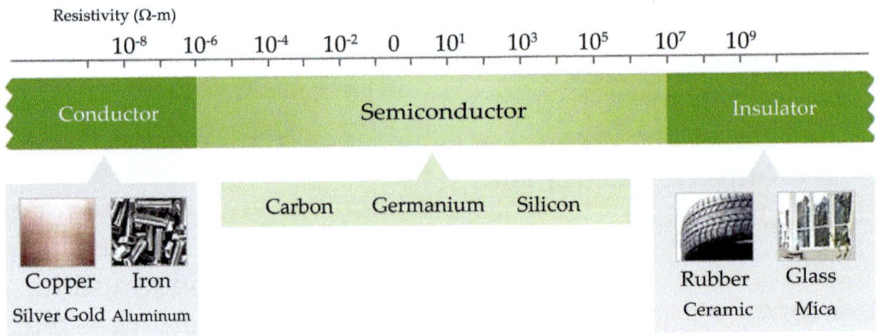

Figure 2.1. Properties of Semiconductors.

Source: https://www.hitachi-hightech.com/global/en/media/about_il01_gif_tcm27–70459.gif

2.2.2.1. Energy Band

"An atom consists of a nucleus, composed of protons and neutrons, and electrons that orbit the nucleus." or "A nucleus, composed of protons and neutrons, and the electrons that orbit it make up an atom." Only a limited number of extremely specific orbits are permitted, and they can only exist in certain discrete levels within the atomic space surrounding the nucleus. Electrons are not permitted to orbit the nucleus at any distance. In a solid, when many atoms come together to form a crystal, their individual energy levels overlap and form continuous bands of energy. This is the band of energy.

Metals, semiconductors, and insulators are distinguished from each other by their band structures. Their band structures are shown in the figure below.Figure 2.2

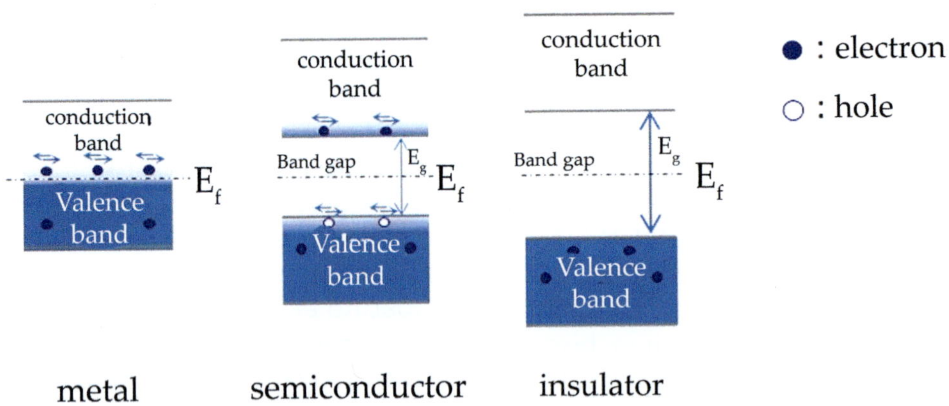

Figure 2.2. Band structures.

Source: https://www.hitachi-hightech.com/global/en/media/properties_01_png_tcm27–70473.png

The Fermi energy in metals is typically within the conduction band or very close to it, not inside the conduction band. This indicates that there are always free-moving electrons in the metal, which enables it to conduct current. We refer to these electrons as free electrons and it is due to them that current passes through a metal. The Fermi energy, or (Ef), is located between the valence and conduction bands in semiconductors and insulators, where they are divided by a forbidden energy gap, or (Eg), of appropriate width. The electron must acquire sufficient energy to cross the band gap and enter the conduction band. After that, it is ready to operate.

Semiconductors at room temperature have a smaller band gap, allowing electrons to easily jump the gap and move to the conduction band. This limited conductivity results in semiconductors conducting less current than metals. At low temperatures, electrons lack the energy to occupy the conduction band, causing an inability for charge movement. At absolute zero, semiconductors act like insulators because there is insufficient thermal energy to excite electrons across the band gap. While the density of electrons in the conduction band is lower than in metals, semiconductors have higher conductivity than insulators, earning them the name "semiconductor" as they conduct halfway between a conductor and insulator.

Since insulators have large band gaps, few electrons are able to cross them. As a result, current flows more slowly through insulators. The band gap energy is what distinguishes semiconductors from insulators. In an insulator with a very large forbidden gap, the electron practically needs a sufficient amount of energy to cross over to the conduction band. Insulators are difficult to conduct electricity through. This indicates that the insulator has very low electrical conductivity.

Semiconductors used in IC are typically high-purity single-crystal silicon, with a purity of 99.999999999%. But when actually making a circuit, impurities are added to control the electrical properties. Depending on the added impurities, they become n-type and p-type semiconductors.

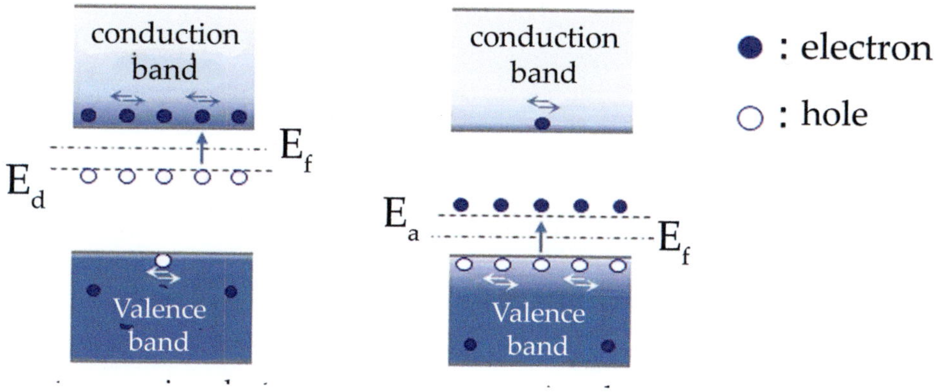

Figure 2.3. Semiconductor crystal.

Source: https://www.hitachi-hightech.com/global/en/media/properties_02_png_tcm27–70474.png

High-purity silicon is combined with pentavalent phosphorus (P) or arsenic (As) to create n-type semiconductors. We refer to these contaminants as donors. The donor's energy level is in close proximity to the conduction band, meaning that the energy gap is minimal. Subsequently, electrons at this energy level contribute to conductivity by being readily excited to the conduction band.

However, trivalent boron (B) and so forth are added to semiconductors of the p-type. We refer to this as an acceptor. The acceptor's energy level is in proximity to the valence band. Here, electrons in the valence band are excited because there are no electrons present. Consequently, the valence band develops holes, which enhances conductivity.

2.3. TYPES OF SEMICONDUCTORS

Semiconductor may be classified as under:

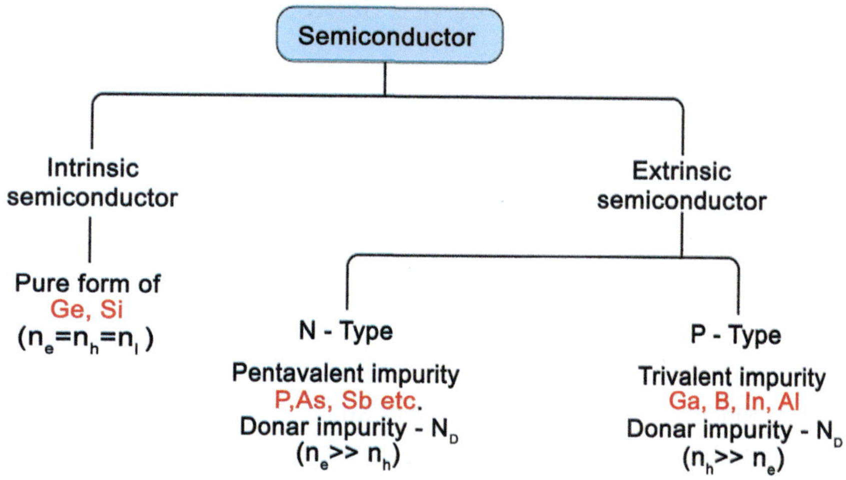

Figure 2.4. Types of Semiconductors.

Source: https://res.cloudinary.com/rs-designspark-live/image/upload/c_limit,w_750/f_auto/v1/article/semi2_585c022dcc2dd3bcd0ccacd27d4dd2146c5a9723

2.3.1. Intrinsic Semiconductors

A semiconductor that is composed of the semiconductor material in its purest form is known as an intrinsic semiconductor. Pure germanium and silicon are two examples of these semiconductors, with forbidden energy gaps of 0.72 eV and 1.1 eV, respectively. There are many electrons with enough energy to cross the tiny energy gap between the valence and conduction bands even at room temperature because the energy gap is so small.

Alternatively, an intrinsic semiconductor may be defined as one in which the number of conduction electrons is equal to the numtber of holes. A schematic energy band diagram of an intrinsic semiconductor at room temperature is shown in Figure 2.5 (Table 2.1).

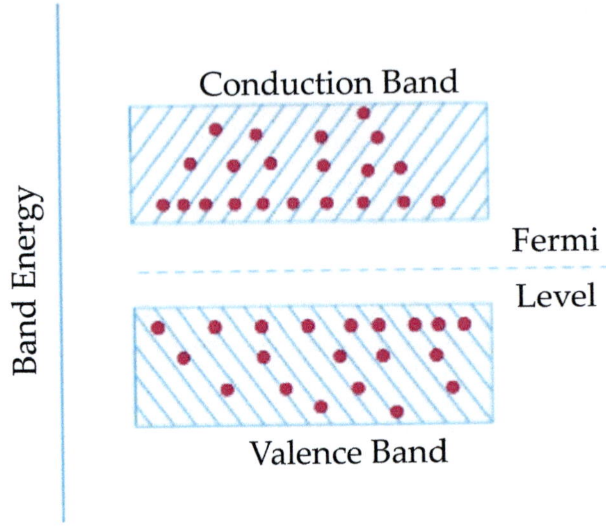

Figure 2.5. Intrinsic semiconductor.

Source: https://uotechnology.edu.iq/dep-laserandoptoelec-eng/branch/lectures/electronic1/%D8%A7%D9%84%D9%85%D8%AD%D8%A7%D8%B6%D8%B1%D8%A9%20%D8%A7%D9%84%D8%A7%D9%88%D9%84%D9%89.pdf

Table 2.1. Working Mechanism of Intrinsic Semiconductors

Electronic Configuration of Silicon and Germanium	
Silicon	1s2 2s22p6 3s2 3p2
Germanium	1s2 2s2 2p6 3s2 3p6 4s2 3d10 4p2

It can be observed from the electron configurations of both elements that both elements possess four electrons in their outermost valence shell. As the temperature of the semiconductor increases, the electrons gain additional thermal energy and are able to break free from their respective shells. This process of ionization within the crystal lattice results in creating vacancies in the atomic bonds. The location where an electron becomes dislodged forms a hole, which effectively carries a positive charge. This hole is then filled by a free electron, causing the previous position to become a hole and the new position to become neutral. This movement of holes, or effective positive charges, occurs within the semiconductor. In an intrinsic semiconductor, the number of free electrons is equal to the number of holes.

Mathematically,

$$ne = nh = ni$$

In this case, the ni represents the total intrinsic carrier concentration, which is equivalent to the total number of electrons or holes. An intrinsic semiconductor behaves like an insulator when its temperature is T=0K. The electrons become excited and shift from the valence band to the conduction band when the temperature rises further (T>0). Due to the partial occupancy" might be misleading. It's more accurate to state that the presence of electrons in the conduction band corresponds to an equal number of holes in the valence band due to the excitation process of the conduction band by these electrons, the valence band has an equal number of holes.

2.3.2. Extrinsic Semiconductors

Those intrinsic semiconductors to which some suitable impurity or doping agent has been added in extremely small amounts (about 1 part in 10^8) are called extrinsic or impurity semiconductors. Depending on the type of doping material used, extrinsic semiconductors can be subdivided into two classes:

i. N-type semiconductors and
ii. P-type semiconductors.

2.3.2.1. N-type Extrinsic Semiconductor

When pure germanium crystal is mixed with a pentavalent substance such as antimony (Sb), a semiconductor of this kind is produced. As depicted in Figure 2.6, each antimony atom uses four of its five electrons to form covalent bonds with the four germanium atoms that surround it. The fifth electron is not needed and is only loosely bound attached to the antimony atom.

As a result, an electric field or an increase in thermal energy can readily excite it from the valence band to the conduction band. The explanation above makes clear that electrons make up the majority of carriers in N-type semiconductors, while holes make up the minority.

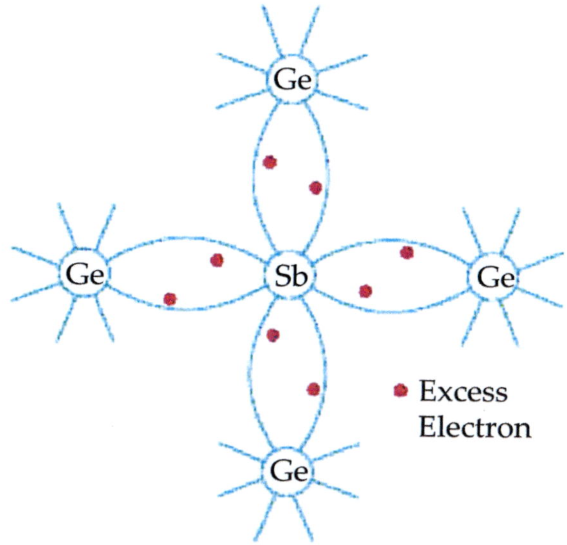

Figure 2.6. N-type semiconductors.

Source: https://uotechnology.edu.iq/dep-laserandoptoelec-eng/branch/lectures/electronic1/%D8%A7%D9%84%D9%85%D8%AD%D8%A7%D8%B6%D8%B1%D8%A9%20%D8%A7%D9%84%D8%A7%D9%88%D9%84%D9%89.pdf

2.3.2.2. P-type Extrinsic Semiconductor

This kind of semiconductor is created when a pure germanium crystal is mixed with traces of a trivalent element, such as boron (B). The boron atom's three valence electrons form covalent bonds with the four surrounding germanium atoms, but as illustrated in the figure, one bond is left incomplete and results in a hole. An extrinsic semiconductor that is P-type (P for positive) is created when boron, an acceptor impurity, creates as many positive holes in a germanium crystal as there are boron atoms. Conduction occurs in this kind of semiconductor when holes in the valence band migrate.

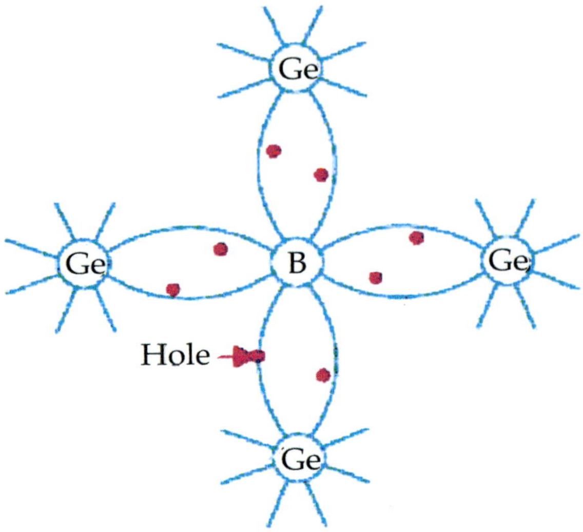

Figure 2.7. P-type Extrinsic Semiconductor.

Source: https://uotechnology.edu.iq/dep-laserandoptoelec-eng/branch/lectures/electronic1/%D8%A7%D9%84%D9%85%D8%AD%D8%A7%D8%B6%D8%B1%D8%A9%20%D8%A7%D9%84%D8%A7%D9%88%D9%84%D9%89.pdf

2.3.2.3. Energy Bands of Extrinsic Semiconductors

A shift in the surrounding temperature causes minority charge carriers to be produced in extrinsic semiconductors. The majority of carriers are also produced by the dopant atoms. Most of these minority carriers are destroyed by the majority carriers during recombination. As a result, the concentration of minority carriers declines.

Therefore, this affects the energy band structure of the semiconductor. In such semiconductors, additional energy states exist:

- Energy state due to donor impurity (E_D); and
- Energy state due to acceptor impurity (E_A).

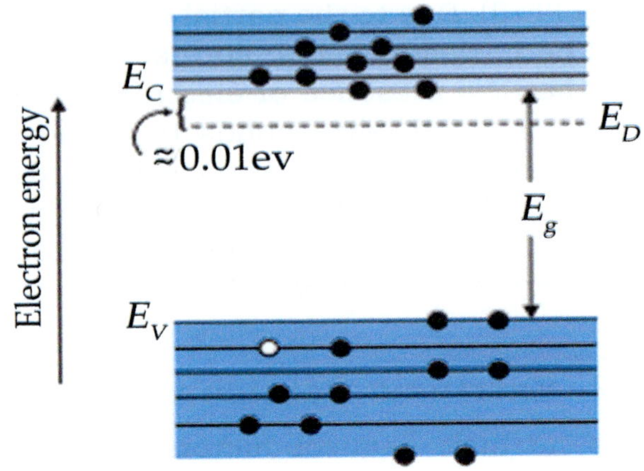

Figure 2.8. Energy band structure of n-type Si semiconductor.

Source: https://d1whtlypfis84e.cloudfront.net/guides/wp-content/uploads/2018/04/05082315/Extrinsic-semiconductor-3.jpg

The above energy band diagram is of n-type Si semiconductor. Here you can see that the energy level of the donor (E_D) is lower than that of the conduction band (E_C). Hence, electrons can move into the conduction band with minimal energy (~0.01 eV). Also, at room temperature, most donor atoms and very few Si atoms get ionized. Hence, the conduction band has most electrons from the donor impurities.

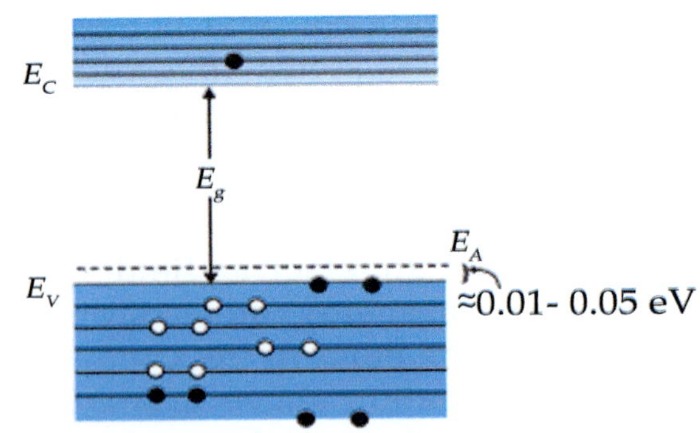

Figure 2.9. Energy band diagram is of p-type Si semiconductor.

Source: https://d1whtlypfis84e.cloudfront.net/guides/wp-content/uploads/2018/04/05082357/Extrinsic-semiconductor-4.jpg

The above energy band diagram is of p-type Si semiconductor. Here you can see that the energy level of the acceptor (E_A) is higher than that of the valence band (E_V). Hence, electrons can move from the valence band to the level EA with minimal energy. Also, at room temperature, most acceptor atoms are ionized. This leaves holes in the valence band. Hence, the valence band has the most holes from the impurities. The electron and hole concentration in a semiconductor in thermal equilibrium is:

$$n_e \times n_h = n_i^2$$

2.3.3. Band Theory of Solids

This theory describes the quantum state that an electron assumes when it is inside a solid metal. Each molecule is made up of multiple distinct energy levels. Electrons in atoms occupy specific energy orbits determined by Pauli's exclusion principle. Two atomic orbitals combine to form a molecular orbit in molecules, which has two distinct energy levels.

It would be better to describe that in a solid, the discrete energy levels of individual atoms merge into continuous bands due to the large number of atoms. As a result, it creates energy bands, a continuum of energy. Plotting all of the available energies for electrons in materials provides us with a very helpful method of visualizing the differences between conductors, insulators, and semiconductors according to this theory.

Thus, in place of discrete energies like in free atoms, the available energy states form bands. In the band theory of solids, there are numerous energy bands. However, the most important energy bands in solids are as follows:

- valence band;
- conduction band; and
- forbidden band.

2.3.3.1. Valence Band

The energy band which comprises valence electrons' energy levels is referred to as the valence band. This band is present below the conduction band. Furthermore, the electrons of this band are loosely bound to the atom's nucleus.

2.3.3.2. Conduction Band

This energy band comprises free electrons' energy level. In order for electrons to be free, external energy must be applied in such a manner that the valence electrons are pushed to the conduction band and become free.

2.3.3.3. Forbidden Band

In essence, this is the energy difference between the conduction and valence band. This is also known as the forbidden gap. We use the forbidden gap to calculate a solid's electrical conductivity. Furthermore, it is possible to classify materials as insulators, semiconductors, and conductors.

2.3.4. Energy Band Inside an Atom

Consider a Sodistaum atom. It comprises 11 electrons. They fill up the energy level following Pauli's exclusion principle.

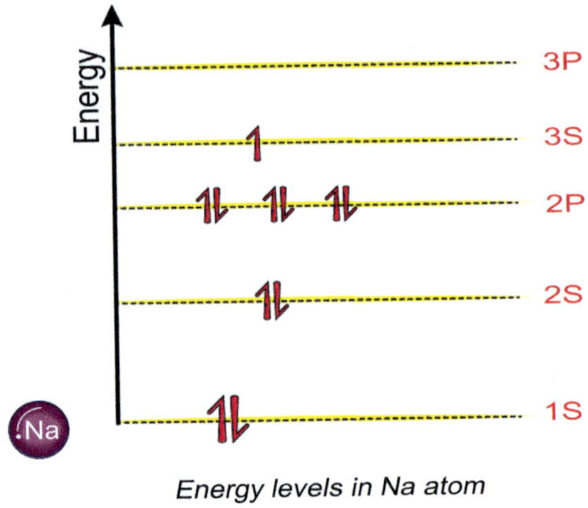

Figure 2.10. Energy level in Na atom.

Source: https://cdn1.byjus.com/wp-content/uploads/2019/02/band-theory-of-solids-1.png

2.3.4.1. Energy Levels Inside a Molecule Made Up of Two Atoms

What happens if two sodium atoms are so close to one another that they nearly form a molecule? Each atom cannot have the configuration that it would have on its own. If they do, many electrons with the same energy levels will result from a violation of Pauli's exclusion principle. What will happen to this system when two atoms get very close to one another? Their individual energy bands will overlap and change into what is known as a molecular orbit. In other words, the 1s molecular orbital is formed when the 1s orbits of individual sodium atoms combine. Two distinct energy levels end up existing in the molecular orbit due to the overlap of two atomic orbitals. An anti-bonding orbital is the higher energy level, and a bonding orbital is the lower energy level.

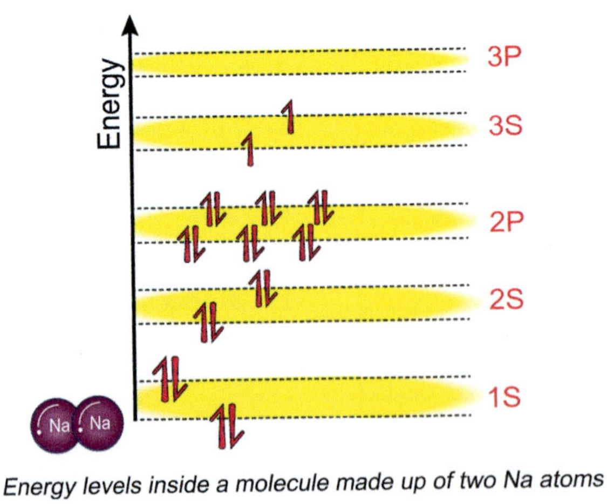

Figure 2.11. Eneergy levels iside a molecule made up of two Na atoms.

Source: https://cdn1.byjus.com/wp-content/uploads/2019/02/band-theory-of-solids-2.png

2.3.4.2. Energy Levels Inside a Molecule Made Up of Three Atoms

Try to imagine what would happen, based on the theory we just learned, if we were to add a third sodium atom to the mixture. This will result in the overlap of three atomic orbitals, creating a single molecular orbital with three distinct energy levels. Here, every molecular orbital will inherit three different energies. Generally speaking, the molecular orbit will have more energy levels the more atoms we add.

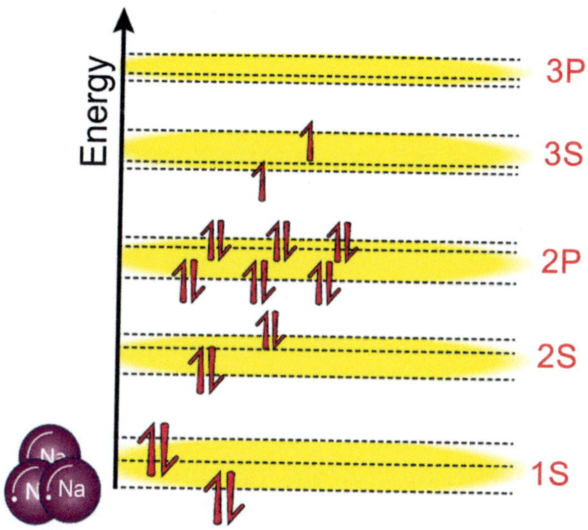

Energy levels inside a molecule made up of three Na atoms

Figure 2.12. Energy levels inside a molecule made up of three Na atoms.

Source: https://cdn1.byjus.com/wp-content/uploads/2019/02/band-theory-of-solids-3.png

2.3.4.3. Energy Levels Inside a Solid Made Up of Avogadro Number of Atoms

At some point, each molecular orbital in a solid composed entirely of sodium with roughly 10^{23} atoms crammed together will have 10^{23} distinct energy levels. Consider drawing the 1s orbital of a sodium solid block, the lower and upper energy levels, and stacking 10^{23} energy levels in between for easier comprehension! There won't be any room for us to distinguish individual energy levels because the spaces between them will be so small. It is therefore convenient to conceptualize it as an energy continuum or continuous energy. When we think about them this way, we can refer to them as energy bands rather than molecular orbits.

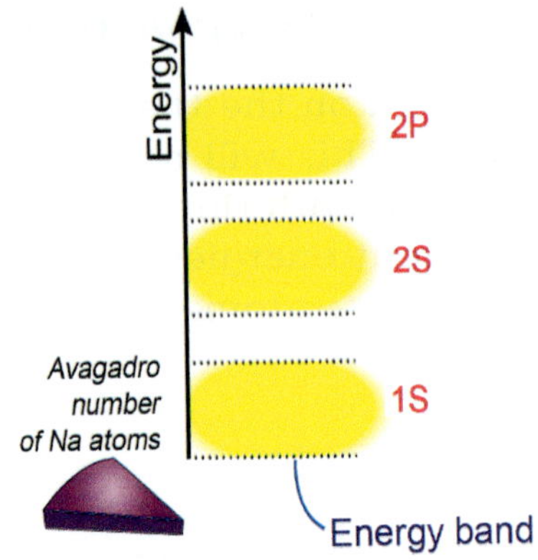

Energy levels inside a molecule made up of Avagadro number of Na atoms

Figure 2.13. Energy levels inside a molecule made up of Avagadro number of Na atoms.

Source: https://cdn1.byjus.com/wp-content/uploads/2019/02/band-theory-of-solids-4.png

2.3.4.4. Energy Levels Inside a Solid Made Up of N-Number of Atoms

Each energy band will typically contain n discrete energy levels if there are n atoms in it. The molecular orbitals in such an n-atom system are referred to as energy bands. Two electrons can fit in a single 1s orbital and a single 2s orbital. Consequently, a 1s and 2s energy band can fit a total of 2n electrons. Six electrons can fit in a single 2p level, which means that a 2p energy band can hold 6n electrons, and so on.

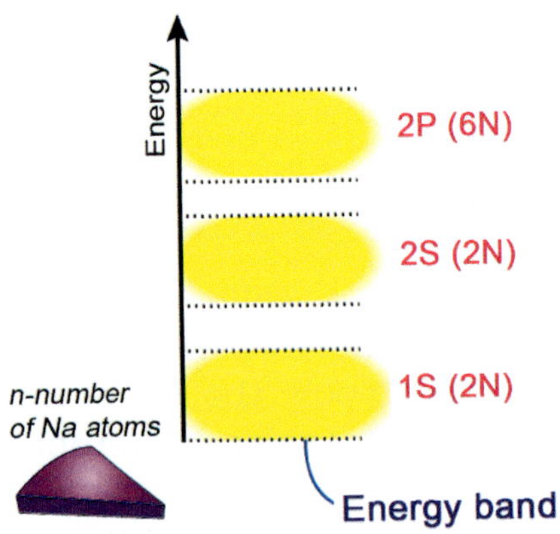

Energy levels inside a molecule made up of n-number of Na atoms

Figure 2.14. Energy band will typically contain n discrete energy levels.

Source: https://cdn1.byjus.com/wp-content/uploads/2019/02/band-theory-of-solids-5.png

2.4. MECHANICAL PROPERTIES OF SEMICONDUCTORS

The physical characteristics that a material displays when forces are applied are known as its mechanical properties. Tensile strength, elongation, hardness, fatigue limit, and modulus of elasticity are a few examples of mechanical properties. Because they are bound by elastic forces, the atoms that make up the solid can deviate from their average position. As a result, the solid is elastic and the atoms vibrate. Since orbitals overlap quantum mechanically, the potential is effectively asymmetric and steeper at short distances. However, a harmonic oscillator is assumed (harmonic approximation) for small amplitudes around the minimum.

2.4.1. Lattice Vibrations

The sound velocity, thermal, elastic, and optical properties of materials can all be explained by lattice vibrations. Atoms in a solid oscillate around their equilibrium position, a phenomenon known as lattice vibration. Due to the atoms' bonds with one another, the equilibrium positions of a crystal form a regular lattice. These nearby atoms' vibrations are not independent of one another. Lattice waves are regular lattices with harmonic forces between atoms and regular modes of vibration.

In the following, we will discuss the dispersion relations for lattice vibrations, i.e., the connection between the frequency ν (or energy $E = h\nu = \hbar\omega$) of the wave and its wavelength λ (or k-vector $k = 2\pi/\lambda$).

2.4.1.1. Monoatomic Linear Chain

The linear chain is a one-dimensional model that provides the best understanding of the fundamental physics of lattice vibrations. Although this term is officially reserved for the quantized lattice vibrations arising from the quantum-mechanical treatment, the mechanical vibrations will also be referred to as phonons.

In the monoatomic linear chain, the atoms of mass M are positioned along a line (x-axis) with a period (lattice constant) a at the positions $x_{n0} = na$. This represents a one-dimensional Bravais lattice. The Brillouin zone of this system is $[-\pi/a, \pi/a]$.

The atoms will interact with a harmonic potential, i.e., the energy is proportional to the displacement $u_n = x_n - x_{n0}$ to the second power. The total potential energy of the system is then:

$$U = \frac{1}{2}C\sum_n (u_n - u_{n+1})^2 . \quad (1)$$

The model assumes that the mass points are connected via massless, ideal springs with a spring constant C. If φ(d) is the interaction energy between two atoms as a function of their distance d, C is given by C = φ" (a). Again, the harmonic approximation is only valid for small displacements, i.e., un ≪ a. The displacement of the atoms can be along the chain (longitudinal wave) or perpendicular to the chain (transverse wave), see Figure. 2.15. We note that for these two types of waves, the elastic constant C must not be the same.

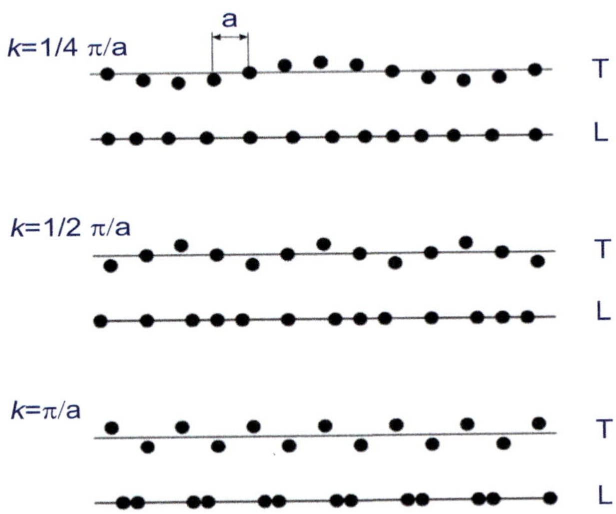

Figure 2.15. Visualization of transverse ('T') and longitudinal ('L') waves in a linear monoatomic chain at different wavevectors.

Source: https://media.springernature.com/lw685/springer-static/image/chp%3A10.1007%2F978-3-319-23880-7_5/MediaObjects/116076_3_En_5_Fig1_HTML.gif

When the sum in (1) has a finite number of terms (n = 0.., N −1), the boundary conditions have to be considered. There are typically two possibilities: The boundary atoms are fixed, i.e., $u_0 = u_{N-1} = 0$, the boundary conditions are periodic, i.e., ui = u_{N+i}. If N ≫ 1, the boundary conditions play no significant role anyway, thus those with the greatest ease for subsequent math are chosen. In solid-state physics, typically periodic boundary conditions are used. Boundary phenomena, such as at surfaces, are then treated separately.

The equations of motion derived from (1) are

$$M\ddot{u}_n = F_n = -\frac{\partial U}{\partial u_n} = -C(2u_n - u_{n-1} - u_{n+1}) \,. \quad (2)$$

We solve for solutions that are periodic in time (harmonic waves), i.e., $u_n(x, t) = u_n \exp(-i\omega t)$. Then the time derivative can be executed immediately as $\ddot{u}_n = -\omega^2 u_n$ and we obtain:

$$M\omega^2 u_n = C(2u_n - u_{n-1} - u_{n+1}) \,. \quad (3)$$

If, also, the solution is periodic in space, i.e., is a (one-dimensional) plane wave, i.e., $u_n(x, t) = v_0 \exp[i(kx-\omega t)]$ with $x = na$, we find from the periodic boundary condition $\exp(ikNa) = 1$ and thus

$$k = \frac{2\pi}{a}\frac{n}{N}, \; n \in \mathbf{N} \,. \quad (4)$$

It is important that, when k is altered by a reciprocal space vector, i.e., k' = k + 2πn/a, the displacements un are unaffected. This property means that there are only N values for k that generate independent solutions. These can be chosen as k = −π/a.., π/a, so that k lies in the Brillouin zone of the lattice. In the Brillouin zone there is a total number of N k-values, i.e., one for each lattice point. The distance between adjacent k-values is $\frac{2\pi}{Na} = \frac{2\pi}{L}$, L being the lateral extension of the system. The displacements at the lattice points n and n + m are now related to each other via

$$\begin{aligned} u_{n+m} &= v_0 \exp(ik(n+m)a) \\ &= v_0 \exp(ikna)\exp(ikma) = \exp(ikma)u_n \,. \end{aligned} \quad (5)$$

Thus, the equation of motion (3) reads

$$M\omega^2 u_n = C(2 - \exp(-ika) - \exp(ika))u_n \,. \quad (6)$$

Using the identity exp(ika)+exp(−ika) = 2 cos(ka), we find the dispersion relation of the monoatomic linear chain (Figure. 2.16):

$$\omega^2(k) = \frac{4C}{M}\frac{1-\cos(ka)}{2} = \frac{4C}{M}\sin^2\left(\frac{ka}{2}\right) \,. \quad (7)$$

The solutions describe plane waves that propagate in the crystal with a phase velocity c = ω/k and a group velocity vg = dω/dk

$$v_g = \left[\frac{4C}{M}\right]^{1/2} \frac{a}{2}\cos\left(\frac{ka}{2}\right) \,. \quad (8)$$

In the vicinity of the Γ point, i.e., k ≪ π/the dispersion relation is linear in k

$$\omega(k) = a\sqrt{C/M}|k| \,. \quad (9)$$

We are used to such linear relations for sound (and also light) waves. The phase and group velocity are the same and do not depend on k. Thus, such solutions are called acoustic. The sound velocity of the medium is given by cs = a C/M.

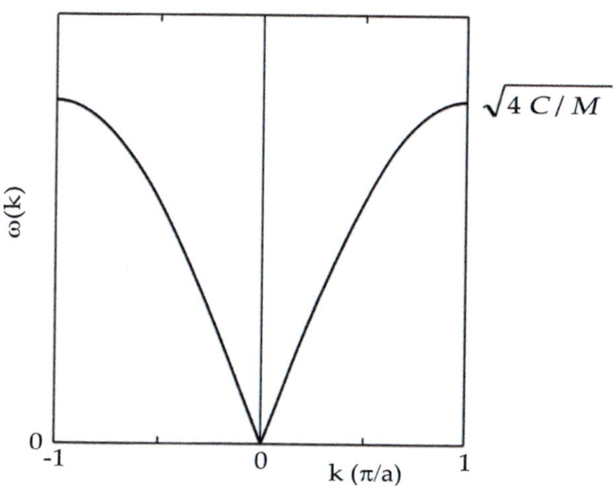

Figure 2.16. Dispersion relation for a monoatomic linear chain.

Source: https://media.springernature.com/lw685/springer-static/image/chp%3A10.1007%2F978-3-030-51569-0_5/MediaObjects/116076_4_En_5_Fig4_HTML.png

It is characteristic of the nonhomogeneous medium that, when k approaches the boundary of the Brillouin zone, the behavior of the wave is altered. For k = π/the wavelength is just λ = 2π/k = 2a, and thus samples the granularity of the medium. The maximum phonon frequency ω_m is

$$\omega_{\rm m} = \sqrt{\frac{4C}{M}}\ . \tag{10}$$

The group velocity is zero at the zone boundary, thus a standing wave is present. The dispersion relations differ because the longitudinal and transverse waves can have different force constants. Unless both of the directions perpendicular to x are equivalent, the transverse branch of the dispersion relation is twofold degenerate.

2.4.1.2. Diatomic Linear Chain

We now examine the scenario in which the system is composed of two distinct types of atoms (Figure. 2.17). This will serve as a model for semiconductors like zincblende that have a diatomic base. It should be noted that even though there are two identical atoms in the base, the diamond structure likewise requires this kind of modeling.

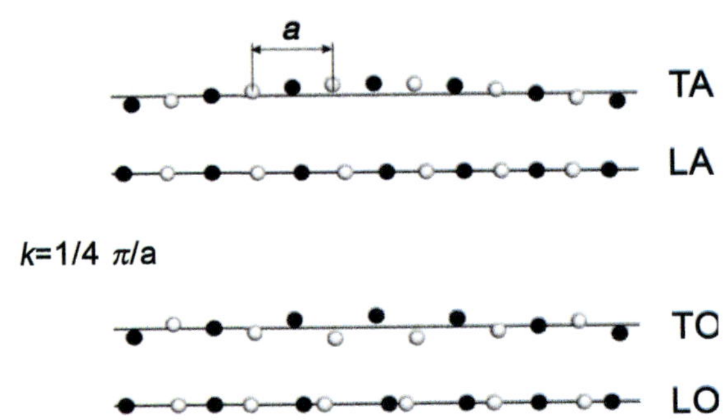

Figure 2.17. Visualization of acoustic and optical waves in a diatomic linear chain.

Source: https://media.springernature.com/lw685/springer-static/image/chp%3A10.1007%2F978-3-319-23880-7_5/MediaObjects/116076_3_En_5_Fig3_HTML.gif

The lattice will be the same and the lattice constant will be a. Alternating atoms of sort 1 and 2 with a relative distance of a/2 are on the chain. The displacements of the two atoms are labeled u_n^1 and u_n^2, both belonging to the lattice point n. The atoms have the masses M1 and M2. The force constants are C1 (for the 1–2 bond within the base) and C2 (for the 2–1 bond between different bases).

The total potential energy of the system is then given as

$$U = \frac{1}{2}C_1 \sum_n \left(u_n^1 - u_n^2\right)^2 + \frac{1}{2}C_2 \sum_n \left(u_n^2 - u_{n+1}^1\right)^2\ . \tag{11}$$

The equations of motion are

$$M_1\ddot{u}_n^1 = -C_1\left(u_n^1 - u_n^2\right) - C_2\left(u_n^1 - u_{n-1}^2\right) \tag{12a}$$

$$M_2\ddot{u}_n^2 = -C_1\left(u_n^2 - u_n^1\right) - C_2\left(u_n^2 - u_{n+1}^1\right) . \tag{12b}$$

With the plane-wave ansatz $u_n^1(x,t) = v_1 \exp\left[\mathrm{i}(kna - \omega t)\right]$ and $u_n^2(x,t) = v_2 \exp\left[\mathrm{i}(kna - \omega t)\right]$ and periodic boundary conditions we find

$$0 = -M_1\omega^2 v_1 + C_1(v_1 - v_2) + C_2(v_1 - \exp(-\mathrm{i}ka)v_2) \tag{13a}$$

$$0 = -M_2\omega^2 v_2 + C_1(v_2 - v_1) + C_2(v_2 - \exp(\mathrm{i}ka)v_1) . \tag{13b}$$

These equations for v1 and v2 can only be solved nontrivially if the determinant vanishes, i.e.

$$0 = \begin{vmatrix} M_1\omega^2 - (C_1 + C_2) & C_1 + \mathrm{e}^{-\mathrm{i}ka}C_2 \\ C_1 + \mathrm{e}^{\mathrm{i}ka}C_2 & M_2\omega^2 - (C_1 + C_2) \end{vmatrix}$$
$$= M_1M_2\omega^4 - (M_1 + M_2)(C_1 + C_2)\omega^2 + 2C_1C_2(1 - \cos(\tag{14}$$

Using the substitutions $C_+ = (C_1 + C_2)/2,\ C_\times = \sqrt{C_1C_2}$, the arithmetic and geometrical averages, and accordingly for M+ and M×, the solution is

$$\omega^2(k) = \frac{2C_\times}{\gamma M_\times}\left[1 \pm \sqrt{1 - \gamma^2\frac{1 - \cos(ka)}{2}}\right] , \tag{15}$$

with

$$\gamma = \frac{C_\times M_\times}{C_+M_+} . \tag{16}$$

The dispersion relation, as shown in Figure. 2.18, now has (for each longitudinal and transverse mode) two branches. The lower branch ('–' sign in (15)) is related to the acoustic mode; neighboring atoms have similar phase (Figure. 2.3). For the acoustic mode ω = 0 at the Γ point and the frequency increases towards the zone boundary. The maximum phonon frequency ω_m is in the upper branch ('+' sign in (15)) at the zone center

$$\omega_\mathrm{m} = \sqrt{\frac{4C_\times}{\gamma M_\times}} = \sqrt{\frac{4C_+M_+}{M_\times^2}} . \tag{17}$$

In the vicinity of the Γ point the dispersion is parabolic with negative curvature:

$$\omega(k) \cong \omega_{\mathrm{m}} \left[1 - \frac{1}{2}\left(\frac{\gamma a}{4}\right)^2 k^2\right] . \qquad (18)$$

The upper branch is called the optical mode (since it can interact strongly with light) and neighboring atoms have opposite phase. Thus, four different vibrations exist that are labeled TA, LA, TO, and LO. Both the TA and TO branches are degenerate.

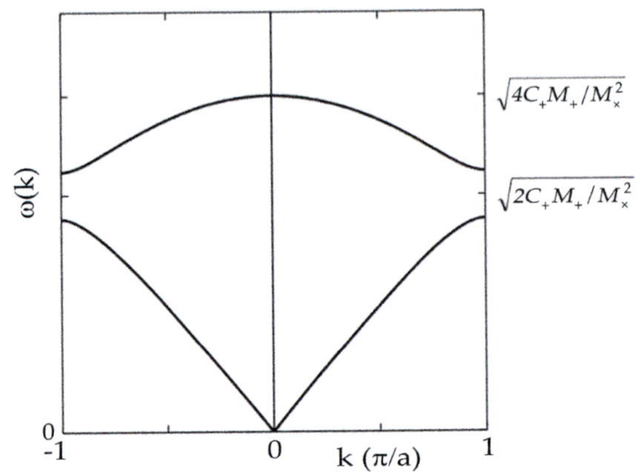

Figure 2.18. Dispersion relation for a diatomic linear chain.

At the zone boundary (X point) a frequency gap exists. The gap center is at

$$\bar{\omega}_{\mathrm{X}} = \frac{\omega_{\mathrm{m}}}{\sqrt{2}} , \qquad (19)$$

and the total width is

$$\Delta\omega_X = \omega_{\mathrm{m}}\sqrt{1-\gamma} = 2\sqrt{\frac{C_+M_+ - C_\times M_\times}{M_\times^2}} . \qquad (20)$$

The group velocity is zero for optical and acoustic phonons at k = π/a and for optical phonons at the Γ point.

Usually two cases are treated explicitly: (i) atoms with equal mass ($M = M_1 = M_2$) and different force constants or (ii) atoms with unequal mass and identical force constants $C = C_1 = C_2$. For the case $C_1 = C_2$ and $M_1 = M_2$, $\gamma = 1$ and thus $\Delta\omega_X = 0$. Then the dispersion relation is the same as for the monoatomic chain, except that the k space has been folded since the actual lattice constant is now a/2.

$$M_1 = M_2$$

In this case, M+ = M× = M and the dispersion relation is

$$\omega^2 = \frac{2C_+}{M}\left[1 \pm \sqrt{1 - \frac{C_\times^2}{C_+^2}\frac{1-\cos(ka)}{2}}\right]. \tag{21}$$

At the zone boundary the frequencies for the acoustic and the optical branch are $\omega_{X,1} = \sqrt{2C_1/M}$ with v1 = v2 and $\omega_{X,2} = \sqrt{2C_2/M}$ with $v_1 = v_2$ and $\omega_{X,2} = \sqrt{2C_2/M}$ with $v_1 = -v_2$, respectively (assuming $C_2 > C_1$). The motion for $k = \pi/a$ is phase shifted by 180° for adjacent bases. Additionally, for the acoustic branch the atoms of the base are in phase, while for the optical branch the atoms of the base are 180° out of phase. The vibration looks as if only one of the springs is strained.

$$C_1 = C_2$$

In this case, $C_+ = C_\times = C$ and the dispersion relation is

$$\omega^2 = \frac{2CM_+}{M_\times^2}\left[1 \pm \sqrt{1 - \frac{M_\times^2}{M_+^2}\frac{1-\cos(ka)}{2}}\right]. \tag{22}$$

At the zone boundary the frequencies for the acoustic and the optical branch are $\omega_{X,1} = \sqrt{2C/M_1}$ with v2 = 0 and $\omega_{X,2} = \sqrt{2C/M_2}$ with $v_2 = 0$ and $\omega_{X,2} = \sqrt{2C/M_2}$ with $v_1 = 0$, respectively (assuming $M_2 < M_1$). In the vibration for $k = \pi/a$ thus only one atom species oscillates, the other does not move. Close to the Γ point the atoms are in phase in the acoustic branch, i.e., $v_1 = v_2$. For the optical branch, the frequency at the Γ point is given by $\omega = \sqrt{2C/M_r}$ (with the reduced mass $M_r^{-1} = M_1^{-1} + M_2^{-1} = 2M_+/M_\times^2$) and the amplitude ratio is given by the mass ratio: $v_2 = -(M_1/M_2)v_1$, i.e., the heavier atom has the smaller amplitude.

2.4.1.2. Lattice Vibrations of a Three-Dimensional Crystal

There are 3N equations of motion when calculations are performed for a three-dimensional crystal with a monoatomic base. These represent the three acoustic branches of the dispersion relation—one LA phonon mode and two TA phonon modes—after being converted to normal coordinates. There are three acoustic branches and three (p-1) optical branches in a crystal containing p atoms in its base. There are three optical phonon branches for a diatomic base (like the zincblende structure): one LO phonon mode and two TO phonon modes. There are three modes in total. It is now necessary to compute the dispersion ω(k) for every direction of k.

In Figure 2.19 and Figure 2.20, the phonon dispersion in silicon and GaAs is shown along particular lines in the Brillouin zone. The main differences are: (i) the degeneracy of the acoustic and optical branch at the X point for the group-IV semiconductor is lifted for the III–V semiconductor due to the different mass of the constituents, (ii) the degeneracy of the LO and TO energies at the Γ point for the group-IV semiconductor is lifted for the III–V semiconductor due to the ionic character of the bond and the macroscopic electric field connected with the long-wavelength LO phonon.

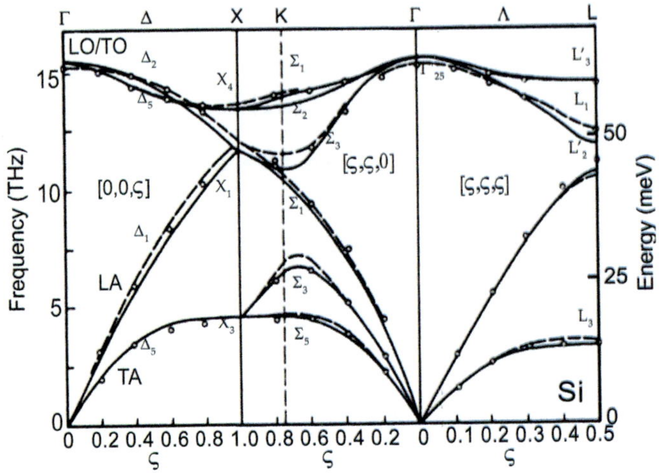

Figure 2.19. Phonon dispersion in Si, experimental data and theory (solid lines: bond charge model, dashed lines: valence force field model).

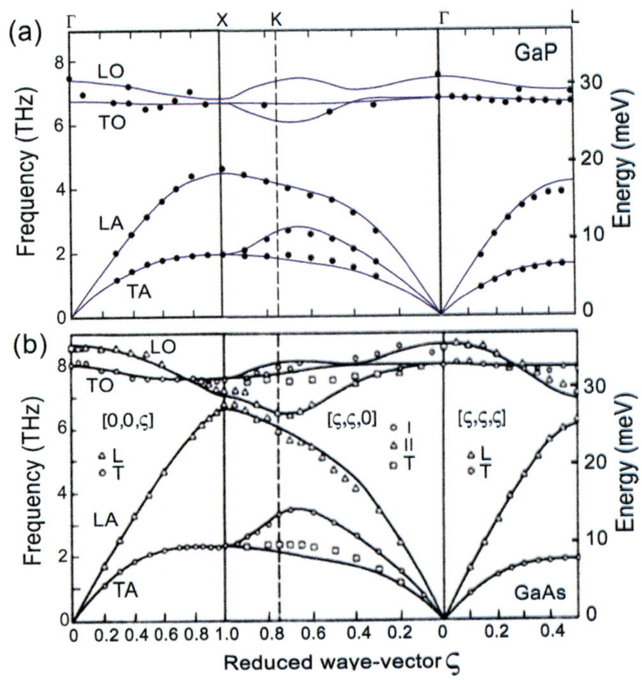

Figure 2.20. Phonon dispersion in (a) GaP and (b) GaAs. Experimental data (symbols) and theory (solid lines, 14-parameter shell model). 'L' and 'T' refer to longitudinal and transverse modes, respectively. 'I' and 'II' (along [ζ, ζ, 0]) are modes whose polarization is in the $(1, \bar{1}, 0)$ plane.

We note that the degeneracy of the TA phonon is lifted for propagation along the ⟨110⟩ directions (Σ) because the two transverse directions ⟨001⟩ and $\langle 1\bar{1}0 \rangle$ are not equivalent.

In boron nitride the masses of the two constituents are so similar that no gap exists between acoustical and optical branches (Fig. 2.21). Also, the density of states (averaged over the entire Brillouin zone) is depicted.

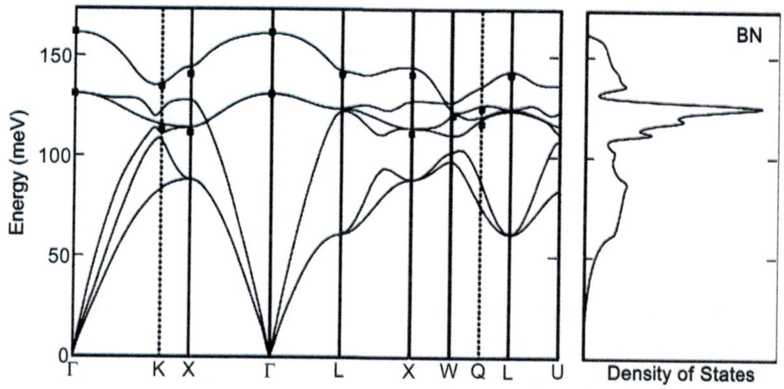

Figure 2.21. Phonon dispersion in BN (left panel), experimental data (symbols) and theory (solid lines, first principles pseudopotential model). In the right panel the density of states is depicted.

The displacement of atoms is shown in Figure. 2.21 for the different phonon modes present in zincblende crystals and in Figure. 2.22 for wurtzite crystals. The modes are labeled with their symmetry (in molecular notation) according to group theory.

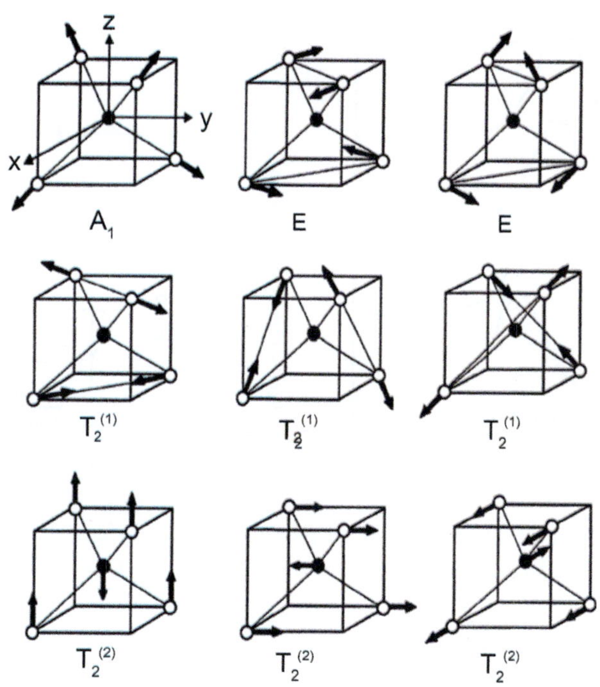

Figure 2.22. Displacement of atoms for various phonon modes in zincblende crystals.

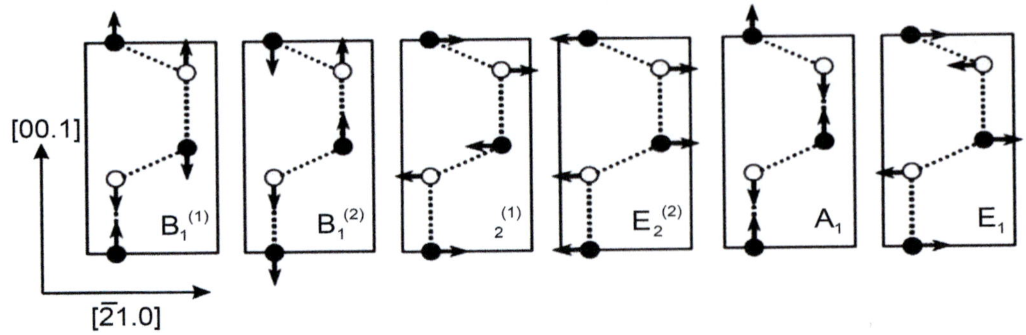

Figure 2.23. Displacement of atoms for various phonon modes in wurtzite crystals.

The dependence of the phonon frequency on the mass of the atoms ($\propto M^{-1/2}$) can be demonstrated with the isotope effect, visualized for GaAs in Figure. 2.24. The dependence of the phonon frequencies on the stiffness of the spring can be seen from Figure. 2.25; the smaller lattice constant provides the stiffer spring.

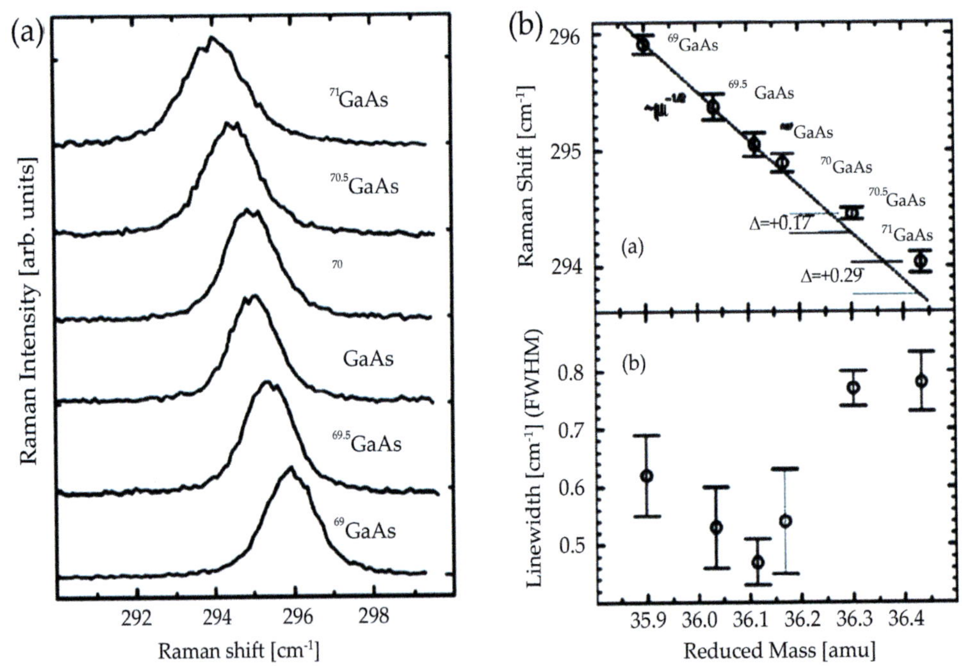

Figure 2.24. (a) Raman spectra of GaAs with different isotope content as labeled. (b) Energy of optical phonons in GaAs with different isotope content [using the Raman spectra shown in (a)].

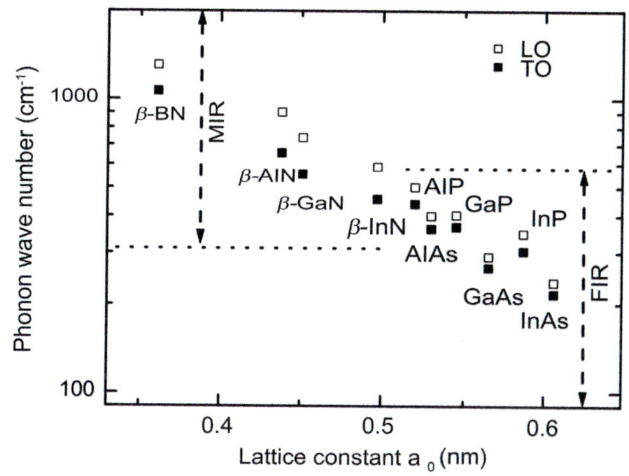

Figure 2.25. Optical phonon frequencies (TO: filled squares, LO: empty squares) for a number of III–V compounds with different lattice constant a0. 1meV corresponds to 8.065 wave numbers (or cm^{-1}).

2.4.2. Density of States

The density of states (DOS) tells how many of the total 3pN modes are in a given energy interval. The states are spaced equally in k-space but not on the energy scale.

For the monoatomic linear chain model, the number of states N(E') from E = 0 up to E = ħω = E'(k') for the dispersion of the acoustic phonons (7) is given as

$$N(E') = k' \frac{N}{\pi/a} = \frac{L}{\pi} k'. \tag{23}$$

Using (7), we find for one polarization ($E_m = \hbar\omega_m$)

$$N(E) = \frac{2N}{\pi} \arcsin\left(\frac{E}{E_{\mathrm{m}}}\right).$$

The DOS D(E) is given by

$$D(E) = \frac{\mathrm{d}N(E)}{\mathrm{d}E} = \frac{2\,N}{\pi\,E_{\mathrm{m}}} \frac{1}{\sqrt{1-(E/E_{\mathrm{m}})^2}}.$$

Often the density of states is scaled by the (irrelevant) system size and given per atom (D/N) or per volume (D/L^3), per area (D/L^2) or per length (D/L) for three-, two- or one-dimensional systems, respectively.

In the diatomic linear chain model, additionally, the optical phonons contribute to the density of states. In Figure. 2.26, the phonon density of states is shown for $\gamma = 0.9$

and for comparison for $\gamma = 1$ (gapless phonon dispersion). For small wavevector, the density of states is $4N/(\pi E_m)$. Within the gap, the density of states vanishes. At the edges of the band gap, the density of states is enhanced. The total number of states for both dispersions is the same.

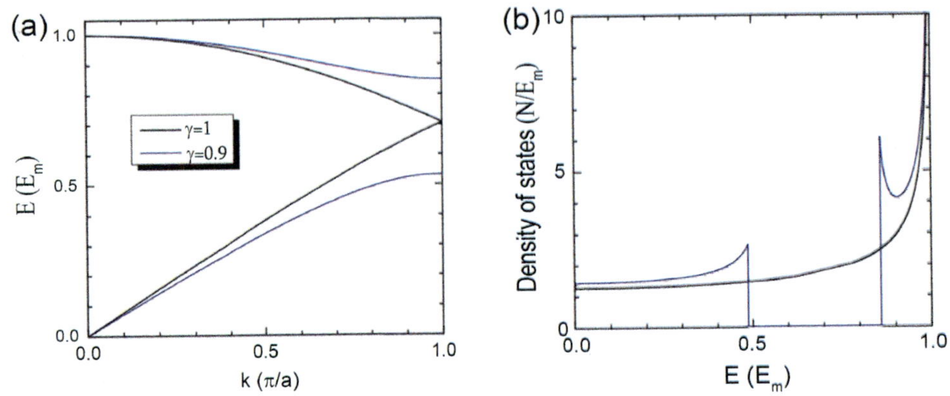

Figure 2.26. (a) Phonon dispersion for the diatomic linear chain model for $\gamma = 1$ (black line) and $\gamma = 0.9$ (blue lines). (b) Corresponding density of states (in units of N/E_m).

In a three-dimensional solid, Eq. (23) is modified to (for three degenerate polarizations)

$$N(E') = \frac{4\pi}{3} \frac{3}{(2\pi/L)^3} k'^3,$$

taking into account all states within a sphere in k-space of radius k.' Assuming a linear dispersion $\omega = v_s k$, we obtain

$$N(E) = \frac{V}{2\pi^2} \frac{E^3}{\hbar^3 v_s^3}.$$

Thus, the density of states is proportional to E_2,

$$D(E) = \frac{3V}{2\pi^2} \frac{E^2}{\hbar^3 v_s^3}.$$

This dependence is the base for Debye's law for the T_3 temperature dependence of the heat capacity.

CASE STUDY

THE SEMICONDUCTOR INDUSTRY

The semiconductor industry is crucial to modern technology, powering devices from smartphones to supercomputers. This case study examines the evolution of the semiconductor industry, its key players, challenges, and future prospects.

Background

Semiconductors are materials with electrical conductivity between conductors (like metals) and insulators (like glass). Silicon is the most common semiconductor material due to its abundance and electrical properties. The invention of the transistor in 1947 marked the beginning of the semiconductor era, leading to the development of integrated circuits (ICs) and microprocessors.

Key Players

1. **Intel Corporation**: Founded in 1968, Intel dominated the semiconductor industry for decades, producing microprocessors for personal computers. However, increased competition and shifts in technology have challenged its market position.
2. **Samsung Electronics:** A major player in both memory and logic semiconductors, Samsung is known for its innovation and vertical integration, controlling various stages of the semiconductor supply chain.
3. **Taiwan Semiconductor Manufacturing Company (TSMC):** TSMC is the world's largest dedicated independent semiconductor foundry, manufacturing chips for fabless companies like Apple, Nvidia, and Qualcomm.
4. **Nvidia Corporation:** Specializing in graphics processing units (GPUs) initially, Nvidia expanded into AI and data centers, becoming a significant player in the semiconductor industry.

Challenges:

1. **Technological Complexity:** Shrinking transistor sizes to increase performance while reducing costs has become increasingly challenging, leading to rising R&D costs and manufacturing complexities.
2. **Supply Chain Disruptions:** The semiconductor industry is susceptible to supply chain disruptions, such as natural disasters and geopolitical tensions, affecting production and global supply.
3. **Intellectual Property (IP) Protection:** Protecting semiconductor IP from theft or infringement is crucial due to the high value of design and manufacturing processes.

Opportunities

1. **Emerging Technologies:** Advancements in AI, 5G, IoT, and electric vehicles create new opportunities for semiconductor companies to develop specialized chips tailored to these applications.
2. **Foundry Services:** The increasing demand for foundry services presents opportunities for companies like TSMC to expand their market share by offering cutting-edge manufacturing technologies to fabless semiconductor firms.
3. **Environmental Sustainability:** The shift towards renewable energy and energy-efficient technologies creates opportunities for semiconductor companies to develop eco-friendly solutions.

Conclusion

The semiconductor industry continues to evolve rapidly, driven by technological advancements, market dynamics, and global trends. Companies must innovate continuously, navigate supply chain challenges, and seize emerging opportunities to maintain competitiveness in this dynamic landscape.

CLASS ACTIVITY

In this class activity, students will engage in a hands-on exploration of semiconductors. Through interactive discussions and demonstrations, they'll delve into the properties and applications of semiconductor materials in electronic devices. The activity aims to reinforce theoretical concepts with practical demonstrations, fostering a deeper understanding of how semiconductors power the digital world. Students will have the opportunity to analyze real-life examples of semiconductor devices and discuss their impact on technology and society. By actively participating in this activity, students will gain valuable insights into the role of semiconductors in shaping our technological landscape.

SUMMARY

- Semiconductors are materials that have electrical conductivity properties between those of conductors (e.g., metals) and insulators (e.g., rubber, glass). They are widely used in the manufacturing of electronic devices such as transistors, diodes, integrated circuits, and solar cells.
- A semiconductor is any of a class of crystalline solids intermediate in electrical conductivity between a conductor and an insulator.
- Semiconductors are employed in the manufacture of various kinds of electronic devices, including diodes, transistors, and integrated circuits.
- Semiconductors possess specific electrical properties. A substance that conducts electricity is called a conductor, and a substance that does not conduct electricity is called an insulator. Semiconductors are substances with properties somewhere between them.
- Semiconductors comprising a single element are called elemental semiconductors, including the famous semiconductor material silicon. On the other hand, semiconductors made up of two or more compounds are called compound semiconductors and are used in semiconductor lasers, light-emitting diodes, etc.
- In atoms, electrons fill their respective energy orbits which follow Pauli's exclusion principle. In molecules, two atomic orbitals combine to form a molecular orbit with two separate energy levels.
- The energy band that comprises valence electrons' energy levels is referred to as the valence band. This band is present below the conduction band. Furthermore, the electrons in this band are loosely bound to the atom's nucleus.
- The atoms that make up the solid have an average position from which they can deviate since they are bound by elastic forces. The atoms thus undergo vibrational motion, and the solid is elastic.
- When calculations are performed for a three-dimensional crystal with a monoatomic base, there are 3N equations of motion.
- The density of states (DOS) indicates the number of 3pN modes within a specific energy range. States are evenly distributed in k-space but not in the energy scale.

REVIEW QUESTIONS

1. What is a semiconductor, and how does it differ from a conductor and an insulator?
2. Describe the role of semiconductors in the field of electronics.
3. What are the most commonly used materials for semiconductor fabrication, and why are they preferred?

4. Explain the significance of transistors in modern electronics and how they function.
5. How do diodes work, and what are their primary applications?
6. Define integrated circuits (ICs) and explain their importance in electronic devices.
7. What is Moore's Law, and how does it impact the semiconductor industry?
8. Discuss the challenges facing the semiconductor industry in sustaining Moore's Law and increasing computing power.
9. How are semiconductors utilized in emerging technologies like artificial intelligence (AI) and the Internet of Things (IoT)?
10. What efforts are being made to address environmental concerns associated with semiconductor manufacturing, and what role can policymakers play in this regard?

REFERENCES

1. Adachi, S. (2004). *Handbook on Physical Properties of Semiconductors (Group IV Elemental Semiconductors)* (Vol. 1). Kluwer.
2. Bernstein, N., Mehl, M. J., Papaconstantopoulos, D. A., Papanicolaou, N. I., Bazant, M. Z., & Kaxiras, E. (2000). Energetic, vibrational, and electronic properties of silicon using a nonorthogonal tight-binding model. *Physical Review B, 62*, 4477–4487.
3. Chelikowsky, J. R., & Cohen, M. L. (2015). Semiconductors: A pillar of pure and applied physics. *Journal of Applied Physics, 117*, 112812(8pp).
4. Chuang, C.-T., Bernstein, K., Joshi, R. V., Puri, R., Kim, K., Nowak, E. J., Ludwig, T., & Aller, I. (2004). Focusing on planar device structures and strained silicon for handling silicon scaling issues in the deep sub-100 nm regime. *IEEE Circuits and Devices Magazine, 20*(1), 6–19.
5. Coleman, J. J. (2012). The development of the semiconductor laser diode after the first demonstration in 1962. *Semiconductor Science and Technology, 27*, 090207(10pp).
6. Orton, J. (2009). *Semiconductors and the Information Revolution: Magic Crystals that Made IT Happen* (pp. 35–36). Academic Press/Elsevier.
7. Phillips, J. C., & Lucovsky, G. (2009). *Bonds and Bands in Semiconductors* (2nd ed.). Momentum Press.
8. Riordan, M. (2007). The silicon dioxide solution: How physicist Jean Hoerni built the bridge from the transistor to the integrated circuit. *IEEE Spectrum, 44*(12), 50–56.
9. Skotnicki, T., Hutchby, J. A., King, T.-J., Wong, H.-S. P., & Boeuf, F. (2005). The end of CMOS scaling. *IEEE Circuits and Devices Magazine, 21*(1), 16–26.

CHAPTER 3

Semiconductor Devices: The P-N Junction

LEARNING OBJECTIVES

After studying this chapter, you will be able to:

- Understand the basic structure of P-N junction in semiconductor
- Define zero applied bias and reverse applied bias
- Explain the junction breakdown
- Discus nonuniformly doped junctions

KEY TERMS FROM THIS CHAPTER

Avalanche breakdown
Charge carriers
Depletion region
Forward bias
Junction capacitance
Built-in potential
Current flow
Electric field
Hyperabrupt junction
Linearly graded junctio

3.1. INTRODUCTION

Semiconductor devices form the backbone of modern electronic technology, playing a crucial role in everything from simple household appliances to advanced computing systems. At the heart of many semiconductor devices lies the p-n junction, a fundamental building block that enables the control and manipulation of electrical currents. This junction is created by joining p-type and n-type semiconductors, materials that have been doped with specific impurities to produce an excess of positive charge carriers (holes) in the p-type and an excess of negative charge carriers (electrons) in the n-type. The interaction between these two types of materials at their interface gives rise to unique electrical properties that are harnessed in various applications.

The p-n junction is characterized by the formation of a depletion region at the boundary between the p-type and n-type materials. In this region, electrons from the n-type material diffuse into the p-type material and recombine with holes, leading to a zone devoid of free charge carriers. This process creates an electric field that opposes further diffusion of charge carriers, establishing a built-in potential barrier. The behavior of this junction under different conditions, such as forward and reverse bias, is pivotal in the operation of many semiconductor devices. For instance, when a forward bias is applied, the potential barrier is reduced, allowing current to flow across the junction. Conversely, a reverse bias increases the barrier, inhibiting current flow and creating a rectifying effect.

The ability to control current flow through the p-n junction is exploited in a wide range of semiconductor devices, including diodes, transistors, and solar cells. In diodes, the p-n junction allows current to pass in one direction while blocking it in the opposite direction, making them essential for converting alternating current (AC) to direct current (DC) in power supplies. Transistors, which are the building blocks of integrated circuits, utilize multiple p-n junctions to amplify and switch electronic signals, enabling the complex operations of modern microprocessors. Solar cells, on the other hand, convert light energy into electrical energy by generating electron-hole pairs at the p-n junction, illustrating the versatility of this fundamental structure.

3.2. BASIC STRUCTURE OF PN JUNCTION IN SEMICONDUCTOR

One or more P-N junctions are used in the majority of semiconductor devices. All semiconductor devices, including switching devices, rectifiers, amplifiers, and linear and digital integrated circuits, are controlled by the P-N junction. Placing a layer of P-type semiconductor next to a layer of N-type semiconductor creates a PN junction in a semiconductor. The term "metallurgical junction" refers to the boundary that divides the N and P regions.

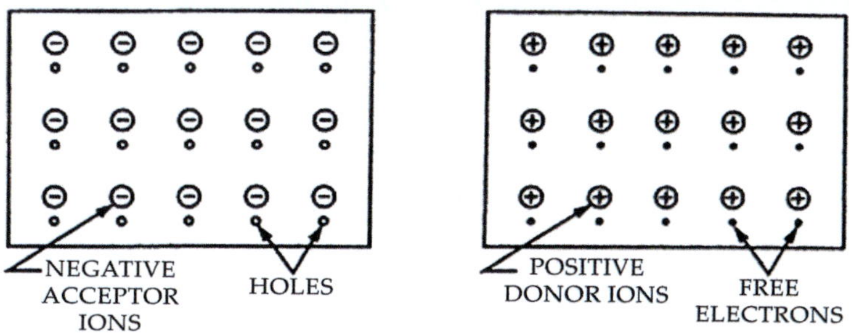

(a) P-Type Semiconductor Material　　*(b) N-Type Semiconductor Material*

Figure 3.1. Two blocks of semiconductor material, one P-type, and the other N-type.

Source: https://www.eeeguide.com/basic-structure-of-pn-junction-in-semiconductor/

Two blocks of semiconductor material, one P-type and the other N-type, are shown in Figure 3.1. The P-type semiconductor block has the same number of fixed negative acceptor ions (represented by an encircled minus sign) and mobile holes (represented by small circles). Similarly, the N-type semiconductor block has the same number of fixed donor positive ions and mobile or free electrons (represented by dots). In P-type materials, the holes, which make up the majority of the charge carriers, are typically evenly distributed throughout the material's volume. In an N-type material, the electrons, which make up the majority of the charge carriers, are equally distributed throughout the material's volume. Because the positive and negative charges in each region are equal, they are all electrically neutral. (Chavez, R., A. Becker, V. Kessler, 2023)

On the formation of PN Junction in Semiconductor, some of the holes from P-type material tend to diffuse across the boundary into N-type material and some of the free electrons similarly diffuse into the P-type material, as illustrated in Figure. 3.2. This happens due to a density gradient (as the concentration of holes is higher on the P-side than that on the N-side and the concentration of electrons is higher on the N-side than that on the P-side). This process is known as diffusion and the current produced because of the diffusion process is known as the diffusion current I_D. The potential distribution diagram is shown in Figure. 3.2. From Figure. 3.2, it is obvious that a potential barrier V_B or V_0 is developed which results in an electric field. This field prevents the respective majority carriers from crossing the barrier region.

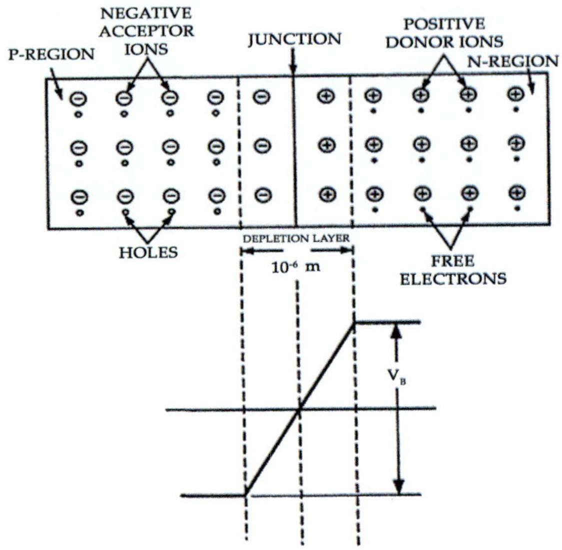

Figure 3.2. The potential distribution diagram.

Source: https://www.eeeguide.com/basic-structure-of-pn-junction-in-semiconductor/

The doping profile of an ideal uniformly doped PN Junction in Semiconductor is depicted in Figure. 3.3.

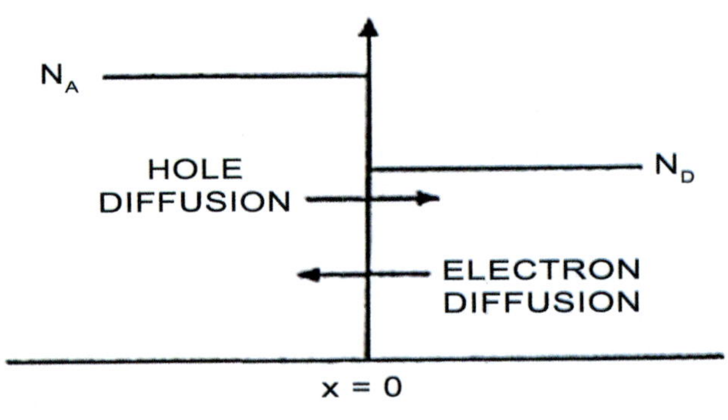

Figure 3.3. Doping profile of an ideal uniformly doped P-N Junction.

Source: https://www.eeeguide.com/basic-structure-of-pn-junction-in-semiconductor/

It is clear that some atoms on the P-side produce negative ions due to the free electrons crossing the junction, which gives them one extra electron above their total number of protons. On the N-side, the electrons also leave behind positive ions, which are atoms with one less electron than protons. The P-side of the junction gains a negative potential when negative ions are produced there. In a similar manner, the N-side generates positive ions and gains a positive potential. No more electrons can migrate from the N-type material to the P-type material due to the negative potential on the P-side. In a similar vein, holes cannot migrate across the boundary any farther due to the positive potential on the N-side. At the junction, a barrier potential is thus created by the initial diffusion of charge carriers. Barrier potential V_0 has a magnitude of a few tenths of a volt, or 0.3 V for Ge and 0.7 V for Si. (Chavez, R., A. Becker, V. Kessler, 2020) The area surrounding the intersection is totally ionized. Consequently, neither the N-side nor the P-side has any free electrons or holes. The depletion region is the area surrounding the junction because the mobile charge carriers (i.e., holes and free electrons) have been removed from this area.

3.3. ZERO APPLIED BIAS

Zero bias PN junction: In this case, we'll look at the step junction's characteristics when it's in thermal equilibrium and there are no external excitations or currents flowing through it. By measuring the width, electric field, and depletion of the space-charge region, we can ascertain the built-in potential barrier.

3.3.1. Built-in Potential Barrier

The junction is in thermal equilibrium when there is no voltage applied across it, meaning that the Fermi energy level remains constant throughout the system. The energy-band diagram for the pn junction in thermal equilibrium is displayed in Figure 3.4. Since the relative positions of the conduction and valence bands with respect to the Fermi energy change between the p and n regions, the conduction and valence band energies must bend as we pass through the space charge region.

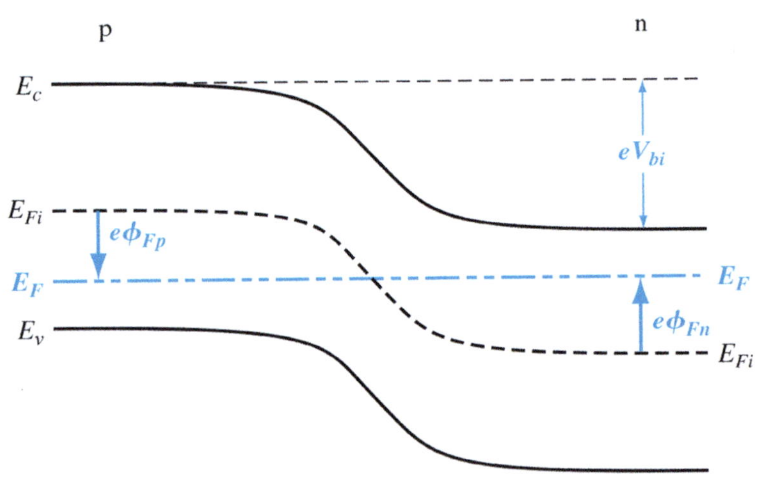

Figure 3.4. Energy-band diagram of a pn junction in thermal equilibrium.

Source: Donald A. Neamen; Semiconductor Physics and Devices Basic Principles Fourth Edition; ISBN 978-0-07-352958-5.

When attempting to enter the conduction band of the p region, electrons in the n region's conduction band encounter a potential obstacle. The built-in potential barrier, or V_{bi}, is the name given to this potential barrier. In addition to maintaining equilibrium between majority carrier holes in the p region and minority carrier holes in the n region, the built-in potential barrier also keeps majority carrier electrons in the n region and minority carrier electrons in the p region. Since new potential barriers will form between the probes and the semiconductor to cancel V_{bi}, it is not possible to measure the potential difference across the junction with a voltmeter. This voltage does not produce any current because the potential V_{bi}, keeps the system in balance.

The intrinsic Fermi level is equidistant from the conduction band edge throughout the junction; thus, the built-in potential barrier can be determined as the difference between the intrinsic Fermi levels in the p and n regions. We can define the potentials ϕ_{Fn} and ϕ_{Fp} as shown in Figure 3.4, so we have

$$V_{bi} = |\phi_{Fn}| + |\phi_{Fp}| \quad (1)$$

In the n region, the electron concentration in the conduction band is given by

$$n_0 = N_c \exp\left[\frac{-(E_c - E_F)}{kT}\right] \quad (2)$$

which can also be written in the form

$$n_0 = n_i \exp\left[\frac{E_F - E_{Fi}}{kT}\right] \quad (3)$$

where n_i and E_{Fi} are the intrinsic carrier concentration and the intrinsic Fermi energy, respectively. We may define the potential ϕ_{Fn} in the n region as

$$e\phi_{Fn} = E_{Fi} - E_F \quad (4)$$

Equation (3) may then be written as

$$n_0 = n_i \exp\left[\frac{-(e\phi_{Fn})}{kT}\right] \quad (5)$$

Taking the natural log of both sides of Equation (5), setting $n_0 = N_d$, and solving for the potential, we obtain

$$\phi_{Fn} = \frac{-kT}{e} \ln\left(\frac{N_d}{n_i}\right) \quad (6)$$

Similarly, in the p region, the hole concentration is given by

$$p_0 = N_a = n_i \exp\left[\frac{E_{Fi} - E_F}{kT}\right] \quad (7)$$

where Na is the acceptor concentration. We can define the potential ϕ_{Fn} in the p region as

$$e\phi_{Fp} = E_{Fi} - E_F \quad (8)$$

Combining Equations (7) and (8), we find that

$$\phi_{Fp} = +\frac{kT}{e} \ln\left(\frac{N_a}{n_i}\right) \quad (9)$$

Finally, the built-in potential barrier for the step junction is found by substituting Equations (6) and (9) into Equation (1), which yields

$$V_{bi} = \frac{kT}{e}\ln\left(\frac{N_a N_d}{n_i^2}\right) = V_t \ln\left(\frac{N_a N_d}{n_i^2}\right) \quad (10)$$

where $V_t = kT/e$ and is defined as the thermal voltage.

At this time, we should note a subtle but important point concerning notation.

3.3.2. Electric Field

Positive and negative space charge densities separate, producing an electric field in the depletion region. The distribution of volume charge density in the pn junction, assuming uniform doping and an abrupt junction approximation, is depicted in Figure 3.5. Assuming that x_p is a positive number, we shall assume that the space charge region abruptly ends in the n region at $x = +x_n$ and in the p region at $x = -x_p$.

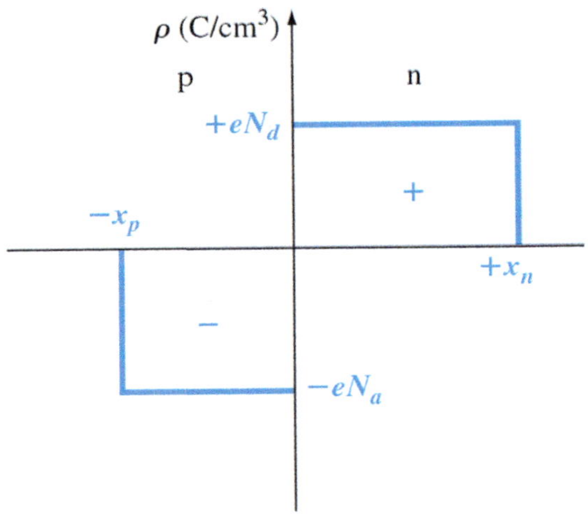

Figure 3.5. The space charge density in a uniformly doped pn junction assuming the abrupt junction approximation.

The electric field is determined from Poisson's equation, which, for a one-dimensional analysis, is

$$\frac{d^2\phi(x)}{dx^2} = \frac{-\rho(x)}{\epsilon_s} = -\frac{dE(x)}{dx} \quad (11)$$

where $\phi(x)$ is the electric potential, $E(x)$ is the electric field, $\rho(x)$ is the volume charge density, and ϵ_s is the permittivity of the semiconductor. From Figure 3.5, the charge densities are

$$\rho(x) = -eN_a \qquad -x_p < x < 0 \quad (12a)$$

And

$$\rho(x) = eN_d \qquad 0 < x < x_n \tag{12b}$$

The electric field in the p region is found by integrating Equation (11). We have

$$E = \int \frac{\rho(x)}{\epsilon_s} dx = -\int \frac{eN_a}{\epsilon_s} dx = \frac{-eN_a}{\epsilon_s} x + C_1 \tag{13}$$

where C_1 is a constant of integration. The electric field is assumed to be zero in the neutral p region for $x < -x_p$ since the currents are zero in thermal equilibrium. Since there are no surface charge densities within the pn junction structure, the electric field is a continuous function. The constant of integration is determined by setting $E = 0$ at $x = -x_p$. The electric field in the p region is then given by

$$E = \frac{-eN_a}{\epsilon_s}(x + x_p) \qquad -x_p \le x \le 0 \tag{14}$$

In the n region, the electric field is determined from

$$E = \int \frac{(eN_d)}{\epsilon_s} dx = \frac{eN_d}{\epsilon_s} x + C_2 \tag{15}$$

where C_2 is again a constant of integration and is determined by setting $E = 0$ at $x = x_n$, since the E-field is assumed to be zero in the n region and is a continuous function. Then

$$E = \frac{-eN_d}{\epsilon_s}(x_n - x) \qquad 0 \le x \le x_n \tag{16}$$

The electric field is also continuous at the metallurgical junction, or at $x = 0$. Setting Equations (14) and (16) equal to each other at $x = 0$ gives

$$N_a x_p = N_d x_n \tag{17}$$

The number of positive charges per unit area in the n region and the number of negative charges per unit area in the p region are equal, according to equation (17).

The electric field in the depletion region is plotted in Figure 3.6. For this geometry, the electric field is oriented in the negative x direction, that is, from the n to the p region. The maximum (magnitude) electric field for the uniformly doped pn junction happens at the metallurgical junction, and the E-field is a linear function of distance through the junction. Even in the absence of any voltage applied between the p and n regions, there is still an electric field in the depletion region.

The potential in the junction is found by integrating the electric field. In the p region then, we have

$$\phi(x) = -\int E(x)dx = \int \frac{eN_a}{\epsilon_s}(x + x_p)dx \tag{18}$$

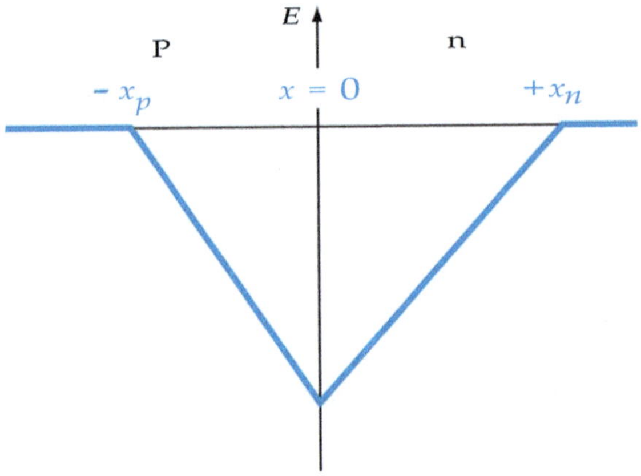

Figure 3.6. Electric field in the space charge region of a uniformly doped pn junction.

Source: Donald A. Neamen; Semiconductor Physics and Devices Basic Principles Fourth Edition; ISBN 978–0-07–352958–5.

Or

$$\phi(x) = \frac{eN_a}{\epsilon_s}\left(\frac{x^2}{2} + x_p \cdot x\right) + C_1'$$

where C_1' is again a constant of integration. The potential difference through the pn junction is the important parameter, rather than the absolute potential, so we may arbitrarily set the potential equal to zero at x = -x_p. The constant of integration is then found as

$$C_1' = \frac{eN_a}{2\epsilon_s}x_p^2 \qquad (20)$$

so that the potential in the p region can now be written as

$$\phi(x) = \frac{eN_a}{2\epsilon_s}(x + x_p)^2 \qquad (-x_p \le x \le 0) \qquad (21)$$

The potential in the n region is determined by integrating the electric field in the n region, or

$$\phi(x) = \int \frac{eN_d}{\epsilon_s}(x_n - x)dx \qquad (22)$$

Then

$$\phi(x) = \frac{eN_d}{\epsilon_s}\left(x_n \cdot x - \frac{x^2}{2}\right) + C_2' \qquad (23)$$

where is another constant of integration. The potential is a continuous function, so setting Equation (21) equal to Equation (23) at the metallurgical junction, or at x = 0, gives

$$C_2' = \frac{eN_a}{2\epsilon_s} x_p^2 \tag{24}$$

The potential in the n region can thus be written as

$$\phi(x) = \frac{eN_d}{\epsilon_s}\left(x_n \cdot x - \frac{x^2}{2}\right) + \frac{eN_a}{2\epsilon_s} x_p^2 \qquad (0 \le x \le x_n) \tag{25}$$

Figure 3.7 is a plot of the potential through the junction and shows the quadratic dependence on distance. The magnitude of the potential at $x = x_n$ is equal to the builtin potential barrier. Then from Equation (25), we have

$$V_{bi} = |\phi(x = x_n)| = \frac{e}{2\epsilon_s}\left(N_d x_n^2 + N_a x_p^2\right) \tag{26}$$

The potential energy of an electron is given by E = -eϕ, which means that the electron potential energy also varies as a quadratic function of distance through the space charge region. The quadratic dependence on distance was shown in the energy band diagram of Figure 3.4, although we did not explicitly know the shape of the curve at that time.

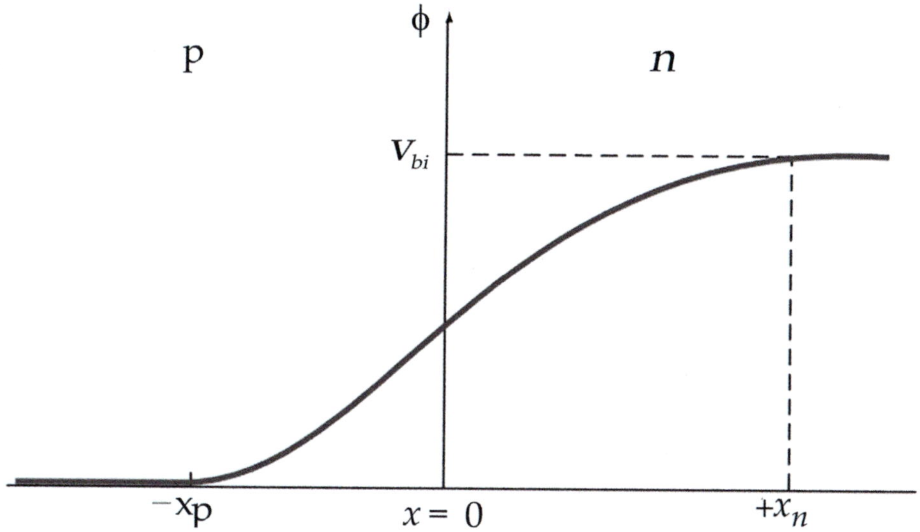

Figure 3.7. Electric potential through the space charge region of a uniformly doped pn junction.

Source: Donald A. Neamen; Semiconductor Physics and Devices Basic Principles Fourth Edition; ISBN 978-0-07-352958-5.

3.3.3. Space Charge Width

$$x_p = \frac{N_d x_n}{N_a} \tag{27}$$

Then, substituting Equation (27) into Equation (26) and solving for x_n, we obtain

$$x_n = \left\{\frac{2\epsilon_s V_{bi}}{e}\left[\frac{N_a}{N_d}\right]\left[\frac{1}{N_a + N_d}\right]\right\}^{1/2} \quad (28)$$

Equation (28) gives the space charge width, or the width of the depletion region, x_n extending into the n-type region for the case of zero applied voltage.

Similarly, if we solve for x_n from Equation (17) and substitute into Equation (26), we find

$$x_p = \left\{\frac{2\epsilon_s V_{bi}}{e}\left[\frac{N_d}{N_a}\right]\left[\frac{1}{N_a + N_d}\right]\right\}^{1/2} \quad (29)$$

where x_p is the width of the depletion region extending into the p region for the case of zero applied voltage.

The total depletion or space charge width W is the sum of the two components, or

$$W = x_n + x_p \quad (30)$$

Using Equations (28) and (29), we obtain

$$W = \left\{\frac{2\epsilon_s V_{bi}}{e}\left[\frac{N_a + N_d}{N_a N_d}\right]\right\}^{1/2} \quad (31)$$

The built-in potential barrier can be determined from Equation (10), and then the total space charge region width is obtained using Equation (31).

3.4. REVERSE APPLIED BIAS

The Fermi energy level in the system will no longer be constant if a potential is applied between the p and n regions, breaking the equilibrium. The energy-band diagram of the pn junction for the scenario in which a positive voltage is applied to the n region relative to the p region is displayed in Figure 3.8. The Fermi level on the n side is lower than the Fermi level on the p side because of the downward positive potential. The applied voltage in energy units is equal to the difference between the two.

The total potential barrier, indicated by V_{total}, has increased. The applied potential is the reverse-biased condition. The total potential barrier is now given by

$$V_{total} = |\phi_{FN}| + |\phi_{FP}| + V_R \tag{32}$$

where VR is the magnitude of the applied reverse-biased voltage. Equation (32) can be rewritten as

$$V_{total} = V_{bi} + V_R \tag{33}$$

where V_{bi} is the same built-in potential barrier we had defined in thermal equilibrium.

3.4.1. Space Charge Width and Electric Field

A pn junction with an applied reverse-biased voltage V_R is depicted in Figure 3.9. The electric field induced by the applied voltage (E_{app}) and the electric field in the space charge region are also depicted in the figure. Since there is no electric field in the neutral p and n regions, or very little electric field there, the applied voltage must cause the electric field in the space charge region to grow larger than the value at thermal equilibrium. Since the electric field begins with a positive charge and ends with a negative charge, an increase in both positive and negative charges is required for the electric field to grow. The only way to increase the number of positive and negative charges in the depletion region for a given impurity doping concentration is to increase the space charge width W. Thus, as the reverse-biased voltage V_R increases, so does the space charge width W. The electric field in the bulk n and p regions is assumed to be zero. When we discuss the current-voltage characteristics in the upcoming chapter, this assumption will become more apparent.

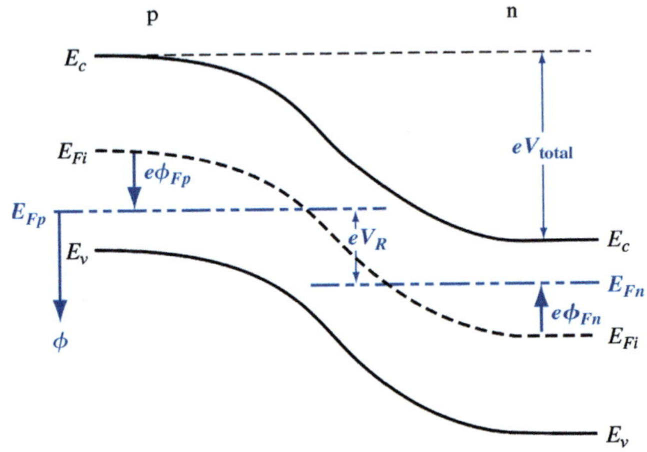

Figure 3.8. Energy-band diagram of a pn junction under reverse bias.

Source: Donald A. Neamen; Semiconductor Physics and Devices Basic Principles Fourth Edition; ISBN 978–0-07–352958–5.

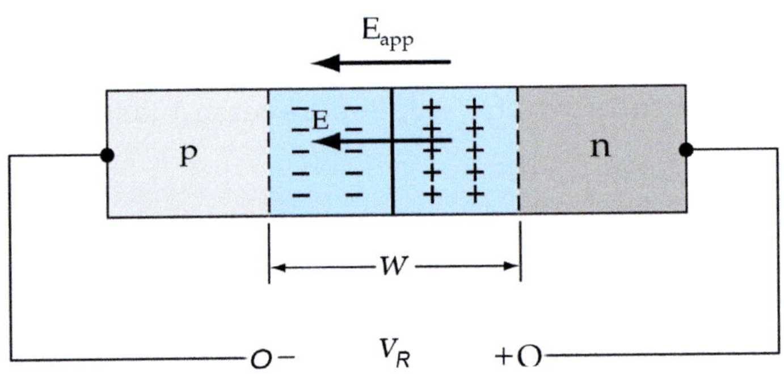

Figure 3.9. A pn junction, with an applied reverse-biased voltage, showing the directions of the electric field induced by V_R and the space charge electric field.

Source: Donald A. Neamen; Semiconductor Physics and Devices Basic Principles Fourth Edition; ISBN 978–0-07–352958–5.

The built-in potential barrier can be replaced by the total potential barrier. The total space charge width can be written from Equation (31) as

$$W = \left\{ \frac{2\epsilon_s (V_{bi} + V_R)}{e} \left[\frac{N_a + N_d}{N_a N_d} \right] \right\}^{1/2} \tag{34}$$

showing that the total space charge width increases as we apply a reverse-biased voltage. By substituting the total potential barrier V_{total} into Equations (28) and (29), the space charge widths in the n and p regions, respectively, can be found as a function of applied reverse-biased voltage.

When a reverse-biased voltage is applied, the depletion region's electric field strength grows. Equations (14) and (16) still determine the electric field, which is still a linear function of distance through the space charge region. The electric field's magnitude increases with reverse-biased voltage because x_n and x_p grow. The metallurgical junction is still the location of the maximum electric field.

The maximum electric field at the metallurgical junction, from Equations (14) and (16), is

$$E_{max} = \frac{-eN_d x_n}{\epsilon_s} = \frac{-eN_a x_p}{\epsilon_s} \quad (35)$$

If we use either Equation (28) or (29) in conjunction with the total potential barrier, $V_{bi} + V_R$, then

$$E_{max} = -\left\{\frac{2e(V_{bi} + V_R)}{\epsilon_s}\left(\frac{N_a N_d}{N_a + N_d}\right)\right\}^{1/2} \quad (36)$$

We can show that the maximum electric field in the pn junction can also be written as

$$E_{max} = \frac{-2(V_{bi} + V_R)}{W} \quad (37)$$

where W is the total space charge width.

3.4.2. Junction Capacitance

The pn junction is connected to a capacitance since the depletion region separates the positive and negative charges. The charge densities in the depletion region for reverse-biased voltages applied to V_R and $V_R + d_{VR}$ are displayed in Figure 3.10. More positive charges in the n region and more negative charges in the p region will be revealed by an increase in the reverse-biased voltage d_{VR}. It is defined as the junction capacitance.

$$C' = \frac{dQ'}{dV_R} \quad (38)$$

where

$$dQ' = eN_d\, dx_n = eN_a\, dx_p \quad (39)$$

The differential charge dQ' is in units of C/cm^2 so that the capacitance C' is in units of farads per square centimeter F/cm^2), or capacitance per unit area.

For the total potential barrier, Equation (28) may be written as

$$x_n = \left\{\frac{2\epsilon_s(V_{bi} + V_R)}{e}\left[\frac{N_a}{N_d}\right]\left[\frac{1}{N_a + N_d}\right]\right\}^{1/2} \quad (40)$$

The junction capacitance can be written as

$$C' = \frac{dQ'}{dV_R} = eN_d \frac{dx_n}{dV_R} \qquad (41)$$

so that

$$C' = \left\{ \frac{e\epsilon_s N_a N_d}{2(V_{bi} + V_R)(N_a + N_d)} \right\}^{1/2} \qquad (42)$$

Exactly the same capacitance expression is obtained by considering the space charge region extending into the p region x_p. The junction capacitance is also referred to as the depletion layer capacitance.

If we compare Equation (34) for the total depletion width W of the space charge region under reverse bias and Equation (42) for the junction capacitance C', we find that we can write

$$C' = \frac{\epsilon_s}{W} \qquad (43)$$

The capacitance per unit area of a parallel plate capacitor is equal to equation (43) in this case. With reference to Figure 3.10, we might have arrived at this same judgment sooner. Remember that since the space charge width depends on the reverse-biased voltage applied to the pn junction, the junction capacitance likewise depends on the reverse-biased voltage.

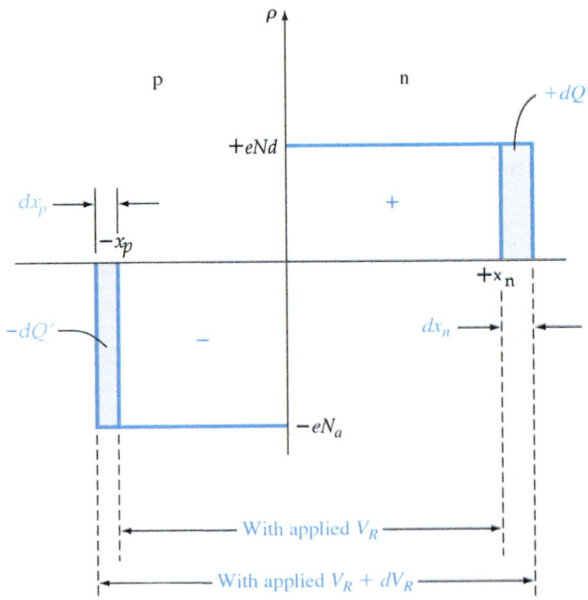

Figure 3.10. Differential change in the space charge width with a differential change in reverse-biased voltage for a uniformly doped pn junction.

Source: Donald A. Neamen; Semiconductor Physics and Devices Basic Principles Fourth Edition; ISBN 978-0-07-352958-5.

3.4.3. One-Sided Junctions

Consider a special pn junction called the one-sided junction. If, for example, $N_a \gg N_d$, this junction is referred to as a p^+n junction. The total space charge width, from Equation (34), reduces to

$$W \approx \left\{ \frac{2\epsilon_s (V_{bi} + V_R)}{eN_d} \right\}^{1/2} \tag{44}$$

Considering the expressions for xn and xp, we have for the p^+n junction

$$x_p \ll x_n \tag{45}$$

and

$$W \approx x_n \tag{46}$$

Almost the entire space charge layer extends into the low-doped region of the junction. This effect can be seen in Figure 3.11.

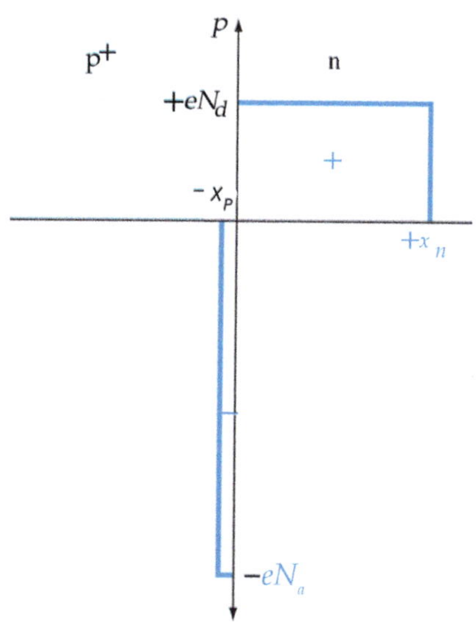

Figure 3.11. Space charge density of a one-sided p+n junction.

Source: Donald A. Neamen; Semiconductor Physics and Devices Basic Principles Fourth Edition; ISBN 978–0-07–352958–5.

The junction capacitance of the p^+n junction reduces to

$$C' \approx \left\{ \frac{e\epsilon_s N_d}{2(V_{bi} + V_R)} \right\}^{1/2} \tag{47}$$

The depletion layer capacitance of a one-sided junction is a function of the doping concentration in the low-doped region. Equation (47) may be manipulated to give

$$\left(\frac{1}{C'}\right)^2 = \frac{2(V_{bi} + V_R)}{e\epsilon_s N_d} \qquad (48)$$

which shows that the inverse capacitance squared is a linear function of applied reverse-biased voltage.

Equation (48) is plotted in Figure 3.12. By extending the curve to the point where $(1/C)^2 = 0$, one can ascertain the junction's inherent potential. The doping concentration of the low-doped region in the junction can be experimentally determined, as the slope of the curve is inversely proportional to it. This capacitance was derived under the following assumptions: a planar junction, the abrupt junction approximation, and uniform doping in both semiconductor regions.

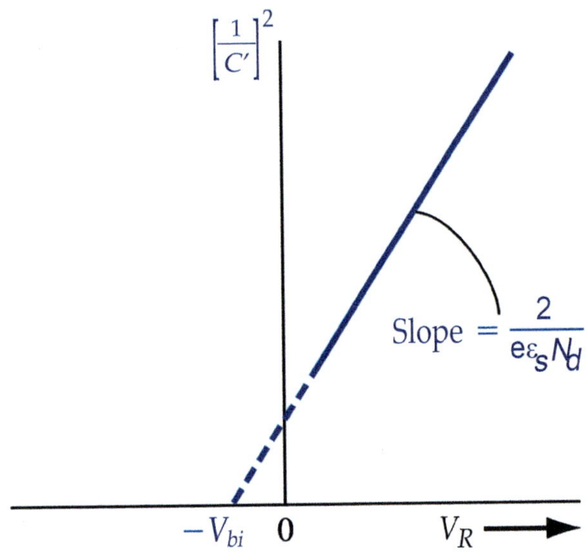

Figure 3.12. $(1/C')^2$ versus V_R of a uniformly doped pn junction.

Source: Donald A. Neamen; Semiconductor Physics and Devices Basic Principles Fourth Edition; ISBN 978–0-07–352958–5.

3.5. JUNCTION BREAKDOWN

The consequences of putting a reverse-biased voltage across the pn junction were ascertained. Reverse-biased voltages, however, might not rise infinitely; at a certain voltage, reverse-biased currents will rise quickly. The breakdown voltage is the applied voltage at this point.

The reverse-biased breakdown in a pn junction is caused by the avalanche and Zener effects, two physical processes. In highly doped pn junctions, Zener breakdown takes place via a tunneling mechanism. When reverse bias is applied to a highly doped junction, electrons may tunnel straight from the valence band on the p side into the conduction band on the n side due to the proximity of the conduction and valence bands on opposite sides of the junction. Figure 3.13a provides a schematic representation of this tunneling process.

When electrons and/or holes travel through the space charge region and gather enough energy from the electric field to collide with atomic electrons in the depletion region to form electron-hole pairs, the process known as avalanche breakdown takes place. Figure 3.13b provides a schematic representation of the avalanche process. A reverse-biased current is produced when the freshly formed electrons and holes travel in opposing directions as a result of the electric field. The avalanche process can also occur if the freshly created electrons and/or holes have enough energy to ionize additional atoms. The avalanche effect will be the main mechanism of breakdown for the majority of pn junctions.

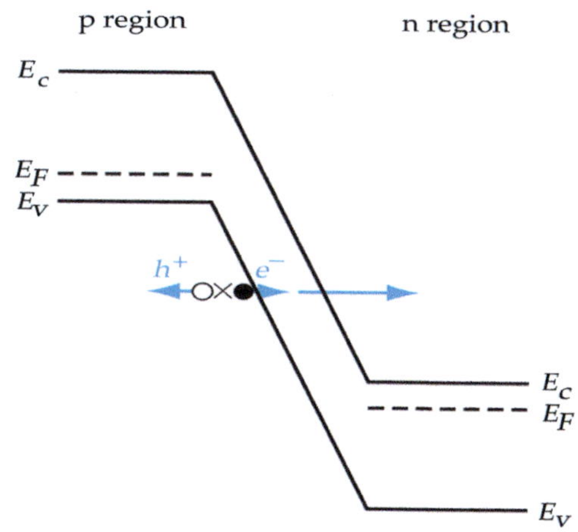

(a)

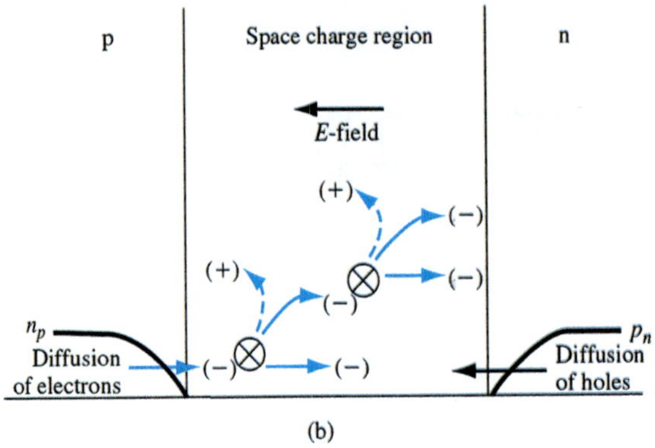

Figure 3.13. (a) Zener breakdown mechanism in a reverse-biased pn junction; (b) avalanche breakdown process in a reverse-biased pn junction.

Source: Donald A. Neamen; Semiconductor Physics and Devices Basic Principles Fourth Edition; ISBN 978-0-07-352958-5.

If we assume that a reverse-biased electron current I_{n0} enters the depletion region at x = 0 as shown in Figure 3.14, the electron current I_n will increase with distance through the depletion region due to the avalanche process. At x = W, the electron current may be written as

$$I_n(W) = M_n I_{n0} \tag{49}$$

where M_n is a multiplication factor. The hole current is increasing through the depletion region from the n to p region and reaches a maximum value at x = 0. The total current is constant through the pn junction in steady state.

We can write an expression for the incremental electron current at some point x as

$$dI_n(x) = I_n(x)\alpha_n\, dx + I_p(x)\alpha_p\, dx \tag{50}$$

where α_n and α_p are the electron and hole ionization rates, respectively. The ionization rates are the number of electrons–hole pairs generated per unit length by an electron (α_n) or by a hole (α_p). Equation (50) may be written as

$$\frac{dI_n(x)}{dx} = I_n(x)\alpha_n + I_p(x)\alpha_p \tag{51}$$

The total current I is given by

$$I = I_n(x) + I_p(x) \tag{52}$$

which is a constant. Solving for $I_p(x)$ from Equation (52) and substituting into Equation (51), we obtain

$$\frac{dI_n(x)}{dx} + (\alpha_p - \alpha_n)I_n(x) = \alpha_p I \tag{53}$$

If we make the assumption that the electron and hole ionization rates are equal so that

$$\alpha_n = \alpha_p \equiv \alpha \tag{54}$$

then Equation (53) may be simplified and integrated through the space charge region. We will obtain

$$I_n(W) - I_n(0) = I \int_0^W \alpha \, dx \tag{55}$$

Using Equation (49), Equation (55) may be written as

$$\frac{M_n I_{n0} - I_n(0)}{I} = \int_0^W \alpha \, dx \tag{56}$$

Since $M_n I_{n0} \approx I$ and since In (0) = In0, Equation (56) becomes

$$1 - \frac{1}{M_n} = \int_0^W \alpha \, dx \tag{57}$$

The avalanche breakdown voltage is defined to be the voltage at which M_n approaches infinity. The avalanche breakdown condition is then given by

$$\int_0^W \alpha dx = 1 \tag{58}$$

The ionization rates are strong functions of electric field and, since the electric field is not constant throughout the space charge region, Equation (58) is not easy to evaluate.

If we consider, for example, a one-sided p⁺n junction, the maximum electric field is given by

$$\mathrm{E}_{\max} = \frac{eN_d x_n}{\epsilon_s} \tag{59}$$

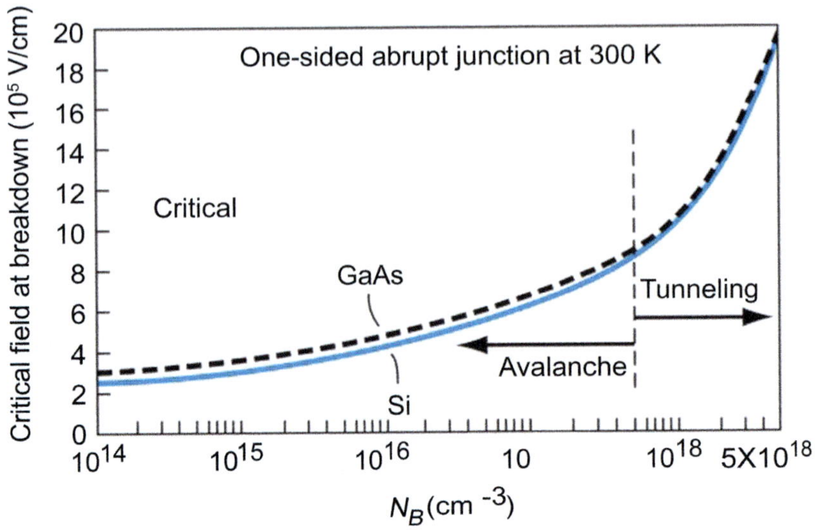

Figure 3.14. Critical electric field at breakdown in a one-sided junction as a function of impurity doping concentrations.

Source: Donald A. Neamen; Semiconductor Physics and Devices Basic Principles Fourth Edition; ISBN 978-0-07-352958-5.

The depletion width x_n is given approximately as

$$x_n \approx \left\{ \frac{2\epsilon_s V_R}{e} \cdot \frac{1}{N_d} \right\}^{1/2} \tag{60}$$

where V_R is the magnitude of the applied reverse-biased voltage. We have neglected the built-in potential V_{bi}.

If we now define V_R to be the breakdown voltage V_B, the maximum electric field, E_{max}, will be defined as a critical electric field, E_{crit}, at breakdown. Combining Equations (59) and (60), we may write

$$V_B = \frac{\epsilon_s E_{crit}^2}{2eN_B} \tag{61}$$

where N_B is the semiconductor doping in the low-doped region of the one-sided junction. The critical electric field, plotted in Figure 3.14, is a slight function of doping.

A planar junction with uniform doping has been under our consideration. A junction with a linear grade will see a drop in breakdown voltage. The breakdown voltage plots for a linearly graded junction and a one-sided abrupt junction are displayed in Figure 3.15. The breakdown voltage will decrease even more if the curvature of a diffused junction is also considered.

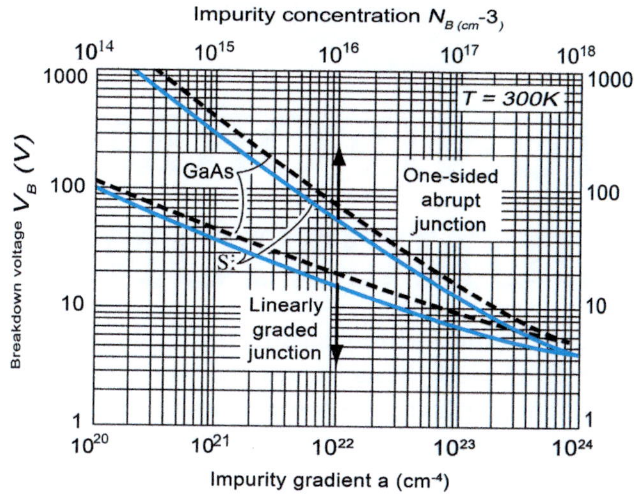

Figure 3.15. Breakdown voltage versus impurity concentration in uniformly doped and linearly graded junctions.

Source: Donald A. Neamen; Semiconductor Physics and Devices Basic Principles Fourth Edition; ISBN 978-0-07-352958-5.

3.6. NONUNIFORMLY DOPED JUNCTIONS

We have assumed that every semiconductor region in the pn junctions we have looked at so far has been uniformly doped. This is rarely the case in real-world pn junction structures. Certain nonuniform doping profiles are used in some electronic applications to achieve unique pn junction capacitance characteristics. (Chavez, R., A. Becker, V. Kessler, 2012)

3.6.1. Linearly Graded Junctions

Starting from an n-type semiconductor that is uniformly doped, for instance, and diffusing acceptor atoms through the surface, the impurity concentrations will typically resemble the ones displayed in Figure 3.16. The metallurgical junction is represented by the point x = x' on the diagram. One can approximate the net p-type doping concentration in the vicinity of the metallurgical junction as a linear function of the distance from the junction. Similarly, the net n-type doping concentration is a linear function of distance extending into the n region from the metallurgical junction as a first approximation. We refer to this effective doping profile as a linearly graded junction.

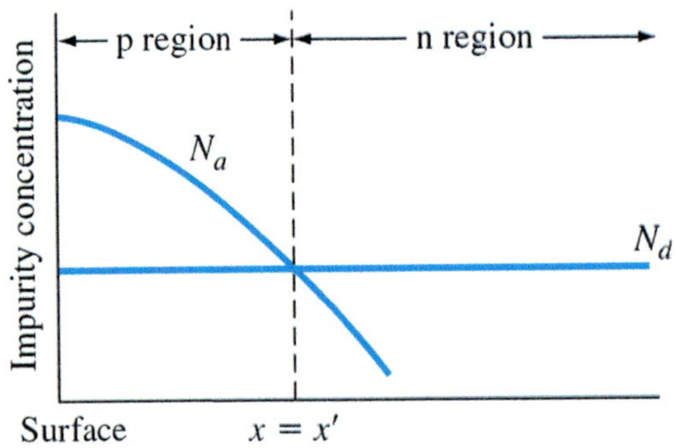

Figure 3.16. Impurity concentrations of a pn junction with a nonuniformly doped p region.

Source: Donald A. Neamen; Semiconductor Physics and Devices Basic Principles Fourth Edition; ISBN 978–0-07–352958–5.

Figure 3.17 shows the space charge density in the depletion region of the linearly graded junction. For convenience, the metallurgical junction is placed at x = 0. The space charge density can be written as

$$\rho(x) = eax \qquad (62)$$

where a is the gradient of the net impurity concentration.

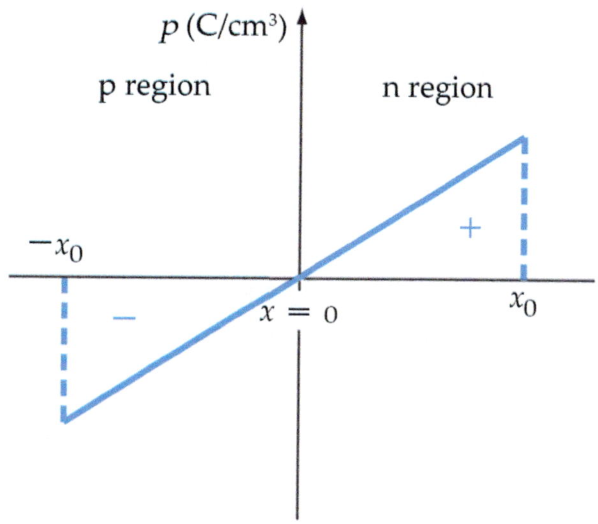

Figure 3.17. Space charge density in a linearly graded pn junction.

Source: Donald A. Neamen; Semiconductor Physics and Devices Basic Principles Fourth Edition; ISBN 978–0-07–352958–5.

The electric field and potential in the space charge region can be determined from Poisson's equation. We can write

$$\frac{dE}{dx} = \frac{\rho(x)}{\epsilon_s} = \frac{eax}{\epsilon_s} \qquad (63)$$

So that the electric field can be found by integration as

$$E = \int \frac{eax}{\epsilon_s}\, dx = \frac{ea}{2\epsilon_s}\left(x^2 - x_0^2\right) \qquad (64)$$

Unlike the linear function observed in the uniformly doped junction, the electric field in the linearly graded junction is a quadratic function of distance. The metallurgical junction is again located where the electric field reaches its maximum value. It is evident that there is no electric field at both $x = +x_0$ and $x = -_{x0.}$ Although the electric field in a semiconductor with nonuniform doping is not exactly zero, it is small, so putting E = 0 in the bulk regions is still a good approximation.

The potential is again found by integrating the electric field as

$$\phi(x) = -\int E\, dx \qquad (65)$$

If we arbitrarily set $\phi = 0$ at $x = -x_0$, then the potential through the junction is

$$\phi(x) = \frac{-ea}{2\epsilon_s}\left(\frac{x^3}{3} - x_0^2 x\right) + \frac{ea}{3\epsilon_s} x_0^3 \tag{66}$$

The magnitude of the potential at $x = +x_0$ will equal the built-in potential barrier for this function. We then have that

$$\phi(x_0) = \frac{2}{3} \cdot \frac{eax_0^3}{\epsilon_s} = V_{bi} \tag{67}$$

Another expression for the built-in potential barrier for a linearly graded junction can be approximated from the expression used for a uniformly doped junction. We can write

$$V_{bi} = V_t \ln\left[\frac{N_d(x_0)N_a(-x_0)}{n_i^2}\right] \tag{68}$$

where $N_d(x_0)$ and $N_a(-x_0)$ are the doping concentrations at the edges of the space charge region. We can relate these doping concentrations to the gradient, so that

$$N_d(x_0) = ax_0 \tag{69a}$$

And

$$N_a(-x_0) = ax_0 \tag{69b}$$

Then the built-in potential barrier for the linearly graded junction becomes

$$V_{bi} = V_t \ln\left(\frac{ax_0}{n_i}\right)^2 \tag{70}$$

There may be situations in which the doping gradient is not the same on either side of the junction, but we will not consider that condition here.

If a reverse-biased voltage is applied to the junction, the potential barrier increases. The built-in potential barrier V_{bi} in the above equations is then replaced by the total potential barrier $V_{bi} + V_R$. Solving for x_0 from Equation (67) and using the total potential barrier, we obtain

$$x_0 = \left\{\frac{3}{2} \cdot \frac{\epsilon_s}{ea}(V_{bi} + V_R)\right\}^{1/3} \tag{71}$$

The junction capacitance per unit area can be determined by the same method that we used for the uniformly doped junction. Figure 3.18 shows the differential charge dQ,' which is uncovered as a differential voltage dV_R is applied. The junction capacitance is then

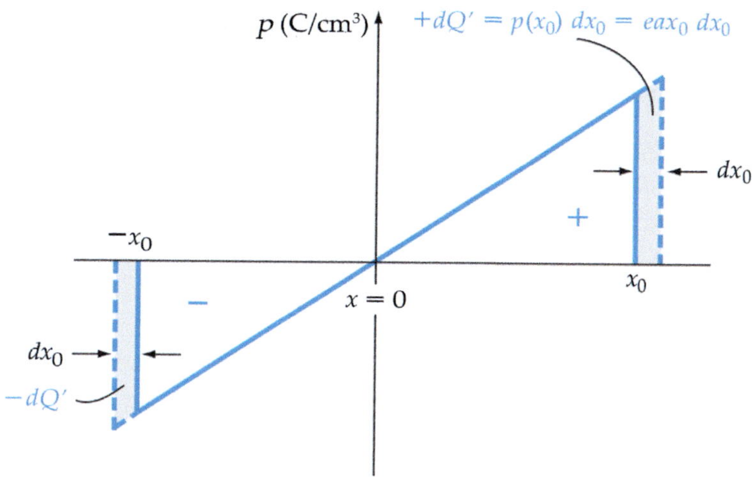

Figure 3.18. Differential change in space charge width with a differential change in reverse-biased voltage for a linearly graded pn junction.

Source: Donald A. Neamen; Semiconductor Physics and Devices Basic Principles Fourth Edition; ISBN 978–0-07–352958–5.

$$C' = \frac{dQ'}{dV_R} = (eax_0)\frac{dx_0}{dV_R} \tag{72}$$

Using Equation (71), we obtain

$$C' = \left\{\frac{ea\epsilon_s^2}{12(V_{bi} + V_R)}\right\}^{1/3} \tag{73}$$

We may note that C′ is proportional to $(V_{bi} + V_R)^{-13}$ for the linearly graded junction as compared to $C'\alpha(V_{bi} + V_R)^{-1/2}$ for the uniformly doped junction. In the linearly graded junction, the capacitance is less dependent on reverse-biased voltage than in the uniformly doped junction.

3.6.2. Hyperabrupt Junctions

There are other possible doping profiles besides the uniformly doped and linearly graded junctions. The generalized one-sided pn junction depicted in Figure 3.19 is where the generalized n-type doping concentration for $x > 0$ is found.

$$N = Bx^m \tag{74}$$

The uniformly doped junction is represented by the case of $m = 0$ and the recently discussed linearly graded junction by $m = +1$. The indicated $m = +2$ and $m = +3$ cases would be similar to an epitaxial n-type layer that is relatively low doped and grown on a much more heavily doped n^+ substrate layer. The term "hyperabrupt junction" describes what happens when the value of m is negative. In this instance, n-type doping is more prevalent close to the metallurgical junction than it is in the semiconductor's bulk. When m is negative, Eq. (74) does not hold at x = 0 and is instead used to approximate the n-type doping over a narrow region close to $x = x_0$.

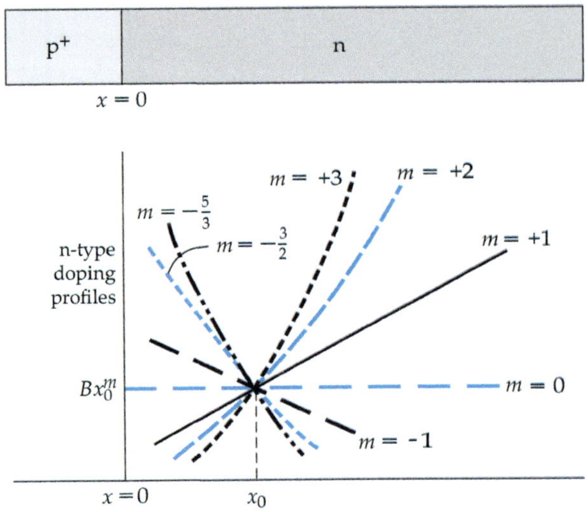

Figure 3.19. Generalized doping profiles of a one-sided p⁺n junction.

Source: Donald A. Neamen; Semiconductor Physics and Devices Basic Principles Fourth Edition; ISBN 978-0-07-352958-5.

The junction capacitance can be derived using the same analysis method as before and is given by

$$C' = \left\{\frac{eB\epsilon_s^{(m+1)}}{(m+2)(V_{bi}+V_R)}\right\}^{1/(m+2)} \tag{75}$$

A desirable property of varactor diodes is that the capacitance becomes a very strong function of reverse-biased voltage when m is negative. A varactor is a device whose reactance can be controlled by bias voltage; the term originates from the word's variable reactor.

If a varactor diode and an inductance are in parallel, the resonant frequency of the LC circuit is

$$f_r = \frac{1}{2\pi\sqrt{LC}} \tag{76}$$

The capacitance of the diode, from Eq. (75), can be written in the form

$$C = C_0(V_{bi}+V_R)^{-1/(m+2)} \tag{77}$$

In a circuit application, we would, in general, like to have the resonant frequency be a linear function of reverse-biased voltage V_R, so we need

$$C \, \alpha \, V^{-2} \tag{78}$$

From Eq. (77), the parameter m required is found from

$$\frac{1}{m+2} = 2 \tag{79}$$

Or

$$m = -\frac{3}{2} \tag{80}$$

A specific doping profile will yield the desired capacitance characteristic.

CASE STUDY

Semiconductor X-Ray Inspection Tool: From Prototype to Production without Alpha & Beta

The Situation

The Bay Area OEM start-up Silicon Valley X-Ray (SVXR – acquired by Bruker) had successfully demonstrated their X-ray inspection technology on their lab tool and needed to rapidly develop a fab-ready product.

Their Asia customer requested a production-worthy tool to go into their semiconductor fab in seven months.

The Challenge

SVXR needed to go directly from a lab tool to a Fab-ready tool on an extremely compressed schedule, bypassing alpha and beta systems. The tool must meet all the SEMI and safety standards, be highly reliable and maintainable, and be built to high-quality requirements. SVXR did not have the time nor the capital to build an internal engineering and manufacturing capability. They chose to outsource to someone with deep experience in designing and building semiconductor tools and scaling manufacturing.

The Solution

SVXR selected Owens Design to be their design and build partner based on their engineering expertise and their ability to scale manufacturing.

Owens Design developed and built the prototype tool in six months. The PO was delivered in June, with the final integration completed by January of the following year. The system was then accepted at a Taiwan semiconductor fab the following June – almost exactly one year from the start.

Owens delivered four additional systems in the following month. Production was set up to support building five systems in parallel while handling the current capacity of building two per month. Capacity is scalable to once a week with four months' notice.

Owens Design also supported a cost reduction roadmap, to reduce cost and increase profitability.

The Results

According to Dr. David Adler, founder and CEO of SVXR, "We wanted to go directly from a prototype to a production product without doing alpha and beta. When we decided that quality and time to market were the most important, we had very few choices because we wanted to go with somebody who knew what they were doing. You don't want to hire a heart surgeon who's doing his first heart surgery on you. You want somebody who's done it many times and knows exactly what they're doing. By going with Owens, we were confident that they would be able to execute on our very accelerated time schedule and produce a high-quality product.

"Today in the market, we're establishing ourselves as the primary company for high-volume inline x-ray manufacturing x-ray inspection and we've established a dominant position in that new market. We're about 100 times faster than our competitors, so there's very little competition for what we do. We're ready to start scaling to much higher volumes so, over the next few years, we plan to scale to tens of units to hundreds of units a year and we will be doing that with Owens.

"I don't think that we would be as successful today as we are without working with Owens. There are a lot of things that can go wrong when moving from a prototype to a product, and because our system is running tens of millions of dollars of product per hour through it, it had to really be reliable and work exactly as we said it would. That required outstanding performance not only from our own team but also from Owens to be able to produce a tool that we and our customers could rely on. They did that and contributed a great deal to our success.

Michael Jupina, VP Engineering of SVXR, added, "SVXR went to Owens for a simple reason: it enabled us to concentrate on our IP—the x-ray imaging—where Owens' many years of experience of building a system was then utilized for us to put together this system that was able to be installed and qualified in IC packaging houses throughout Southeast Asia."

CLASS ACTIVITY

EXPLORING THE P-N JUNCTION

Objective:

To understand the formation, behavior, and applications of P-N junctions in semiconductor devices.

Materials Needed

- Whiteboard and markers
- Semiconductor samples (if available)
- Multimeter (if available)
- Computers or tablets with internet access
- Worksheets

Assessment:

- Provide students with worksheets that include diagrams of P-N junctions in both forward and reverse bias.
- Ask students to label the diagrams and explain the movement of charge carriers.
- Draw and explain the process of forming a P-N junction by joining P-type and N-type materials.

SUMMARY

- The P-N junction is the control element for the performance of all semiconductor devices such as rectifiers, amplifiers, switching devices, linear and digital integrated circuits. The PN Junction in Semiconductor is produced by placing a layer of P-type semiconductor next to the layer of N-type semiconductor.
- The P-type semiconductor block has mobile holes (shown by small circles) and the same number of fixed negative acceptor ions (shown by encircled minus sign).
- A potential difference exists across the space charge region. Under zero applied bias, this potential difference, known as the built-in potential barrier, maintains thermal equilibrium and holds back the majority of carrier electrons in the n region and the majority of carrier holes in the p region.
- Zero bias PN junction – Here we will examine the properties of the step junction in thermal equilibrium, where no currents exist and no external excitation is applied. We will determine the built-in potential barrier through the depletion or space-charge region, electric field, and width of the space-charge region.
- Electrons in the conduction band of the n region see a potential barrier in trying to move into the conduction band of the p region. This potential barrier is referred to as the built-in potential barrier and is denoted by $V_{bi.}$ The built-in potential barrier maintains equilibrium between majority carrier electrons in the n region and minority carrier electrons in the p region, and also between majority carrier holes in the p region and minority carrier holes in the n region.
- The electric fields in the neutral p and n regions are essentially zero, or at least very small, which means that the magnitude of the electric field in the space charge region must increase above the thermal equilibrium value due to the applied voltage.
- The avalanche breakdown process occurs when electrons and/or holes, moving across the space charge region, acquire sufficient energy from the electric field to create electron-hole pairs by colliding with atomic electrons within the depletion region.
- Avalanche breakdown occurs when a sufficiently large reverse-biased voltage is applied to the pn junction. A large reverse-biased current may then be induced in the pn junction. The breakdown voltage, as a function of the doping concentrations in the pn junction, is derived. In a one-sided pn junction, the breakdown voltage is a function of the doping concentration in the low-doped region.

REVIEW QUESTIONS

1. Why is an electric field formed in the space charge region? Why is the electric field a linear function of distance in a uniformly doped pn junction?
2. Where does the maximum electric field occur in the space charge region?

3. Why is the space charge width larger on the lower doped side of a pn junction?
4. What is the functional dependence of the space charge width on reverse-biased voltage?
5. Why does the space charge width increase with reverse-biased voltage?
6. Why does a capacitance exist in a reverse-biased pn junction? Why does the capacitance decrease with increasing reverse-biased voltage?
7. What is a one-sided pn junction? What parameters can be determined in a one-sided pn junction?
8. Why does the breakdown voltage of a pn junction decrease as the doping concentration increases?

REFERENCES

1. Bowers, D. F. (2014). A fast precision operational amplifier featuring two separate control loops. In *2014 IEEE Bipolar/BiCMOS Circuits and Technology Meeting (BCTM)* (pp. 72–75). IEEE. https://doi.org/10.1109/BCTM.2014.6940028.
2. Chavez, R., Becker, A., Kessler, V., Engenhorst, M., Petermann, N., Wiggers, H., Schierning, G., & Schmechel, R. (2013). A new thermoelectric concept using large area PN junctions. *MRS Proceedings, 1543*, 3–8. http://dx.doi.org/10.1557/opl.2013.954.
3. De Langen, K.-J., & Huijsing, J. H. (1998). Compact low-voltage power-efficient operational amplifier cells for VLSI. *IEEE Journal of Solid-State Circuits, 33*(10), 1482–1496. https://doi.org/10.1109/4.720403.
4. Gilasgar, M., Barlabé, A., & Pradell, L. (2020). High-efficiency reconfigurable dual-band class-F power amplifier with harmonic control network using MEMS. *IEEE Microwave and Wireless Components Letters, 30*(7), 677–680. https://doi.org/10.1109/LMWC.2020.2995063.
5. Hosseini, S. E., & Dehrizi, H. G. (2012). A new BJT-transistor with ability of controlling current gain. In *International Multi-Conference on Systems, Signals & Devices* (pp. 1–4). IEEE. https://doi.org/10.1109/SSD.2012.6198032.
6. Kong, L., Liu, H., Zhu, X., Boon, C. C., Li, C., Liu, Z., & Yeo, K. S. (2020). Design of a wideband variable-gain amplifier with self-compensated transistor for accurate dB-linear characteristic in 65 nm CMOS technology. *IEEE Transactions on Circuits and Systems I: Regular Papers, 67*(12), 4187–4198. https://doi.org/10.1109/TCSI.2020.3006354.
7. Mishra, U. (2008). *Semiconductor Device Physics and Design* (p. 155). Springer. https://doi.org/10.1007/978-1-4020-6481-4.
8. Nam, H., Nguyen, D.-A., Kim, Y., & Seo, C. (2023). Design of 6 GHz variable-gain low-noise amplifier using adaptive bias circuit for radar receiver front end. *Electronics, 12*(9), 2036. https://doi.org/10.3390/electronics12092036.

9. Perez-Verdu, B., Huertas, J. L., & Rodriguez-Vazquez, A. (1988). A new nonlinear time-domain op-amp macromodel using threshold functions and digitally controlled network elements. *IEEE Journal of Solid-State Circuits, 23*(4), 959–971. https://doi.org/10.1109/JSSC.1988.11994.

10. Wen, S., & Chung, D. D. L. (2001). Rectifying and thermocouple junctions based on Portland cement. *Journal of Materials Research, 16*(7), 1989–1993. http://dx.doi.org/10.1557/jmr.2001.0272.

CHAPTER 4 Photovoltaic Devices

LEARNING OBJECTIVES

After studying this chapter, you will be able to:

- Describe the photovoltaic cell generations
- Explain photodetectors and solar cells

KEY TERMS FROM THIS CHAPTER

Back surface field
Light-induced degradation
Passivated Emitter and Rear Cell
Photocurrent
Photolithography
Selenization
Solar energy
Electrodeposition
Monocrystalline
Photoconductor
Photodetector
Photovoltaic cells
Solar cell
Transparent conductive oxide

4.1. INTRODUCTION

Solar cells, also referred to as photovoltaic (PV) devices, are technological innovations that use the photovoltaic effect to directly convert sunlight into electricity. Utilized extensively in renewable energy systems to capture solar energy, a clean and limitless resource, these devices are essential parts of solar panels. Photons from sunlight are absorbed by semiconductor materials used to make photovoltaic devices, mainly silicon. Through the excitation of electrons during the absorption process, electron-hole pairs are created that are divided by an internal electric field, producing a direct current (DC).

Over time, there has been a notable improvement in the efficiency of photovoltaic devices, which dictates the percentage of sunlight converted into electrical power that can be used. Improvements in materials science have resulted in lower production costs and increased efficiency, as demonstrated by the creation of multi-junction and thin-film solar cells. PV technology is used in both large-scale installations like solar farms that supply electricity to the grid and small-scale applications like powering calculators and streetlights.

PV devices have several advantages for the environment, including energy independence and the ability to be installed in remote areas without access to conventional power sources. The adoption of photovoltaic technology is anticipated to increase as international efforts to combat climate change intensify, and it will play a critical role in the switch to sustainable energy systems.

4.2. PHOTOVOLTAIC CELL GENERATIONS

Shifting away from traditional energy production methods is imperative due to concerns about climate change and the rising demand for electricity brought on by, among other factors, an ever-increasing population. One of the things contributing to the continuous change in climate is the increase in atmospheric carbon dioxide levels brought on by the burning of fossil fuels. When compared to fossil fuel sources, switching to renewable energy will result in energy with a lower environmental impact. By fully utilizing solar energy, we are able to create the most advanced energy harvesting technologies that can turn solar energy into electrical power.

The currently used solar energy is very marginal—0.015% is used for electricity production, 0.3% for heating, and 11% is used in the natural photosynthesis of biomass. In contrast, fossil fuels provide between 80 and 85% of the world's energy needs. Fossil fuels present a challenge because of their finite resources and environmental harm caused by CO2 emissions. For example, one ton of carbon dioxide is released into the atmosphere for every ton of coal burned. This carbon dioxide is harmful to the environment and is the main contributor to climate change, the greenhouse effect, global warming, and ozone depletion.

Today, there is a pressing need to discover new sources of renewable energy. To ensure a clean and sustainable future, humanity must discover alternate energy sources. Solar energy is the most advantageous alternative renewable energy source in this situation because of its wide accessibility, adaptability, and environmental friendliness.

Conversion efficiency, or the ratio of solar energy input to electrical energy output, is the most widely used metric to assess the performance of photovoltaic technologies. The systems short-circuit current, open-circuit voltage, and fill factor are just a few of the component characteristics that make up the efficiency. These characteristics are all dependent on fundamental material properties and manufacturing flaws.

The material used to create a photovoltaic cell determines both its efficiency and cost-effectiveness. Extensive research has been conducted in this domain to determine the most economical and efficient material for the construction of photovoltaic cells. The specifications for an ideal material for PV solar cells include the following:

- The cells are expected to have a band gap between 1.1 and 1.7 eV;
- Should have a direct band structure;
- Need to be easily accessible and non-toxic; and
- Should have high photovoltaic conversion efficiency.

The development of techniques to achieve the highest efficiency at the lowest production cost is a major issue in the field of photovoltaic cell development. It is feasible to increase the efficiency of solar cells by employing practical strategies to lower the internal losses of the device. There exist three fundamental categories of losses: optical, quantum, and electrical, each with distinct origins. Reducing losses of any kind necessitates using various, frequently cutting-edge manufacturing techniques for solar modules and cells. Taking into consideration the balance between photogeneration and radiative recombination, the well-known Shockley–Queisser (SQ) limit establishes an upper efficiency limit for technologies that are commercially available.

But since quantum losses are closely related to the internal structure of the cell and its material characteristics, reducing them holds the most potential. The idea of band gap, which describes the lowest energy needed for a photon to incident onto the cell surface and participate in the photovoltaic conversion process, is pertinent in this context. The band gap, which is largely influenced by the material used to make the photovoltaic cell, has a relationship with the efficiency of the cell. With a band gap of roughly 1.12 eV, silicon is the fundamental and most widely used material for solar cells. However, the material's physical properties, particularly the band gap width, can be changed by making changes to its crystal structure.

The inability to absorb photons below the band gap and the thermalization of solar photons with energies above the band gap energy are the two main loss mechanisms in conventional photovoltaic cells. In an effort to enhance solar cell performance, concepts for third-generation solar cells have been put forth to address these two loss mechanisms. The objective of these solutions is to utilize the complete spectrum by introducing innovative mechanisms that generate new pairs of electrons.

Among these ideas, the concept of cells whose fundamental characteristic is an extra intermediate band in the silicon band gap model exhibits significant development potential for raising the power generation efficiency of solar cells made of silicon. There are currently many avenues for research and development on the introduction of intermediate bands in semiconductors. It is situated between the valence band and the conduction band, and its purpose is to permit the absorption of photons with energies below the width of the energy gap, leading to higher quantum efficiency (a higher number of excited electrons in relation to the number of photons incident onto the surface of the cell). One of them is the use of ion implantation, where two techniques can be distinguished: implantation of high-dose metal ions into the silicon layer and introduction of dopants at extremely high concentrations to the semiconductor substrate.

Reducing different kinds of losses that have an impact on the final cell efficiency is necessary to increase solar cell efficiency. The highest verified research cell conversion efficiencies for various photovoltaic technologies are compiled by the National Renewable Energy Laboratory (NREL) and are current as of 1976. Results for cell efficiency are provided for the following semiconductor families: crystalline silicon cells, thin film technologies, multi-junction cells, single-junction gallium arsenide cells, and emerging photovoltaic technologies. A flag with the efficiency and technology symbol across the right edge indicates the most recent world record for a specific technology.

There are four primary generations of photovoltaic cells: first, second, third, and fourth. An increasingly significant factor in the ongoing energy transition is photovoltaics. Developments in materials science and manufacturing techniques have played a major part in that evolution. Still, a lot of work needs to be done before photovoltaics can produce cheaper, greener energy. Research in this area focuses on printable solar cell materials like quantum dots, graphene, or intermediate band gap cells, and efficient photovoltaic devices like multi-junction cells.

A photovoltaic cell's main function is to convert solar radiation, which is known as the photovoltaic effect, from pure light into electrical energy. The production of photovoltaic cells involves a number of technologies, including material modification with various cell component photoelectric conversion efficiencies. Photovoltaic technologies can be categorized into four main generations as a result of the development of numerous unconventional manufacturing techniques for producing functional solar cells (Almosni et al., 2018) (Figure 4.1).

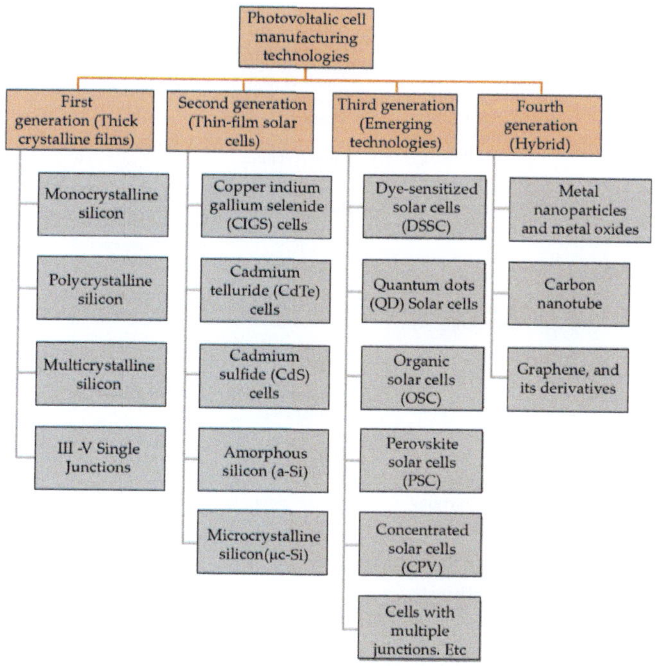

Figure 4.1. Various solar cell types and current developments within this field.

Source. https://www.ncbi.nlm.nih.gov/pmc/articles/PMC9414585/bin/materials-15–05542-g002.jpg.

The generations of various photovoltaic cells essentially tell the story of the stages of their past evolution. There are four main categories that are described as the generations of photovoltaic technology for the last few decades, since the invention of solar cells:

1. First Generation: This category includes photovoltaic cell technologies based on monocrystalline and polycrystalline silicon and gallium arsenide (GaAs).
2. Second Generation: This generation includes the development of first-generation photovoltaic cell technology, as well as the development of thin film photovoltaic

cell technology from "microcrystalline silicon (μc-Si) and amorphous silicon (a-Si), copper indium gallium selenide (CIGS) and cadmium telluride/cadmium sulfide (CdTe/CdS) photovoltaic cells."

3. Third Generation: This generation counts photovoltaic technologies that are based on more recent chemical compounds. In addition, technologies using nanocrystalline "films," quantum dots, dye-sensitized solar cells, solar cells based on organic polymers, etc., also belong to this generation.
4. Fourth Generation: This generation includes the low flexibility or low cost of thin film polymers along with the durability of "innovative inorganic nanostructures such as metal oxides and metal nanoparticles or organic-based nanomaterials such as graphene, carbon nanotubes and graphene derivatives" (Luque et al., 2011).

Examples of solar cell types for each generation along with average efficiencies are shown in Figure 4.2.

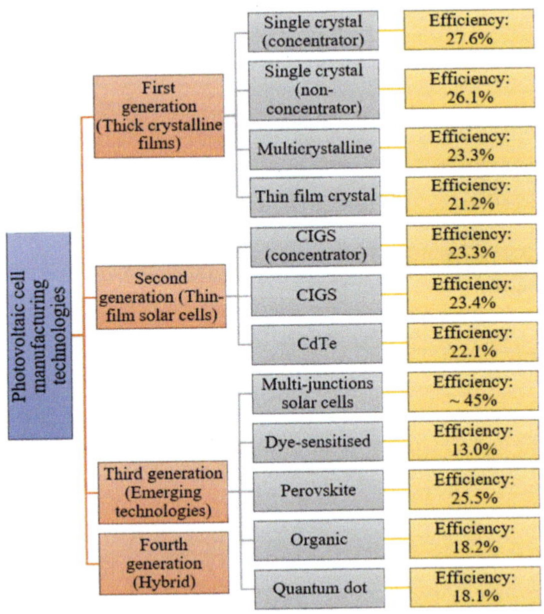

Figure 4.2. Examples of photovoltaic cell efficiencies.

Source. https://www.ncbi.nlm.nih.gov/pmc/articles/PMC9414585/bin/materials-15–05542-g003.jpg

4.2.1. First Generation of Photovoltaic Cells

The first category of photovoltaics to hit the market was silicon-based PV cells, which drew raw materials and processing knowledge from the microelectronics industry. Nowadays, silicon-based solar cells account for 90% of the global market share and more than 80% of installed capacity. They are the most widely used cells because of their comparatively high efficiency. Materials based on thick crystalline layers of silicon (Si) are used in the first generation of photovoltaic cells. This generation is based on single III-V junctions (GaAs) and mono-, poly-, and multicrystalline silicon.

Comparison of first-generation photovoltaic cells:

- Solar Cells Based on Monocrystalline Silicon (M-Si)

Efficiency: 15 ÷ 24%; Band gap: ~1.1 eV; Life span: 25 years; Advantages: Stability, high performance, long service life; Restrictions: High manufacturing cost, more temperature sensitivity, absorption problem, material loss.

- o Solar cells based on polycrystalline silicon (p-si)

Efficiency: 10 ÷ 18%; Band gap: ~1.7 eV; Life span: 14 years; Advantages: Manufacturing procedure is simple, profitable, decreases the waste of silicon, higher absorption compared to m-si; Restrictions: Lower efficiency, higher temperature sensitivity.

- o Solar cells based on GaAs

Efficiency: 28 ÷ 30%; Band gap: ~1.43 eV; Life span: 18 years; Advantages: High stability, lower temperature sensitivity, better absorption than m-si, high efficiency; Restrictions: Extremely expensive. First-generation photovoltaic cells use p-n junction technology, primarily in the form of silicon photovoltaic cells based on mono- or polycrystalline wafers. The Czochralski process is used to fabricate monocrystalline silicon wafers, which are the building blocks for monocrystalline silicon solar cells. The Si blocks are grown from tiny monocrystalline silicon seeds and then cut. Because it is more efficient than multicrystalline material, monocrystalline material is used extensively. Strict specifications for material purity, high material consumption during cell production, cell manufacturing processes, and constrained module sizes made up of these cells are some of the major technological challenges related to monocrystalline silicon (Crabtree et al., 2018) (Figure 4.3).

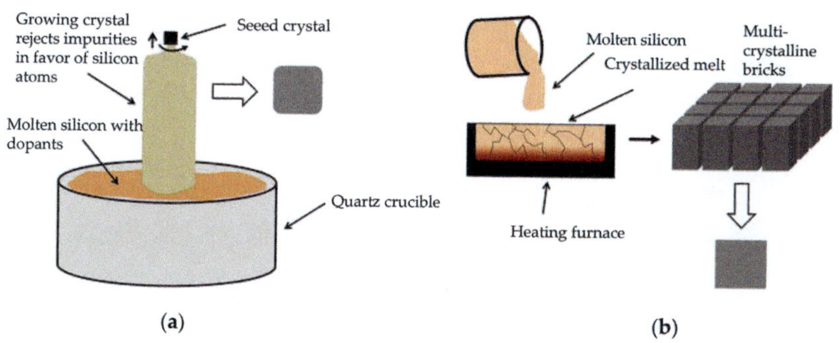

Figure 4.3. A picture showing (a) the Czochralski process for monocrystalline blocks and (b) the process of directional solidification for multicrystalline blocks

Source. https://www.ncbi.nlm.nih.gov/pmc/articles/PMC9414585/bin/materials-15–05542-g004.jpg

High-purity silicon is melted and then crystallized in a large crucible using a directional solidification process to create multicrystalline silicon blocks. Similar to the Czochralski process, this method lacks a reference crystal orientation, leading to the production of silicon material with various orientations. The most widely used base material for solar cells is boron-doped p-type Si substrates. High-efficiency solar cells are also made using n-type silicon substrates, although they come with more technical difficulties than p-type substrates, such as the need to achieve uniform doping throughout the silicon block.

At least six steps must be completed in order to produce crystalline solar cells. Surface texturing, doping, diffusion, oxide removal, metallization, anti-reflective coating, and firing are a few examples of these. Cell efficiency and other parameters are measured at the end of the process (using standard test conditions). The quality of the materials used in the production of photovoltaic cells determines their efficiency. It looks like 29.4% is the theoretical efficiency threshold for first-generation photovoltaic cells, and as early as 20 years ago, a value that was close enough was attained. As early as 1999, 25% efficiency was reached at the laboratory scale; after that, very little progress in efficiency values has been made. Crystalline silicon photovoltaic cells have been around for a while, and their efficiency has risen from 6% when they were first discovered to a record 26.1% efficiency since then. Cell efficiency is restricted by certain elements, such as volume defects. The development of Passivated Emitter and Rear Cell (PERC) technology, which further lowers the recombination rate on the back surface, or the introduction of an aluminum back surface field (Al-BSF), are examples of breakthroughs in the production of these cells.

4.2.1.1. Al-BSF Photovoltaic Cells

Distributed p-n junction silicon solar cells were developed as early as the 1950s, not long after the invention of semiconductor diodes. Phosphorus diffusion in boron-doped wafers is the industry standard for forming p-n junctions; originally, boron diffusion in arsenic-doped wafers was employed. Recombination on the back side could be decreased following the switch in the 1960s from n-type to p-type wafers thanks to the application of an aluminum back-surface field (Al-BSF) created by fusing the back contact to the substrate. For the past few decades, this relatively straightforward contact screen printing design has dominated the market, accounting for 70–90% of the share (Parida B. et al.,. 2011) (Figure 4.4).

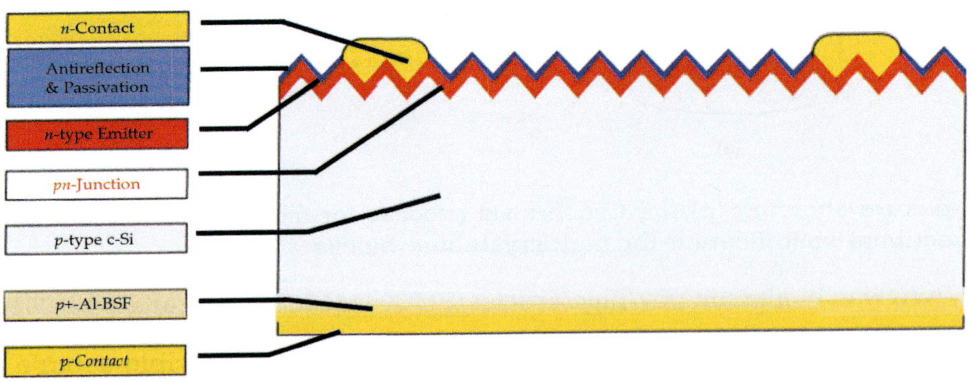

Figure 4.4. Silicon solar cell structure: Al-BSF.

Source: https://www.ncbi.nlm.nih.gov/pmc/articles/PMC9414585/bin/materials-15-05542-g005.jpg

Standard aluminum back surface field (Al-BSF) technology is one of the most widely used solar cell technologies due to its relatively simple manufacturing process. It works by completely depositing Al on the entire rear side (RS) of the p-type substrate during the screen-printing process, creating a p+ BSF that helps repel electrons from the substrate's rear and enhances cell performance. Figure 4.5 depicts the process flow for fabricating Al-BSF solar cells. Commercial solar cells are typically designed with a grid on the front side and full area contacts on the back side (Petrova-Koch V. and so forth, 2009).

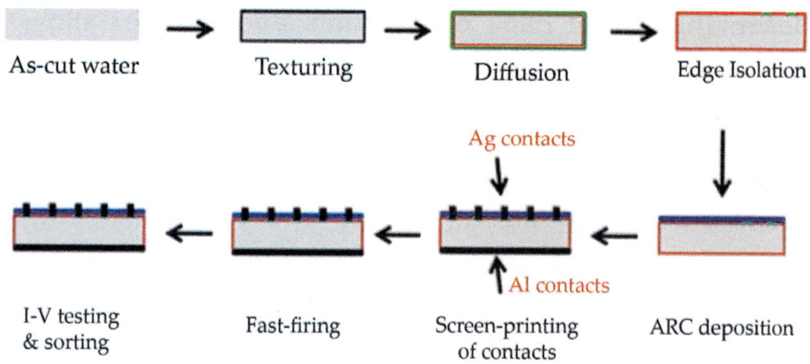

Figure 4.5. Al-BSF solar cell manufacturing process.

Source. https://www.ncbi.nlm.nih.gov/pmc/articles/PMC9414585/bin/materials-15–05542-g006.jpg

4.2.1.2. PERC Photovoltaic Cells

However, around 2013, the industrial Al-BSF cell's efficiency increased to about 20%. To obtain improved electrical and optical qualities, it has therefore become appealing to swap out the fully contacted Al-BSF cell for a PERC (Passivated Emitter and Rear Cell) structure with local back contacts. By adding a passivation layer to the rear side to enhance passivation and internal reflection, the passivated emitter and rear contact (PERC) solar cell enhances the Al-BSF architecture. It has been discovered that aluminum oxide works well as a material for rear side passivation (Metz et al., 2014) (Figure 4.6).

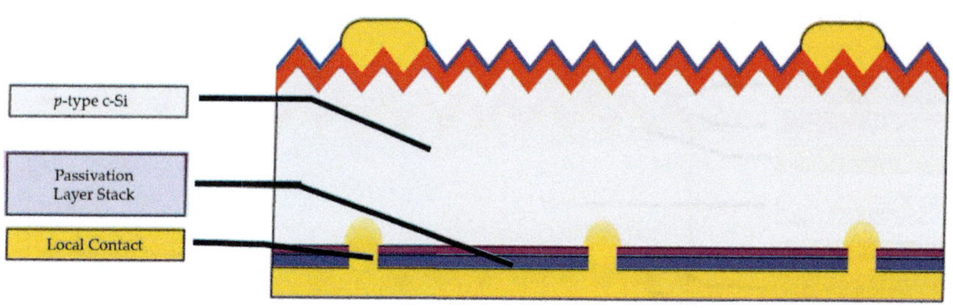

Figure 4.6. Silicon solar cell structure: PERC.

Source. https://www.ncbi.nlm.nih.gov/pmc/articles/PMC9414585/bin/materials-15–05542-g007.jpg

This cell structure's potential was shown as early as the 1980s, but due to its high cost in comparison to yield gain, it could only be processed in a laboratory. In theory, there was a relatively low industry threshold for implementing the PERC technology in mass industrial production because the Al-BSF process only required two new steps, i.e., passivation of the rear surface and accurate local back contact calibration. However, it took decades to develop a PERC process that was profitable. A number of reasons led to the implementation of PERC in low-cost, high-volume production, and the increase in productivity to levels ranging from 22% to 23.4%:

- Introduction of aluminum oxide back surface passivation by plasma-enhanced chemical vapor deposition (PECVD) and formation of local back surface field (BSF) by laser ablation of back passivation layer and Al alloy;
- Introduction of a selective emitter process in low-cost manufacturing, a "back-etching" process, or through a laser doping process;
- Reducing the width of front metallization fingers from about 100 µm to less than 30 µm in high-volume production while reducing contact resistance for lightly phosphorus-doped silicon;
- Adding a low-cost hydrogenation step at the end of the cell formation process to passivate volume defects and inactivate boron–oxygen complexes responsible for light-induced degradation (LID); and
- Reappearance of monocrystalline silicon wafers as a result of cost reduction in silicon ingot production by the Czochralski method and the introduction of diamond wire cutting.

4.2.1.3. SHJ-Type Photovoltaic Cells

Other high-performance cell designs, like heterojunction solar cells (SHJ) and interdigitated back contact (IBC) solar cells, have been mass-produced in tandem with PERC cells. HIT cells, also known as silicon heterojunction solar cells (SHJ), use passivating contacts made of a stack of layers of intrinsic and doped amorphous silicon. One of the main technological obstacles linked to this intriguing cell configuration is the inability to employ processes above 200 °C once the amorphous silicon layer has been deposited. This means that other approaches involving low-temperature pastes or galvanic contacts—which exclude the well-known burned-in screen-printed metal contacts—are required (Huang et al., 2017) (Figure 4.7).

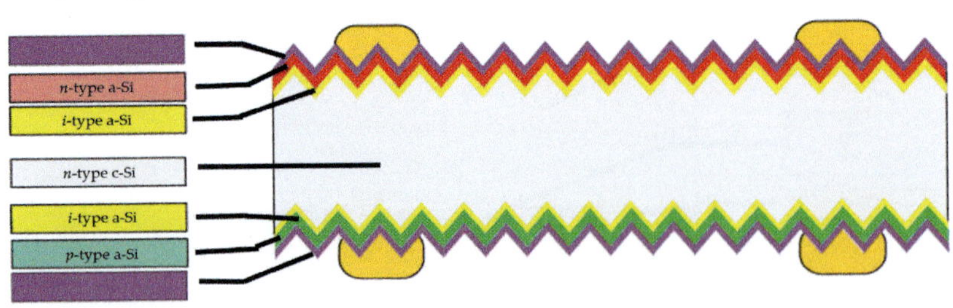

Figure 4.7. Silicon solar cell structures: heterojunction (SHJ) in rear junction configuration.

Source. https://www.ncbi.nlm.nih.gov/pmc/articles/PMC9414585/bin/materials-15–05542-g008.jpg

Strong efforts are being made right now to create high-capacity production lines that could rival current standard production lines in terms of performance. The adoption of SHJ technology will require overcoming several obstacles, including rising cell manufacturing tool costs, decreasing the amount of silver used or substituting it with copper through the development of Cu electroplating technology, and lowering the amount of indium used in the transparent conductive oxide (TCO) layer.

Additionally, the symmetric structure of the HIT solar cell offers two benefits. The structure is less stressed, which is crucial when processing thinner wafers, and the cell can be utilized in what is known as a bifacial module, which can generate more electricity than a regular module (Taguchi et al., 2013) (Figure 4.8).

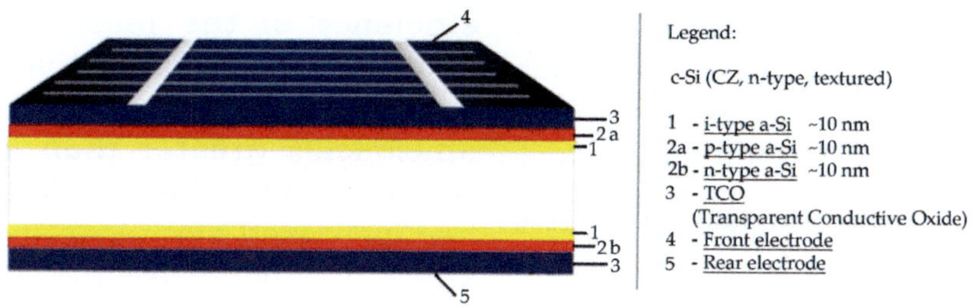

Figure 4.8. Structure of an HIT solar cell.

Source. https://www.ncbi.nlm.nih.gov/pmc/articles/PMC9414585/bin/materials-15–05542-g009.jpg

4.2.1.4. Photovoltaic Cells Based on Single III-V Junctions

This section concludes with a review of GaAs-based single III-V junctions. The highest photovoltaic conversion efficiency is achieved by III-V materials, which reach 29.1% with a GaAs single junction in sunlight and 47.1% with a six-junction device in concentrated sunlight. Because the absorption layers in these devices are usually between 2 and 5 µm thick, they can also be made to be lightweight, flexible, and able to be placed on curved surfaces. The III-V devices have a track record of excellent performance in demanding applications like space and high stability.

Nearly all III-V device performance records are the result of the prevalent III-V layer deposition technique, metal-organic vapor phase epitaxy (MOVPE). However, because of the high cost of the precursors, the relatively low usage of these precursors, and the lengthy batch growth cycles, this process has historically been regarded as an expensive growth technique. Recent research has shown that both MOVPE and hydrogen vapor phase epitaxy (HVPE) techniques can greatly increase the growth rate and demonstrate a much greater use of precursor chemicals; HVPE can also solve the issue of precursor cost. Currently, finishing involves a lot of labor-intensive, expensive, and relatively inefficient process steps, such as contact alignment, photolithography, manual spin coating application, metal evaporation, and lifting.

4.2.2. Second Generation of Photovoltaic Cells

Amorphous silicon or CdTe, gallium selenide, and copper (CIGS) thin film photovoltaic cells are intended to be a less expensive alternative to crystalline silicon cells. Although they have better mechanical qualities that are perfect for flexible applications, there is a chance that their efficiency will suffer as a result. While the first solar cell generation was an example of microelectronics, the development of thin films necessitated new growth techniques and allowed the industry to expand into other domains, such as electrochemistry.

The second-generation photovoltaic cell comparison:

- Solar cells based on amorphous silicon (a-Si)

Efficiency: 5 ÷ 12%; Band gap: ~1.7 eV; Lifespan: 15 years; Advantages: Less expensive, available in large quantities, non-toxic, high absorption coefficient; Restrictions: Lower efficiency, difficulty in selecting dopant materials, poor minority carrier lifetime.

- Solar cells based on cadmium telluride/cadmium sulfide (CdTe/CdS)

Efficiency: 15 ÷ 16%; Band gap: ~1.45 eV; Lifespan: 20 years; Advantages: High absorption rate, less material required for production; Restrictions: Lower efficiency, Cd being extremely toxic, Te being limited, more temperature-sensitive.

- Solar cells based on copper indium gallium selenide (CIGS)

Efficiency: 20%; Band gap: ~1.7 eV; Life span: 12 years; Advantages: Less material required for production; Restrictions: Very high-priced, not stable, more temperature-sensitive, highly unreliable.

4.2.2.1. CIGS Photovoltaic Cells

One important area that required attention was lowering the excessive reliance on semiconductor materials. This served as the catalyst for the development of CIGS and other thin-film photovoltaic cells in the second generation. The record value for CIGS efficiency is 23.4%, which is on par with the highest efficiencies of silicon cells. It should be highlighted, though, that because of the nature of large-scale processing, the efficiency of the research cells does not equate to an efficiency that can be achieved on an industrial scale. However, module efficiencies greater than 20% are already commonplace. The efficiency of CIGS cells has increased significantly in recent years, and more increases are anticipated in the future due to things like increased research into alkaline treatment after deposition.

Semiconducting chalcopyrite alloys of groups I–III–VI, also referred to as CIGS $(Ag,Cu)(In,Ga)(S,Se)_2$, are especially good absorber materials for solar cells. They are intrinsically stable in operation and can be used to create solar cells with direct band gaps ranging from ~1 to 2.6 eV, high absorption coefficients, and favorable internal defect parameters that permit high minority carrier lifetimes. In a monocrystalline device, the first yield to be recorded was 12% in the middle of the 1970s. Following significant advancements in CIGS thin film absorbers, processing, and contacts, thin film cells with a small area and a 23.4% efficiency were produced. The current record module efficiencies for glass are 17.5%, and for flexible steel they are 18.6%.

Although CIGS solar cells have been developed with a standard substrate configuration, flexible solar products can also be formed by deposition of CIGS at relatively low temperatures on metal or polymer substrates. Co-evaporation/devaporation and sputtering are the primary methods used to deposit CIGS thin films; electrochemical and ion beam-assisted deposition are used to a lesser degree. Because these are quaternary compounds, it is imperative to maintain the thin film's stoichiometry throughout the fabrication process. Additionally, efforts are being made to create fully or partially solution-deposited CIGS solar cells, which some believe may be the final step toward the development of ultra-thin, flexible, coiled PV modules.

The following procedures can be used to explain how to increase the efficiency of CIGS cells: (1) CIS compound evaporation; (2) reactive elemental bilayer deposition; (3) selenization of sputtered metal precursors; (4) chemical bath deposition of CdS with ZnO:Al as emitter; (5) gallium alloying; (6) sodium alkali incorporation; (7) three-step co-deposition; (8) heavy alkali ion exchange post-deposition treatment; and (9) sulfurization after selenization (SAS). Development is not linear; there is full potential for optimizing the intricate relationships between those methods and others that are in the works (e.g., silver alloys), which are still unrealized. Many scientists with expertise in CIGS believe that efficiencies of up to 25% can be attained.

CIGS is a multipurpose material that can be made using various methods and applied in different ways. Reactive co-deposition, electrodeposition, metal precursor deposition followed by sulfo-selenization, and solution processing are the four primary depositing methods currently utilized to create CIGS films. The greatest commercial successes and all recent world records have been attained by reactive co-deposition or two-step sulfo-selenization of metal precursors. Glass, metal films, and polymers are just a few of the substrates on which CIGS can be deposited. Glass can be used to create rigid modules, but polymer films and metal can be used for applications that call for lighter or more flexible modules. The next competitive advantage for photovoltaic technologies is becoming CIGS's relatively benign environmental impact (especially without CdS) when compared to other technologies, as the global energy markets evolve toward an appreciation of aspects of the circular economy and greenhouse gas reduction.

Based on CIGS technology, photovoltaic cells consist of a stack of thin films deposited by magnetron sputtering onto a glass substrate: a zinc-doped oxide (ZnO:Al) top electrode, a CdS buffer layer, a CIGS absorbing layer, and a bottom molybdenum (Mo) electrode. The CIGS active layer is deposited by co-evaporation and the CdS buffer layer using a chemical bath in a consistent process (Salhi 2022) (Figure 4.9).

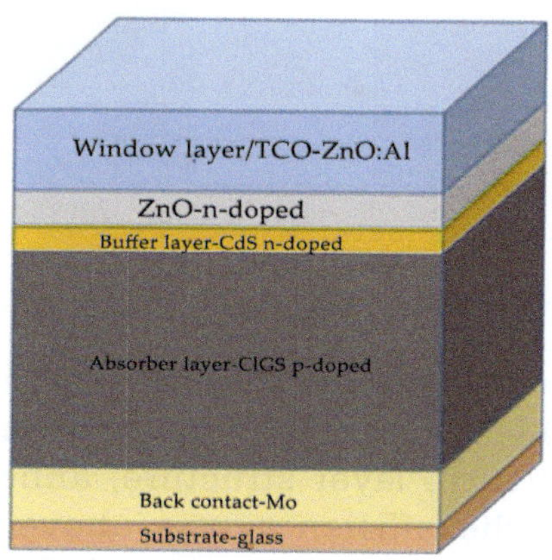

Figure 4.9. Demonstration of the CIGS-based standard solar cell stack.

Source. https://www.ncbi.nlm.nih.gov/pmc/articles/PMC9414585/bin/materials-15-05542-g010.jpg

4.2.2.2. CdTe Photovoltaic Cells

CdTe-based solar cells are also classified as second-generation photovoltaic cells.

One intriguing characteristic of CdTe is its ability to reduce cell size. Its high spectral efficiency allows the absorber thickness to be lowered to approximately 1 µm without significantly compromising efficiency, though more research is required. Because of their lighter weight, super-thin cells are especially appealing for flexible applications, especially in building-integrated photovoltaics (BIPV). Additionally, because transparent coatings are available, transparent photovoltaic panels with CdTe can be developed. Their levels of transparency range from roughly 10% to 50%, with the drawback that greater transparency inevitably results in lower efficiency. However, since most transparent panels are covered in double glass, they could eventually replace window panels in buildings. In addition to producing electricity that the panel could use to power itself, they would also help with thermal insulation and noise reduction (Wu X. 2004) Figure 4.10

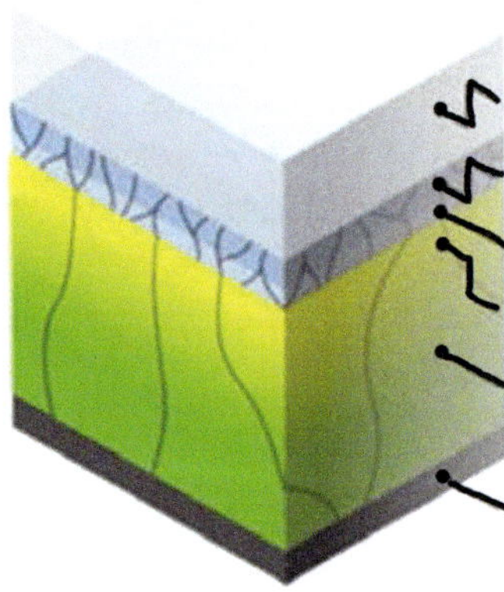

Figure 4.10. Schematic of a CdTe solar cell.

Source. https://www.ncbi.nlm.nih.gov/pmc/articles/PMC9414585/bin/materials-15–05542-g011.jpg.

With the passage of time, CdTe solar cell technology has advanced significantly. The efficiency of certified cells reached 10% in the 1980s, and with the use of a glass/ SnO_2/CdS/CdTe layer structure, annealing in a $CdCl_2$ environment, and subsequent Cu diffusion, the efficiency was above 15% in the 1990s. Transparent conductive oxide (TCO) layers made of sputtered Cd_2SnO_4 and Zn_2SnO_4 allowed the cells' efficiency to reach 16.7% by the 2000s. New records for cell efficiency have been reached over the last ten years, reaching 22.1%. Building-integrated photovoltaics and rooftop systems are using more and more CdTe technology.

A benchmark cell with an efficiency of 16.5% was created by NREL in 2001, and it stood as the standard for almost a decade. First Solar and GE Global Research have made multiple improvements to the record efficiency during the last two years. Less than 10% of the global PV market is currently made up of CdTe thin films, though capacity is anticipated to rise. First Solar, which has record cell efficiencies of 22.1% and average commercial module efficiencies of 17.5–28%, produces the majority of commercial CdTe cells.

Research, development, and manufacturing of CdTe-based photovoltaic cells date back several decades, surpassing Bell Labs' (Murray Hill, NJ, USA) initial investigations on Si crystalline cells in the 1950s. The leading businesses, Solar Cells Inc., BP Solar (Madrid, Spain), and Matsushita (Kadoma, Osaka, Japan), have been focusing on the commercialization of the underlying technology — forerunner to GE PrimeStar (Denver, CO, USA), Abound Solar (Loveland, CO, USA), and First Solar (Tempe, AZ, USA). First Solar (Tempe, AZ, USA) is presently the leading producer of thin-film CdTe PV, having produced 25 GW of PV modules since 2002.

Solar cells with an efficiency of between 10 and 16% have been produced using a variety of reasonably simple and affordable techniques. Several cost-effective deposition techniques have shown promise, such as close-space sublimation, spray deposition, electrodeposition, screen printing, and sputtering.

A CdS (0.4 μm)/CdTe (3.5 μm) thin film solar cell, in which the CdS and CdTe layers are deposited using metal–organic CVD (MOCVD) and CSS deposition techniques, respectively, has recently been reported to have a record efficiency of 16%. The majority of high-performance solar cells have a superstrate type device configuration, in which CdTe is deposited on top of a CdS window layer. Typically, the structure of the device is composed of glass/CdS/CdTe/Cu-C/Ag. To maximize device performance, post-deposition heat treatment of the CdTe layer in the presence of $CdCl_2$ is typically required.

The nearly maximum photocurrent achieved by enhancing the optical characteristics of the cell, eliminating parasitically absorbing CdS, and adding CdSexTe1−x with a smaller band gap is partially responsible for the most recent efficiency gain. By increasing the carrier lifetime and extending the absorber's bandwidth from approximately 1.4 to 1.5 eV, $CdSe_xTe_{1-x}$ improves photocurrent collection without causing a proportionate loss of photocurrent. ZnTe significantly increases the contact ohmicity and, consequently, the efficiency in the rear contact.

4.2.2.3. Kesterite Photovoltaic Cells

Compared to CdTe and CIGS chalcogenide materials, kesterite thin film materials have garnered greater attention recently. $Cu_2ZnSnS_xSe_{4-x}$ (CZTSSe) thin film photovoltaic material is garnering global interest due to its remarkable efficiency and Earth-derived composition. A lot of work is being done on new architectural designs or material engineering to produce high-performance CZTSSe thin film solar cells. Up until recently, the most sophisticated thin film CZTSSe solar cells could only achieve a power conversion efficiency (PCE) of 11.1% by employing the hydrazine suspension method. Additional vacuum and non-vacuum deposition methods also demonstrated efficacy in generating CZTSSe solar cells with a PCE greater than 8%. Even so, the record equipment with a PCE of 11% is still far below the physical limit, also known as the Shockley–Queisser (SQ) limit, which is approximately 31% efficiency in Earth's environment.

We prepare CZTSSe layers using a hydrazine-based pure solution method, starting with a Cu-poor and Zn-rich stoichiometry (Cu/ (Zn + Sn) = 0.8 and Zn/ Sn = 1.1). The layers are spin-coated onto Mo-coated soda-lime glass and annealed at temperatures above 500 °C. For device fabrication, CZTSSe layers are deposited on Mo-coated glass substrates, followed by 25 nm CdS deposition in a chemical bath and sputtering with 10 nm ZnO/50 nm ITO. A 2 µm thick Ni/Al top metal contact and 110 nm MgF_2 are then deposited on top of the devices using electron beam evaporation. The device area is determined by mechanical scribing.

4.2.2.4. Photovoltaic Cells Based on Amorphous Silicon

The second generation of cells classified as using amorphous silicon are the most common thin film technology. These cells have an efficiency of 5% to 7%, which can increase to 8–10% for double and triple junction structures. Varieties include amorphous silicon carbide (a-SiC), amorphous germanium silicon (a-SiGe), microcrystalline silicon (µ-Si), and amorphous silicon nitride (a-SiN). Hydrogen is necessary to dope the material for hydrogenated amorphous silicon (a-Si:H). The gas-phase deposition technique is typically used to create these photovoltaic cells using either metal or gas as the substrate material.

Roll-to-roll is a commonly used manufacturing process for a-Si:H cells. To be utilized as a deposition surface, a cylindrical sheet—typically made of stainless steel—must first be rolled out. After washing and cutting to size, an insulating layer is applied to the sheet. The silicon layer is then covered with a transparent conductive oxide (TCO) following the application of a-Si:H to the reflector. Ultimately, the module is closed after the various layers are joined by laser cuts.

Typically, amorphous silicon is created using PECVD at low substrate temperatures ranging from 150 to 300°C. A 300 nm a-Si:H layer can absorb nearly 90% of photons beyond the passband in one go, enabling the production of solar cells that are lighter and more flexible.

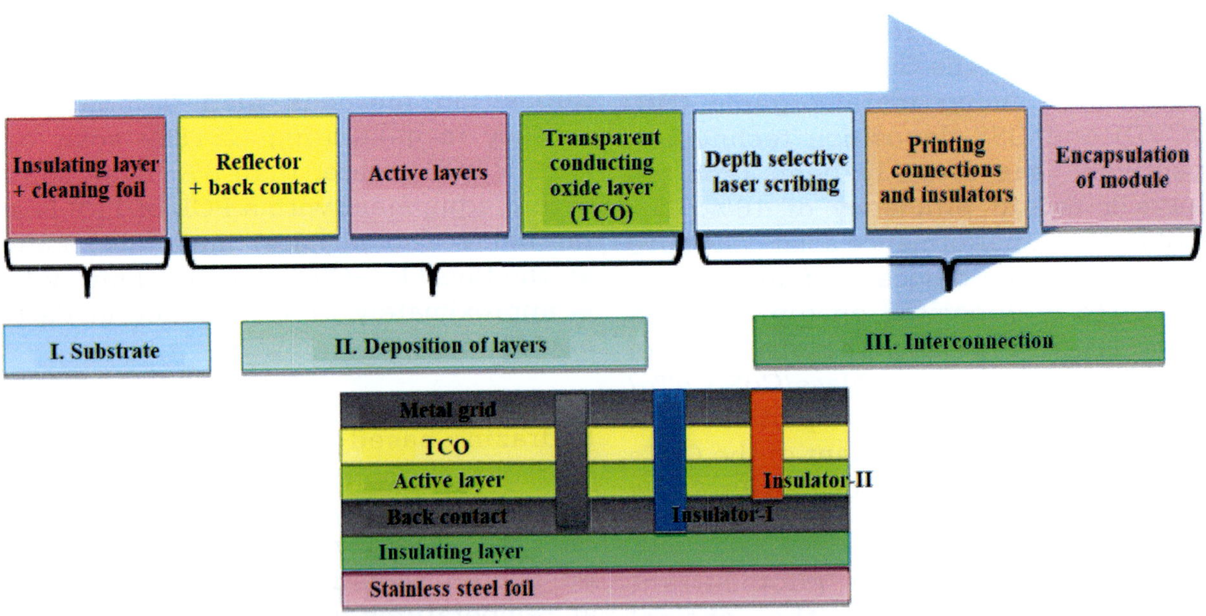

Figure 4.11. Manufacturing process of a-Si-based solar PV cell.

Source. https://www.ncbi.nlm.nih.gov/pmc/articles/PMC9414585/bin/materials-15–05542-g012.jpg.

The detailed manufacturing process of an a-Si-based photovoltaic cell is depicted in Figure 4.11. In comparison to first-generation solar cells, thin-film photovoltaic cells are more affordable, thinner, and flexible. In first-generation photovoltaic cells, the light-absorbing layer had a thickness of 200–300 μm; in second-generation cells, it is 10 μm. Thin films of photovoltaic cells employ semiconductor materials that range from "micromorphic and amorphous silicon" to quaternary or binary semiconductors like "cadmium telluride (CdTe) and copper indium gallium selenide (CIGS)" (Garcia-Barrientos et al., 2021).

4.2.3. Third Generation of Photovoltaic Cells

The third generation of solar cells encompasses tandem, perovskite, dye-sensitized, organic, and emerging concepts. These cells offer a variety of approaches, ranging from high-end, expensive III-V multi-junction cells for space applications to low-cost, low-efficiency dye-sensitized organic solar cells. Even though some third-generation photovoltaic cells have been studied for more than 25 years, they are sometimes referred to as "emerging concepts" due to their low market penetration.

New techniques in silicon photovoltaic cell innovation focus on creating extra energy levels in the semiconductor's band structure. The cutting-edge research in manufacturing technology and efficiency enhancements is currently centered on third-generation solar cells. Adding more energy levels to the semiconductor's band gap (IBSC and IPV cells) and using more ion implantation during the manufacturing process are two current strategies to boost the efficiency of PV cells. Other innovative third-generation cells that are lesser-known commercial "emerging" technologies include:

1. Organic materials (OSC) photovoltaic cells;
2. Perovskites (PSC) photovoltaic cells;
3. Dye-sensitized (DSSC) photovoltaic cells;
4. Quantum dots (QD) photovoltaic cells; and
5. Multi-junction photovoltaic cells.

Third-generation photovoltaic cell comparison:

- Solar cells based on dye-sensitized photovoltaic cells

Efficiency: 5 - 20%; Advantages: Lower cost, low light and wider-angle operation, lower internal temperature operation, robustness, and extended lifetime; Restrictions: Problems with temperature stability, poisonous and volatile substances.

- Solar cells based on quantum dots

Efficiency: 11 - 17%; Advantages: Low production cost, low energy consumption; Restrictions: High toxicity in nature, degradation.

- Solar cells based on organic and polymeric photovoltaic cells

Efficiency: 9 - 11%; Advantages: Low processing cost, lighter weight, flexibility, thermal stability; Restrictions: Low efficiency.

- Solar cells based on perovskite

Efficiency: 21%; Advantages: Low-cost and simplified structure, light weight, flexibility, high efficiency, low manufacturing cost; Restrictions: Unstable.

- Multi-junction solar cells

Efficiency: 36% and higher; Advantages: High performance; Restrictions: Complex, expensive.

4.2.3.1. Organic and Polymeric Materials Photovoltaic Cells (OSC)

Since organic solar cells (OSCs) have the potential to be used in a variety of scenarios due to the special advantages of organic semiconductors, such as their application to large-scale roll-to-roll processing, flexibility, semi-transparency, low weight, low cost, and ability to be processed in solution, they are advantageous in solar energy applications. The potential for donor: acceptor bulk heterojunction (BHJ) compounds has been investigated globally in solution-processed organic solar cells (OSCs) that absorb near-infrared (NIR) radiation. Furthermore, due to their industrial potential, NIR-absorbing OSCs have gained interest as top-tier components in next-generation optoelectronic devices, like NIR photodetectors and translucent solar cells. The value of OSC is rising as a result of the development of non-fullerene acceptors (NFAs) that absorb light in the near-infrared (NIR) range. However, organic donor materials with NIR-absorbing capabilities have not yet been thoroughly investigated in comparison to acceptor materials that do the same.

The most sophisticated BHJ structure offers great promise for lightweight, inexpensive organic solar cells by combining organic donor and acceptor materials. With the development of new NIR photoactive materials with low bandwidth, significant progress has been made over the last ten years, with power conversion efficiencies reaching more than 14% for a single-junction device and more than 17% for a tandem device. Low-band donor and non-fullerene acceptor materials with wide-range solar coverage extended to the NIR region generally show lower exciton binding energy, stronger dipole moment, easier delocalization of π electrons, more tightly superimposed electronic orbitals, and higher dielectric constant than wide-band organic photovoltaic materials. Due to these characteristics, low-bandwidth photovoltaic materials—including single-junction and tandem devices—play a significant part in high-performance organic solar cells. Optimizing the weight ratio of donor to acceptor materials, adding ultra-low band gap materials as a third component to increase NIR light utilization efficiency, and varying the active layer's thickness to strike a balance between charge accumulation and photon collection can all be considered as creative approaches to active layer design. A great deal of work has gone into optimizing the translucent top electrode, with the goals of better compatibility with active layers, enhanced reflectance in the NIR or ultraviolet (UV) light range, and balanced conductivity and transmittance in the visible light range. In terms of device engineering, selective transmission and reflection have been realized for the simultaneous improvement of power conversion efficiency and average transmission of translucent OSC visible light by means of photon crystal, anti-reflection coating, optical microcavity, and dielectric/metal/dielectric (DMD) structures.

4.2.3.2. Dye-Sensitized Photovoltaic Cells (DSSC)

In the current generation of photovoltaic cells, conjugated polymers and organic semiconductors are regarded as advanced materials due to their success in flat panel displays and LEDs. Figure 4.12 depicts a schematic of dye-sensitized organic photovoltaic cells (DSSCs). Polymer/organic photovoltaic cells can also be further subdivided into devices that operate differently, such as plastic (polymer) and organic photovoltaic devices (OPVDs), dye-sensitized organic photovoltaic cells (DSSCs), and photoelectrochemical photovoltaic cells (Keis, et al., 2002).

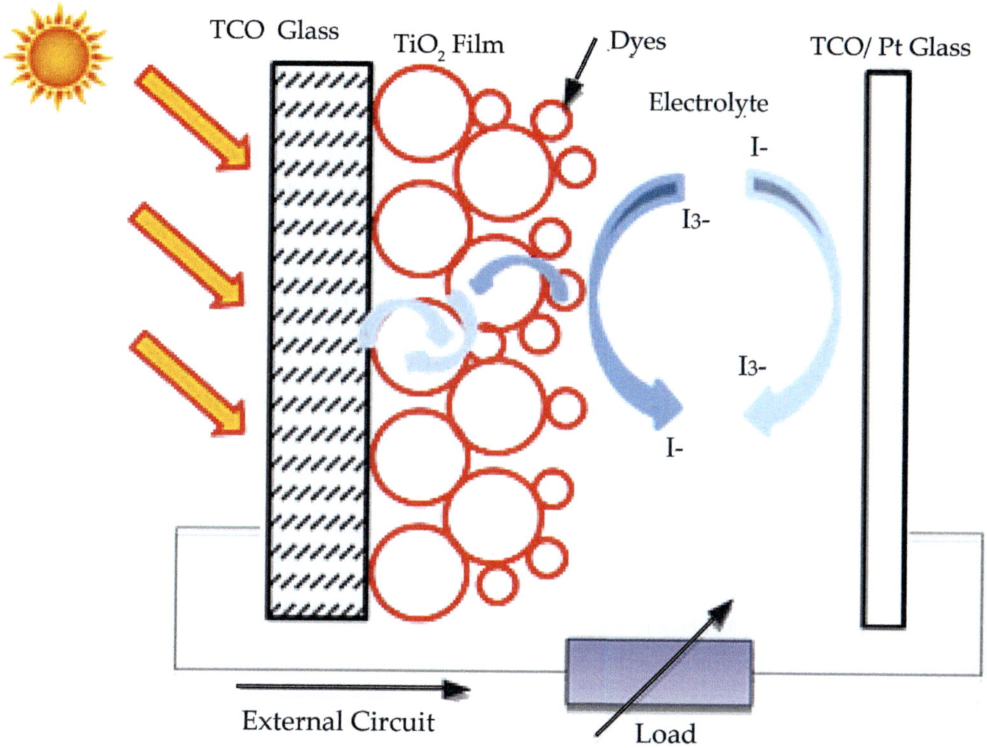

Figure 4.12. Schematic representation of a DSSCs.

Source. https://www.ncbi.nlm.nih.gov/pmc/articles/PMC9414585/bin/materials-15–05542-g013.jpg.

Dye-sensitized solar cells (DSSCs) are considered one of the most effective materials in nanotechnology for capturing energy in photovoltaic technologies. These cells have a unique structure that combines organic and inorganic components, with a porous layer of nanocrystalline titanium dioxide (TiO_2) serving as an electron conductor in contact with an electrolyte solution containing light-absorbing organic dyes. When light is absorbed, a charge transfer occurs at the interface, allowing for the transport of holes in the electrolyte.

With a power conversion efficiency of approximately 11%, efforts are being made to commercialize dye-sensitized photovoltaic modules. A key feature of DSSCs is the photosensitization of nanosized TiO_2 coatings with optically active dyes, leading to an efficiency of over 10%.

DSSCs hold promise as photovoltaic devices because of their simple fabrication, low material costs, and their benefits in transparence, color capability, and mechanical flexibility. Cell stability and low photoelectric conversion efficiency are the key obstacles to the commercialization of DSSCs. For DSSCs, the maximum theoretically achievable energy conversion efficiency was calculated to be 32%; nevertheless, the highest efficiency that has been documented to date is only 13%.

Extensive efforts are being made to enhance the efficiency of the DSSC by comprehending its governing parameters. Many attempts have been made to create a counter electrode (CE), modify a wide band gap semiconductor to serve as a working electrode, and optimize the dye's redox pair and absorbance.

Apart from enhancing the efficiency of DSSCs, a significant problem that requires resolution in subsequent research is the material cost.

4.2.3.3. Perovskite Photovoltaic Cells

Metal halide perovskites (MHPs) are the basis of the groundbreaking new photovoltaic cell concept known as perovskite solar cells (PSCs), e.g., methylammonium iodide ($MAPbI_3$) and formamidine lead iodide ($FAPbI_3$), in that order.

A direct band gap with a high absorption coefficient, a long carrier lifetime and diffusion length, a low defect density, and ease of composition and band gap tuning are just a few of the features that make MHPs desirable as photovoltaic absorbers. The first description of MHP as a sensitizer in a dye cell based on liquid electrolyte conducting holes was published in 2009. . In 2012, MHP demonstrating ~10% efficiency of PSCs based on a solid-state hole conductor sparked an explosion of PSC studies. The efficiency of a single PSC junction rose to a certified level of 25.2% after roughly ten years of research.

Improvements in material quality achieved by a variety of synthetic approaches developed with a basic understanding of MHP growth mechanisms have had a significant impact on the development of PSCs. Understanding the interconnected, intricate processes involved in perovskite growth (e.g., a wide variety of high-efficiency growth modes such as single-step growth, sequential growth, dissolution process, vapor process, post-deposition processing, non-stoichiometric growth, additive-assisted growth, and fine-tuning of structure dimensions) have been developed with the help of nucleation, grain growth, and microstructure evolution. Recent efforts have been focused on interface engineering, with a particular emphasis on the introduction of a two-dimensional perovskite surface layer in order to improve stability and reduce open-circuit voltage losses.

The composition of perovskite is getting simpler due to advancements in synthetic control, primarily in the direction of FAPbI3. This will surely help simplify scale deposition techniques and provide a fundamental knowledge of these cells' characteristics.

4.2.3.4. Quantum Dots Photovoltaic Cells

Solar cells made from quantum dots (QDs) are also referred to as nanocrystalline solar cells. These cells are created through epitaxial growth on a crystal substrate, forming three-dimensional quantum dots surrounded by high potential barriers. The electrons and electron holes within a quantum dot have discrete energy levels due to their confinement within a small space. The ground state energy of these particles within a quantum dot is determined by the size of the dot (Tian, et al., 2013) Figure 4.13

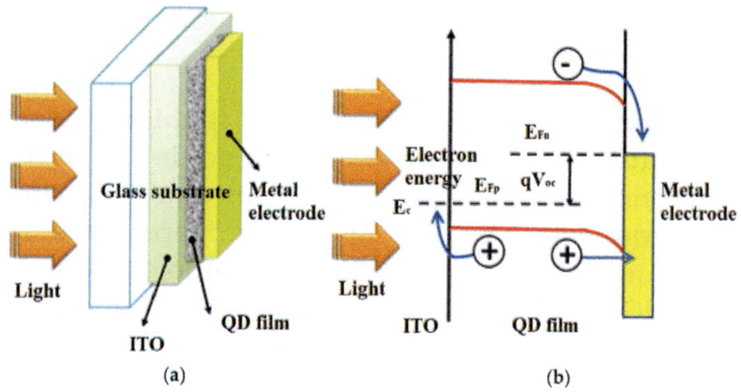

Figure 4.13. (a) A scheme of a solar cell based on quantum dots, (b) solar cell band diagram.

Source. https://www.ncbi.nlm.nih.gov/pmc/articles/PMC9414585/bin/materials-15–05542-g014.jpg.

Their absorption coefficients are comparatively high for nanocrystalline cells. In a solar cell, light absorption and exciton formation are followed by two processes that happen in succession: charge separation and charge transport. The short lifetime and poor mobility of excitons in conducting polymers lead to small exciton diffusion lengths (10–20 nm) in organic compounds. Stated differently, the conversion efficiency decreases when excitons that form far from the electrode or carrier transport layer recombine.

The corresponding nanocrystals (NCs) or quantum dots (QDs) have received a great deal of attention as a result of the development of thin film solar cells using metal halide perovskites. Today, by combining mixed colloidal QDs with perovskites, the record efficiency of QD solar cells was increased to 16.6%. A multitude of possibilities for photovoltaics are made possible by the new nanomaterials' universality in terms of fabrication ease, band gap tuning, and surface chemistry control. These include single-junction, elastic, translucent, controlled cells with heterostructures, as well as multi-junction tandem solar cells that would advance the field even further. On the other hand, a smaller size distribution could improve QD solar cells' performance in a number of ways. First, since larger QDs act as a band tail or shallow trap that complicates transport, electron transport may be easier in smaller QDs. Second, the largest size QD (smallest band gap) near the contacts may be the limit of the open-circuit voltage (VOC) of QD solar cells. By minimizing these losses, increasing the homogeneity and uniformity of QD size would also enhance PV performance. It is possible that more controlled synthesis could benefit QD cells, even though controlled experiments like these have not yet been reported.

4.2.3.5. Multi-Junction Photovoltaic Cells

Multi-junction (MJ) solar cells are made up of multiple p-n junctions made of different semiconductor materials. Each junction responds to light of a different wavelength by generating an electric current, increasing the device's efficiency and ability to convert incident sunlight into electricity. The term "tandem solar cell" refers to the idea of using different materials with different band gaps in order to harness as many photons as possible. Many different materials could be used to construct an entire cell, opening up a wide range of possible designs.

The cells are typically integrated in a monolithic fashion and linked in series via a tunnel junction. Current matching among the cells is achieved by varying the band gap and thickness of each individual cell. According to an analysis, the theoretical viability of utilizing multiple band gaps is 44% for two band gaps, 49% for three band gaps, 54% for four band gaps, and 66% for an infinite number of gaps. Figure 4.14 shows the design of an InGaP/(In)GaAs/Ge triple solar cell and highlights key technologies to improve conversion efficiency (Gao, et al., 2020).

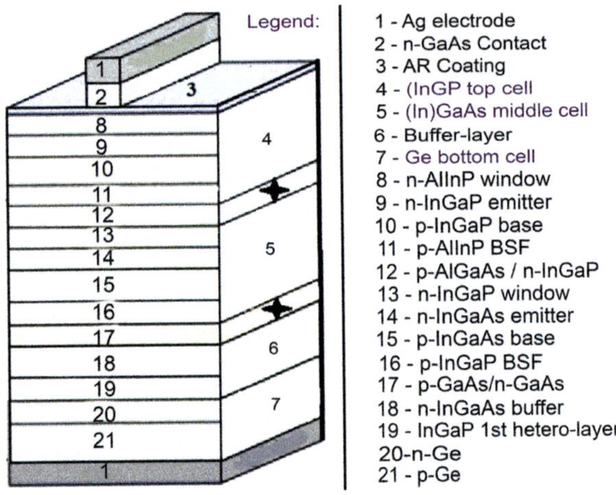

Figure 4.14. Schematic illustration of a triple-junction cell and approaches for improving efficiency of the cell.

Source. https://www.ncbi.nlm.nih.gov/pmc/articles/PMC9414585/bin/materials-15–05542-g015.jpg.

In space photovoltaics, grid-matched InGaP/(In)GaAs/Ge triple solar cells have found widespread application and have demonstrated the highest true efficiency of more than 36%. The space environment's heavy radiation bombardment of different energetic particles damages solar cells inevitably and leads to the creation of more non-radiative recombination centers, which shorten minority carrier diffusion lengths and lower solar cell efficiency. In multi-junction solar cells, the sub-cells are connected in series, and the sub-cell with the highest radiation degradation reduces the multi-junction solar cell's efficiency. Reducing the dopant concentration and base region thickness are two ways to improve the radiation resistance of (In)GaAs sub-cells.

4.2.3.6. Photovoltaic Cells with Additional Intermediate Band

The National Renewable Energy Laboratory (NREL) estimates that multi-junction and IBSC photovoltaic cells have the highest efficiency under experimental conditions (47.1%). The extra intermediate band in silicon's band gap is precisely what makes these cells unique. The international literature currently lists two varieties of these cells: IPV (Impurity Photovoltaic Effect) and IBSC (Intermediate Band Solar Cells).

One method for improving the infrared response of PV cells and, consequently, the efficiency of solar-to-electric energy conversion is the Impurity Photovoltaic Effect (IPV). The introduction of deep radiation defects in the semiconductor crystal structure serves as the basis for the IPV effect theory. Photons with energies lower than the band gap width are guaranteed a multi-step absorption mechanism by means of these defects. Under some circumstances, adding IPV dopants to the structure of silicon solar cells improves the conversion efficiency, short circuit current density, and spectral response.

Intermediate Band Solar Cells are a significant area of research with enormous development potential (IBSCs). They are an example of a third-generation solar cell concept that incorporates materials other than silicon. The goal of the intermediate band gap solar cell (IBSC) concept is to absorb photons whose energy matches the cell structure's sub-band width. These photons are absorbed by a material that resembles a semiconductor and has an intermediate band (IB) in the band gap of a conventional semiconductor in addition to the conduction and valence bands. To add a new energy level, extremely high doses of metal ions are injected into the silicon layers of IBSCs (López, et al., 2020) Figure 4.15

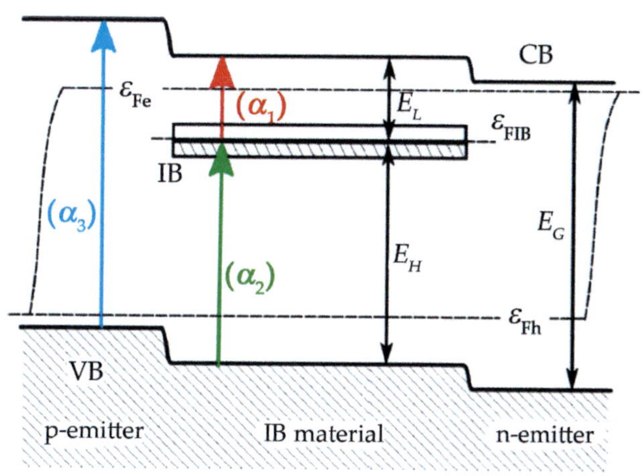

Figure 4.15. Energy band diagram of an intermediate band solar cell (IBSC).

Source. https://www.ncbi.nlm.nih.gov/pmc/articles/PMC9414585/bin/materials-15–05542-g016.jpg.

A model was created based on studies done on the impact of defects incorporated into the silicon structure, which states that increasing the number of specific deep defects in the charge carrier capture region enhances the efficiency of photovoltaic cells. Defects that promote the movement of majority carriers and those that prevent the accumulation of minority carriers are of special interest. This has a major impact on lessening the charge carrier capture site's recombination process. Lastly, we combine effective surface passivation with simultaneous reduction of optical losses by introducing defects into the structure of the silicon underpinning the solar cell.

There are two ways to introduce intermediate bands into semiconductors through ion implantation: either high-dose metal ions are implanted into the silicon layer, or very concentrated dopants are introduced into the semiconductor substrate. Ion implantation is being used more often in the production of photovoltaic cells, which could lower deployment costs and make silicon cells more affordable by raising their efficiency. Ion implantation technology allows for more accurate doping of the silicon layer, the creation of extra energy levels in the band gap, and the reduction of the time required for each stage of the cell fabrication process. These benefits ultimately result in higher quality and lower production costs.

Recently, ion implantation has been gaining traction in the solar industry, replacing the diffusion technique that has been utilized for a long time. With advancements in technology, cell performance is anticipated to enhance further, achieving higher efficiencies. This technology offers precise control over the amount and distribution of dopant doses, resulting in improved uniformity, repeatability, and efficiency above 19%, with a narrower distribution of cell performance (Figure 4.16).

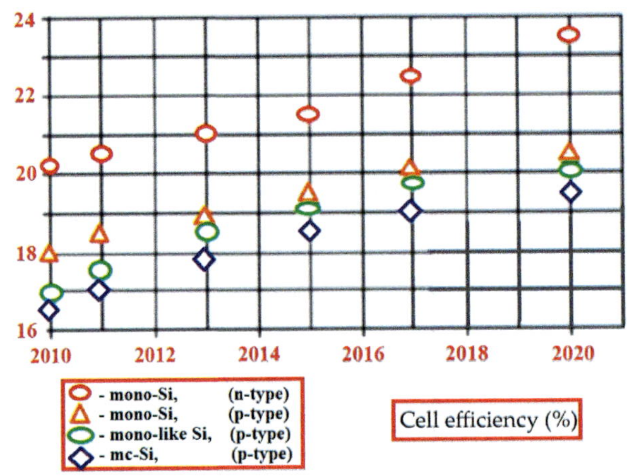

Figure 4.16. Stabilized cell efficiency trend curves.

Source. https://www.ncbi.nlm.nih.gov/pmc/articles/PMC9414585/bin/materials-15–05542-g017.jpg.

By accelerating the impurity ions to a high energy level and implanting the ions into the semiconductor, specific ions containing the necessary impurity are introduced into the semiconductor using the ion implantation method. The depth of ion implantation is determined by the energy imparted to the impurity ions. A controllable dose of impurity ions can be injected deeply into the semiconductor using the ion implantation technique, in contrast to diffusion technology, which only introduces the impurity ion dose at the surface.

4.3. FOURTH GENERATION OF PHOTOVOLTAIC CELLS

Because they combine the affordability and adaptability of polymer thin films with the stability of organic nanostructures like metal nanoparticles and oxides, carbon nanotubes, graphene, and their derivatives, fourth-generation photovoltaic cells are also referred to as hybrid inorganic cells. These gadgets—often referred to as "nanophotovoltaics"—may represent photovoltaics' bright future.

4.3.1. Graphene-Based Photovoltaic Cells

Graphene, carbon nanotubes, metal nanoparticles, thin polymer layers, and their derivatives are used in the fourth generation, which offers superior flexibility and affordability. Graphene was singled out for special attention as the nanomaterial of the future. Graphene-based materials are being investigated for use in PV devices in place of currently used conventional materials because of their special qualities, which include high carrier mobility, low resistivity and transmittance, and 2D lattice packing. However, the synthesis of graphene materials with the right structure and properties is essential to achieving adequate device performance and practical applications.

A careful selection of techniques is necessary for specific applications because the properties of graphene are inherently linked to the process of fabricating it. Specifically, highly conductive graphene can be used in flexible photovoltaic devices. It can also be used as an electrode interlayer material and selective charge-taking element due to its high compatibility with metal oxides, metallic compounds, and conductive polymers.

Over the last twenty years, graphene has become an integral part of photovoltaic technology, serving as a transparent electrode, hole/electron transport material, and interfacial buffer layer in solar cell devices. Various types of graphene-based solar cells have been developed, including organic bulk heterojunction (BHJ) cells, dye-sensitized cells, and perovskite cells. Graphene has been instrumental in achieving energy conversion efficiencies of over 20.3% in perovskite solar cells and 10% in BHJ organic solar cells. Beyond its role in charge extraction and transport, graphene also provides protection against environmental degradation due to its unique 2D lattice structure, ensuring the long-term stability of photovoltaic devices.

Silicon semiconductors are combined with semi-metallic graphene that has a zero-band gap to form Schottky junction solar cells. The first graphene-silicon solar cell was not identified as an n-silicon cell until 2010, despite the fact that graphene was first discovered in 2004. A schematic of a graphene-silicon solar cell with a Schottky junction is presented in Figure 4.17. Graphene sheets (GS) with an effective area of 0.1–0.5 cm^2 were wet deposited on pre-patterned Si/SiO_2 substrates after being grown by chemical vapor deposition (CVD) on nickel films. A Schottky junction is formed when the exposed n-Si substrate is coated with graphene. Au electrodes were used to establish contact with the graphene sheet (Mahmoudi, et al., 2018).

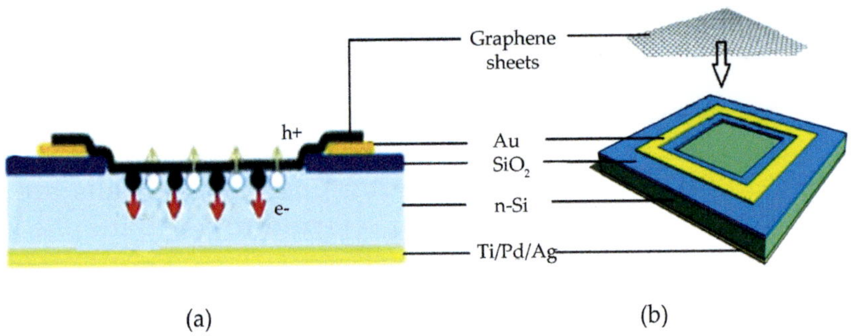

Figure 4.17. Graphene–silicon Schottky junction solar cell. (a) Cross-sectional view, (b) schematic illustration of the device configuration.

Source. https://www.ncbi.nlm.nih.gov/pmc/articles/PMC9414585/bin/materials-15–05542-g018.jpg.

The two primary approaches used in graphene synthesis are top-down and bottom-up. The objective of the top-down method is to intercalate and exfoliate graphite into graphene sheets using solid, liquid, or electrochemical exfoliation. The exfoliation of graphite oxide into graphene oxide (GO), followed by chemical or thermal reduction, is another method that falls under this classification. Making graphene from molecular precursors via chemical vapor deposition (CVD) or epitaxial growth is a bottom-up method. The manufacturing process determines the structure, morphology, and properties of the resultant graphene, such as the number of layers, degree of defects, electrical and thermal conductivity, solubility, and hydrophilicity or hydrophobicity.

Graphene, despite only having one atom in thickness, has the ability to absorb 23% of incident white light. Because graphene has a strong interaction with light and satisfies both the optical (high transmittance) and electrical (low layer resistance) requirements of a typical transparent conductive electrode, graphene integration into silicon solar cells is a promising platform. It is significant to remember that the number of layers affects graphene's transmittance and layer resistance. The optical transparency decreases in tandem with the layer resistance as the number of graphene layers rises.

Graphene provides numerous benefits for PV technology, including flexibility, environmental stability, low electrical resistivity, and photocatalytic properties. However, it must be meticulously engineered to meet the specific needs and standards of the intended applications.

One issue with graphene applications is the lack of an easier, more dependable method for depositing a low-cost, well-ordered monolayer of flakes on target substrates with different surface characteristics. The adhesion of the deposited graphene thin film is the other issue, which has not yet received enough attention. CVD can be used to create large-area continuous graphene layers with high electrical conductivity and optical transparency. Because of its excellent conductivity and transparency properties, as well as its naturally low manufacturing cost, graphene holds great promise as an anode in organic photovoltaic devices, potentially replacing indium tin oxide (ITO).

The main drawback of graphene is its low hydrophilicity, which has an adverse effect on the design of devices that are processed in solution. However, this drawback can be addressed by surface modification through non-covalent chemical functionalization. It is expected that new applications involving this new class of CVD graphene materials will soon emerge in plastic electronics and optoelectronics, given graphene's excellent conductivity properties, mechanical strength, and flexibility. The finding opens the door for inexpensive graphene layers to take the place of ITO in electroluminescent and photovoltaic devices.

4.4. PHOTODETECTORS AND SOLAR CELLS

4.4.1. Photodetectors

Semiconductor devices called photodetectors have the ability to transform optical signals into electrical signals. Three processes are involved in the operation of a photodetector: the generation of carriers by incident light, the transport and/or multiplication of carriers by any available current-gain mechanism, and the interaction of the current with the external circuit to produce the output signal.

Photodetectors are used in many different applications, such as optical-fiber communications detectors and infrared sensors in optoisolators. Photodetectors with high sensitivity at operating wavelengths, fast response times, and low noise levels are required for these applications. The photodetector should also be small, dependable under the necessary operating conditions, and use low biasing voltages or currents.

4.4.1.1. Photoconductor

A photoconductor is made up of a semiconductor slab with ohmic contacts at both ends, as seen in Fig. 4.18a, and a layout of interdigitated contacts, as shown in Fig. 4.18b. When light strikes the surface of the photoconductor, it generates electron-hole pairs through band-to-band transition (intrinsic) or transitions involving forbidden-gap energy levels (extrinsic), leading to an increase in conductivity.

For the intrinsic photoconductor, the conductivity is given by

$$\sigma = q\left(\mu_n n + \mu_p p\right), \qquad (1)$$

and the increase in conductivity under illumination is due mainly to the increase in the number of carriers.

The intrinsic photoconductor has a specific long-wavelength cutoff, while for the extrinsic photoconductor, photoexcitation can happen between the band edge and an energy level within the energy gap. In this scenario, the long-wavelength cutoff is dictated by the depth of the forbidden-gap energy level.

$$n \quad n_0 \exp\left(\frac{t}{\tau}\right), \tag{2}$$

where τ is the carrier lifetime. From Eq. 2 the recombination rate is

$$\left|\frac{dn}{dt}\right| \quad \frac{1}{\tau} n_0 \exp\left(\frac{t}{\tau}\right) \quad \frac{n}{\tau}. \tag{3}$$

If we assume a steady flow of photon flux impinging uniformly on the surface of a photoconductor with an area A = WL, the total number of photons arriving at the surface is (P_{opt}/hν) per unit time, where P_{opt} is the incident optical power and hν is the photon energy. At steady state, the carrier-generation rate G must be equal to the recombination rate n/τ. If the detector thickness D is much larger than the light penetration depth 1/α, the total steady-state carrier-generation rate per unit volume is

$$G \quad \frac{n}{\tau} \quad \frac{\eta(P_{opt} / h\nu)}{WLD}, \tag{4}$$

where η is the quantum efficiency, the number of carriers generated per photon, and n is the carrier density, the number of carriers per unit volume. The photocurrent flowing between the electrodes is

$$I_p = (\sigma \mathscr{E})WD = (q\mu_n n\mathscr{E})WD = (qn\upsilon_d)WD, \tag{5}$$

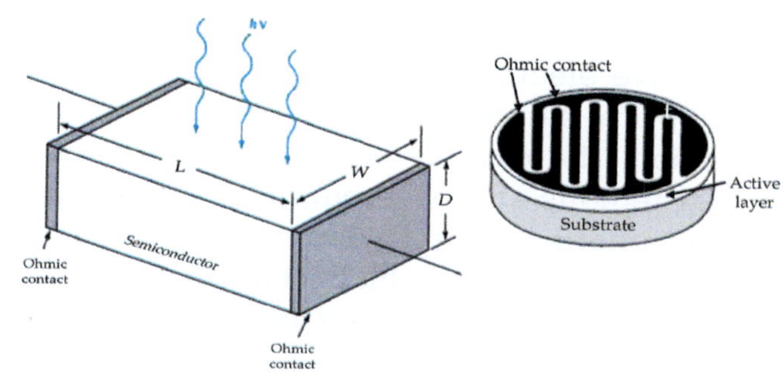

(a) (b)

Figure 4.18. (a) Schematic diagram of a photoconductor that consists of a slab of semiconductor and a contact at each end. (b) Typical layout consists of interdigitated contacts with a small gap.

Source. https://pd-zdh.xaut.edu.cn/__local/A/F0/D3/EA002AB461DFDFC92BCFAD35BC4_C62E9AA9_1335DE9.pdf

where $\mathscr{E}$ is the electric field inside the photocoductor and υd is the carrier drift velocity. Substituting n in Eq. 4 into Eq. 5 gives

$$I_p \quad q\left(\eta \frac{P_{opt}}{h\nu}\right)\cdot\left(\frac{\mu_n \tau \mathcal{E}}{L}\right). \tag{6}$$

If we define the primary photocurrent as

$$I_{ph} \quad q\left(\eta \frac{P_{opt}}{h\nu}\right), \tag{7}$$

the photocurrent gain from Eq. 6 is

$$\boxed{\text{Gain} \equiv \frac{I_p}{I_{ph}} \quad \frac{\mu_n \tau \mathcal{E}}{L} \quad \frac{\tau}{t_r},} \tag{8}$$

where tr ≡ L/ υd = L/μn$\mathcal{E}$ is the carrier transit time. The gain depends on the ratio of carrier lifetime to the transit time.

4.4.2. Photodiode

A photodiode is a type of p-n junction that is operated under reverse bias. Note that the electric-field distribution is nonuniform and the maximum field is at the junction. When an optical signal enters the depletion region of the photodiode, the electric field within this region helps to separate the electron-hole pairs that are generated by the light, leading to the flow of photocurrent Ip through the external circuit. The photogenerated holes move through the depletion region, diffuse into the neutral p region, and combine with electrons from the negative electrode. Similarly, photogenerated electrons move in the opposite direction. If an optical signal enters the photodiode within a certain distance outside of the depletion region, the carriers will diffuse into the depletion region and drift across it. These neutral regions can be thought of as extensions of the electrodes to the depletion region. The photocurrent is dependent on the number of generated electron-hole pairs and the velocities at which the carriers drift. It is important to note that the photocurrent in the external circuit is due to the flow of electrons, even though both electrons and holes are moving within the depletion region. To ensure high-frequency operation, the depletion region should be thin to minimize transit time. Conversely, the depletion layer needs to be thick to enhance quantum efficiency by allowing more incident light absorption. Therefore, a balancing act is needed between response speed and quantum efficiency.

4.4.2.1. Quantum Efficiency

The quantum efficiency, as mentioned above, is the number of EHPs generated for each incident photon:

$$\eta = \left(\frac{I_p}{q}\right) \cdot \left(\frac{P_{opt}}{hv}\right)^{-1},$$ (9)

where I_p, also referred to as the external quantum efficiency, is the photogenerated current that results from the absorption of incident optical power Popt at a wavelength of λ (corresponding to a photon energy of hν). The photogenerated number of EHPs per absorbed photon is the definition of internal quantum efficiency. The absorption coefficient α is one of the main factors that determines η. There is a limited wavelength range in which significant photocurrent can be generated because α is a strong function of wavelength. The bandgap determines the long-wavelength cutoff λ_c, which is approximately 1.8 μm for germanium and 1.1 μm for silicon. The values of α are too small to provide significant band-to-band absorption at wavelengths longer than λ_c. Since the values of α are too large (~10^5 cm^{-1}) for wavelengths much shorter than λc, the radiation is mostly absorbed very close to the surface where recombination time is short. As a result, before they can gather in the p-n junction depletion region, the photocarriers can recombine. The photogenerated carriers in the depletion region may disappear by recombination or by trapping without contributing to the photocurrent. The quantum efficiency is never 100% and is influenced by the absorption coefficient and device design. To improve quantum efficiency, it is important to reduce surface reflections on the device to enhance absorption in the depletion region and to prevent carrier recombination or trapping by enhancing material and device quality.

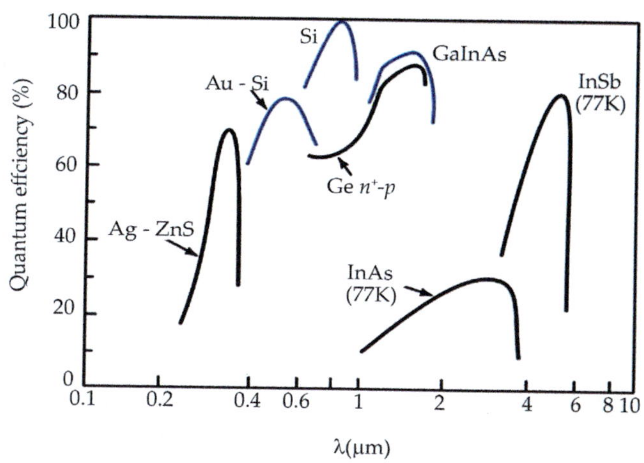

Figure 4.19. Quantum efficiency versus wavelength for various photodetectors.

Source. https://pd-zdh.xaut.edu.cn/__local/A/F0/D3/EA002AB461DFDFC92BCFAD35BC4_C62E9AA9_1335DE9.pdf.

Figure 4.19 illustrates quantum efficiency versus wavelength for various high-speed photodiodes. Metal-semiconductor photodiodes perform well in the ultraviolet and visible regions, while silicon photodiodes with antireflection coatings can achieve 100% quantum efficiency near the 0.8- to 0.9-μm range in the near-infrared region. In the 1.0- to 1.6-μm range, germanium photodiodes and Group III-V photodiodes like GaInAs exhibit high quantum efficiencies. For wavelengths beyond this range, photodiodes require cooling (e.g., to 77 K) for efficient operation.

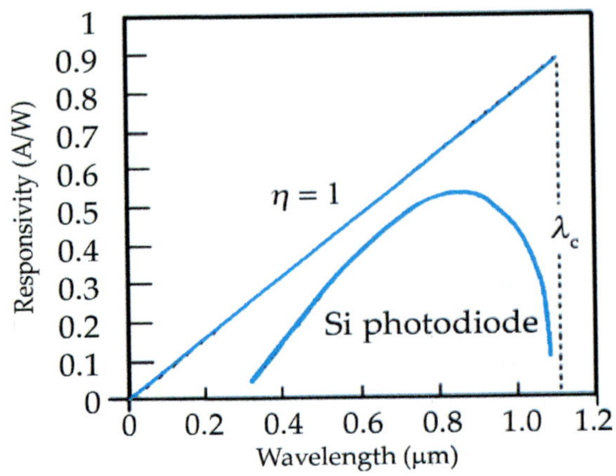

Figure 4.20. Responsivity vs. wavelength for an ideal photodiode with η = 1 and for a typical commercial Si photodiode.

Source. https://pd-zdh.xaut.edu.cn/__local/A/F0/D3/EA002AB461DFDFC92BCFAD35BC4_C62E9AA9_1335DE9.pdf.

4.4.2.2. Responsivity

The responsivity $\mathscr{R}$ of a photodiode is defined as the generated photocurrent (I_p) per incident optical power (P_{opt}). $\mathscr{R}$ is also called the spectral responsivity or radiant sensivity:

$$\mathscr{R} = I_p / P_{opt} \quad (10)$$

From the definition of quantum efficiency, we have

$$\mathscr{R} = I_p / P_{opt} = \eta q / h\nu = \eta q \lambda / hc \quad (11)$$

If a photodiode has an ideal quantum efficiency of 100%, then R should be linearly proportional to the wavelength. In practice, the relationship of R and λ is shown in Figure. 4.20. The quantum efficiency limits the responsivity below the ideal photodiode.

4.4.2.3. Response Speed

Three factors limit the response speed: (1) the depletion region's capacitance; (2) the depletion region's drift time; and (3) the diffusion of carriers. There will be a significant delay as carriers produced outside the depletion region must diffuse to the junction. The junction should form very close to the surface in order to minimize the diffusion effect. When the depletion region is broad, the lightest will be absorbed. The frequency response will be limited by transit time effects if the depletion layer is too wide. Additionally, it shouldn't be overly thin because a high capacitance C, where R is the load resistance, will lead to a high RC time constant. The width at which the modulation period is roughly half of the depletion layer transit time is the ideal compromise. For instance, the ideal depletion-layer thickness in silicon (with a saturation velocity of 10^7 cm/s) is roughly 25 µm at a modulation frequency of 2 GHz.

4.4.3. Solar Cells

4.4.3.1. A pn Junction Solar Cell

Photodiodes can be used as solar cells to convert solar energy to electrical energy. Consider the solar cell connected in a circuit, as shown below Figure 4.21

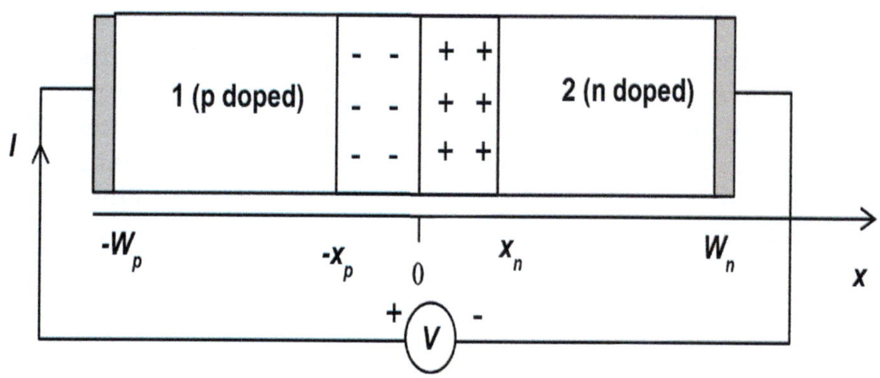

Figure 4.21. A pn Junction Solar Cell.

Source. https://courses.cit.cornell.edu/ece533/Lectures/handout5.pdf.

The relevant equations for the cell are,

$$I = I_O\left(e^{\frac{qV}{KT}} - 1\right) - I_L \quad (12)$$

$$IR + V = 0 \quad (13)$$

The following Figure 4.22 shows graphic solutions to these equations for various resistor R values. The operating points of the cell are represented by the solutions, which correspond to the intersection of the curves. Note that the pn junction in a solar cell is always forward biased.

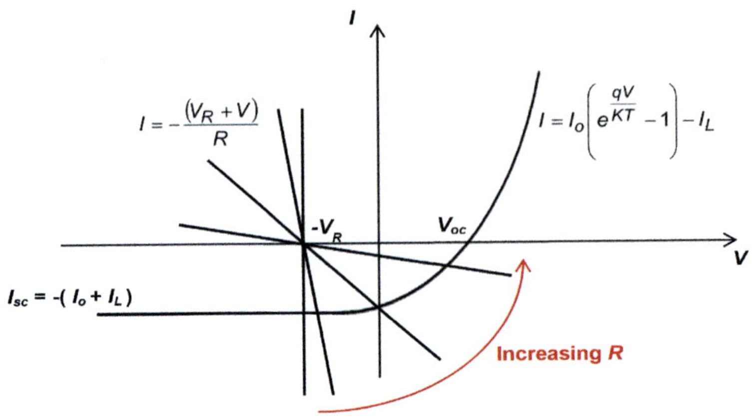

Figure 4.22. The pn junction in a solar cell is always forward biased.

Source. https://courses.cit.cornell.edu/ece533/Lectures/handout5.pdf.

If the external resistor R is too high, the voltage output from the cell will be high, but the current will be low, resulting in a low power delivery to the resistor. Conversely, if the external resistor R is too low, the voltage output will be low, but the current will be high, also leading to a low power delivery. The optimal value of R is one that maximizes power transfer from the cell to the external resistor. This optimal value can be determined by finding the balance between voltage and current in the circuit.

First, we maximize the power delivered to the resistor,

$$\frac{d}{dV}\left\{\left[I_o\left(e^{\frac{qV}{KT}}-1\right)-I_L\right](-V)\right\}=0$$

$$\Rightarrow e^{\frac{qV_m}{KT}}\left(1+\frac{qV_m}{KT}\right)=\frac{I_L}{I_o}+1$$

Here, V_m is the voltage that maximizes the output power of the cell and the corresponding current is,

$$I_m=I_o\left(e^{\frac{qV_m}{KT}}-1\right)-I_L$$

The power delivered to the resistor is equal to the area of the small lightly shaded rectangle in the Figure above. Once we know the optimal operating point, we can choose the resistor R such that the optimal operating point is the solution of Equations (12) and (13). The optimal value of R is,

$$\frac{1}{R} = \frac{qI_o}{KT} e^{\frac{qV_m}{KT}} = \frac{1}{R_{dm}}$$

Therefore, the optimal value of R is equal to the differential resistance R_{dm} of the diode at the optimal operating point. Note that the power output of the solar cell satisfies,

$$P_{out} = -I_m V_m < -I_{sc} V_{oc}$$

The product $-I_{sc}V_{oc}$ corresponds to the area of the large dark shaded rectangle in the Figure above. The fill factor FF of a solar cell is defined as the ratio of the areas of the two rectangles,

$$FF = \frac{-I_m V_m}{-I_{sc} V_{oc}}$$

Needless to say, a fill factor unity would be highly desirable. The sharper the diode turn-on behavior the larger is the fill factor.

CASE STUDY

PHOTOVOLTAIC SOLAR SYSTEM

Here we have studied two actual scenarios of a user having the same plant capacity, same estimated generation but different O&M Agencies. Both the O&M agencies' scope is the same, but one of the two agencies understood the scope right and followed the correct practices, which are depicted in the following example.

Here, we don't want to point out any of the O&M agencies, but just want to focus on the right practices to be followed.

Carry out Compliance-based Condition Monitoring/Testing

Current Practice

Normally, compliance-based testing is carried out to match the laws and regulations in an appropriate manner. Companies follow it or get it done based on the timeline given.

It is necessary to carry out compliance-based testing but not at the cost of losses the required downtime is incurring. Companies fail to perform analysis before going for compliance-based testing.

Corrected Practice

Before opting for the compliance-based testing, it is crucial to analyze some factors. Let's understand it in a simple way -

Parameters	**Agency 1**	**Scenario 2**
Plant Capacity DC	250 kWp	250 kWp
Estimated Generation	3,00,863 kWh	3,00,863 kWh
Downtime	100 Hours	80 Hours
Number of Faults	25	3
Generation Loss/Year	30,000 Units	37,000 Units
Replacement Expense	4.5 Lacs	12 Lacs
Water Usage	2.5 Lac Litre	2.3 Lac Litre
Cost Of Generation	4.30 Rs/kWh	4.30 Rs/kWh
Cost Of Water	0.033 Rs/Litre	0.033 Rs/Litre
Total Expense	5,87,500	13,66,590

Analysis

From the above-mentioned data, one can select scenario 2 over scenario 1 thinking of less downtime and a smaller number of faults.

But from the calculations performed in scenario 1 and scenario 2, it is clearly visible that the other expense incurring in scenario 2 is much higher than what is incurring in scenario 1.

Conclusion

So, practically, one should go ahead with scenario 1 and not with scenario 2 just by thinking of lesser downtime and faults.

In the above-mentioned scenario, agency 1 carried out the root cause analysis, hence resulting in more faults but forgot to consider the MMTR and MTTF factors.

We normally get such work done from the O&M Contractor whose approach is to keep the system running rather than to minimize/fix the fault completely, which can result in the failure of the equipment. Normally, the company doesn't get a replacement for such losses because of the poor management of the assets with respect to the guidelines given by the original equipment manufacturer or system design basis.

CLASS ACTIVITY

Students design and build a model city powered by the sun! They learn about the benefits of solar power and how architectural and building engineers integrate photovoltaic panels into the design of buildings. Students should understand the steps of the engineering design process. They should also have a general understanding of what electricity is and how it is used. Students should be able to use a ruler to make measurements and be able to connect a circuit in series.

SUMMARY

- Photovoltaic (PV) devices, commonly known as solar cells, are technology that converts sunlight directly into electricity through the photovoltaic effect.
- Silicon-based PV cells were the first sector of photovoltaics to enter the market, using processing information and raw materials supplied by the industry of microelectronics.
- Standard aluminum back surface field (Al-BSF) technology is one of the most widely used solar cell technologies due to its relatively simple manufacturing process.
- The thin film photovoltaic cells based on CdTe, gallium selenide, and copper (CIGS) or amorphous silicon have been designed to be a lower-cost replacement for crystalline silicon cells.
- Second-generation photovoltaic cells also include CdTe-based solar cells. An interesting property of CdTe is the reduction in cell size—due to its high spectral efficiency, the absorber thickness can be reduced to about 1 μm without much loss in efficiency, although further work is needed.
- The third generation of solar cells (including tandem, perovskite, dye-sensitized, organic, and emerging concepts) represent a wide range of approaches, from inexpensive low-efficiency systems (dye-sensitized, organic solar cells) to expensive high-efficiency systems (III-V multi-junction cells) for applications that range from building integration to space applications.
- Solar cells made from these materials are called quantum dots (QDs) and are also known as nanocrystalline solar cells.
- Semi-metallic graphene having a zero band gap creates Schottky junction solar cells with silicon semiconductors.
- Photodetectors are semiconductor devices that can convert optical signals into electrical signals.
- A photodiode is basically a p-n junction operated under reverse bias. Note that the electric-field distribution is nonuniform and the maximum field is at the junction.
- The response speed is limited by three factors: (1) diffusion of carriers, (2) drift time in the depletion region, and (3) capacitance of the depletion region.

REVIEW QUESTIONS

1. Discuss about the first and second generation of photovoltaic cells.
2. What are the organic and polymeric materials photovoltaic cells (OSC)?
3. What do you understand by photodetectors?
4. What is photodiode?
5. Explain about solar cells.

REFERENCES

1. Alaaeddin, M. H., Sapuan, S. M., Zuhri, M. Y. M., Zainudin, E. S., & Al-Oqla, F. M. (2019). Photovoltaic applications: Status and manufacturing prospects. Renewable and Sustainable Energy Reviews, 102, 318–332. https://doi.org/10.1016/j.rser.2018.12.026.
2. Alami, A. H., Ramadan, M., Abdelkareem, M. A., Alghawi, J. J., Alhattawi, N. T., Mohamad, H. A., & Olabi, A. G. (2022). Novel and practical photovoltaic applications. Thermal Science and Engineering Progress, 29, 101208. https://doi.org/10.1016/j.tsep.2022.101208.
3. Almosni, S., Delamarre, A., Jehl, Z., Suchet, D., Cojocaru, L., Giteau, M., Behaghel, B., Julian, A., Ibrahim, C., & Tatry, L. (2018). Material challenges for solar cells in the twenty-first century: Directions in emerging technologies. *Science and Technology of Advanced Materials, 19*, 336–369. https://doi.org/10.1080/14686996.2018.1433439.
4. Azzouzi, G., & Tazibt, W. (2013). Improving silicon solar cell efficiency by using the impurity photovoltaic effect. Energy Procedia, 41, 40–49. https://doi.org/10.1016/j.egypro.2013.09.005.
5. Bera, D., Qian, L., Tseng, T.-K., & Holloway, P. H. (2010). Quantum dots and their multimodal applications: A review. Materials, 3, 2260–2345. https://doi.org/10.3390/ma3042260.
6. Billewicz, P., & Węgierek, P. (2016). Laboratory stand for examining the influence of environmental conditions on electrical parameters of photovoltaic cells. *Przegl⊠d Elektrotechniczny, 92*, 176–179. https://doi.org/10.15199/48.2016.08.48.
7. Billewicz, P., Węgierek, P., Grudniewski, T., & Turek, M. (2017). Application of ion implantation for intermediate energy levels formation in the silicon-based structures dedicated for photovoltaic purposes. Acta Physica Polonica A, 132, 274–277. https://doi.org/10.12693/APhysPolA.132.274.
8. Cai, J., Ruffieux, P., Jaafar, R., Bieri, M., Braun, T., Blankenburg, S., Muoth, M., Seitsonen, A. P., Saleh, M., Feng, X., et al., (2010). Atomically precise bottom-up fabrication of graphene nanoribbons. Nature, 466, 470–473. https://doi.org/10.1038/nature09211.
9. Crabtree, G. W., & Lewis, N. S. (2018). Physics of sustainable energy, using energy efficiently and producing it renewably. In *Proceedings of the AIP Conference*. Berkeley, CA, USA. March 1–2, 2008. American Institute of Physics.
10. Dambhare, M. V., Butey, B., & Moharil, S. V. (2021). Solar photovoltaic technology: A review of different types of solar cells and its future trends. *Journal of Physics: Conference Series, 1913*, 012053. https://doi.org/10.1088/1742-6596/1913/1/012053.
11. Das, S., Pandey, D., Thomas, J., & Roy, T. (2019). The role of graphene and other 2D materials in solar photovoltaics. Advanced Materials, 31, 1802722. https://doi.org/10.1002/adma.201802722.
12. Dunlap-Shohl, W. A., Zhou, Y., Padture, N. P., & Mitzi, D. B. (2019). Synthetic approaches for halide perovskite thin films. Chemical Reviews, 119(5), 3193–3295. https://doi.org/10.1021/acs.chemrev.8b00318.

13. Eswaraiah, V., Aravind, S. S. J., & Ramaprabhu, S. (2011). Top-down method for synthesis of highly conducting graphene by exfoliation of graphite oxide using focused solar radiation. Journal of Materials Chemistry, 21, 6800–6803. https://doi.org/10.1039/c1jm10808e.

14. Fernández, S., Gandía, J. J., Saugar, E., Gómez-Mancebo, M. B., Canteli, D., & Molpeceres, C. (2021). Sputtered non-hydrogenated amorphous silicon as alternative absorber for silicon photovoltaic technology. Materials, 14(21), 6550. https://doi.org/10.3390/ma14216550.

15. Fthenakis, V., Athias, C., Blumenthal, A., Kulur, A., Magliozzo, J., & Ng, D. (2020). Sustainability evaluation of CdTe PV: An update. Renewable and Sustainable Energy Reviews, 123, 109776. https://doi.org/10.1016/j.rser.2020.109776.

16. Gangopadhyay, U., Jana, S., & Das, S. (2013). State of art of solar photovoltaic technology. In Proceedings of the International Conference on Solar Energy Photovoltaics (pp. 19–21). Hindawi Limited.

17. Gao, H., Yang, R., & Zhang, Y. (2020). Improving radiation resistance of GaInP/GaInAs/Ge triple-junction solar cells using GaInP back-surface field in the middle subcell. Materials, 13(8), 1958. https://doi.org/10.3390/ma13081958.

18. Garcia-Barrientos, A., Bernal-Ponce, J. L., Plaza-Castillo, J., Cuevas-Salgado, A., Medina-Flores, A., Garcia-Monterrosas, M. S., & Torres-Jacome, A. (2021). Analysis, synthesis and characterization of thin films of a-Si (n-type and p-type) deposited by PECVD for solar cell applications. Materials, 14(21), 6349. https://doi.org/10.3390/ma14216349.

19. Geim, A., & Novoselov, K. (2007). The rise of graphene. Nature Materials, 6, 183–191. https://doi.org/10.1038/nmat1849.

20. Ghosh, S., & Yadav, R. (2021). Future of photovoltaic technologies: A comprehensive review. Sustainable Energy Technologies and Assessments, 47, 101410. https://doi.org/10.1016/j.seta.2021.101410.

21. Goetzberger, A., Hebling, C., & Schock, H. W. (2003). Photovoltaic materials, history, status and outlook. *Materials Science and Engineering R: Reports, 40*, 1–46. https://doi.org/10.1016/S0927-796X(02)00092-X.

22. Green, M. A., & Bremner, S. P. (2016). Energy conversion approaches and materials for high-efficiency photovoltaics. Nature Materials, 16(1), 23–34. https://doi.org/10.1038/nmat4676.

23. Hayat, M. B., Ali, D., Monyake, K. C., Alagha, L., & Ahmed, N. (2019). Solar energy—A look into power generation, challenges, and a solar-powered future. *International Journal of Energy Research, 43*, 1049–1067. https://doi.org/10.1002/er.4252.

24. Hu, Z., Wang, J., Ma, X., Gao, J., Xu, C., Yang, K., & Zhang, F. (2020). A critical review on semitransparent organic solar cells. Nano Energy, 78, 105376. https://doi.org/10.1016/j.nanoen.2020.105376.

25. Huang, H., Lv, J., Bao, Y., Xuan, R., Sun, S., Sneck, S., Li, S., Modanese, C., Savin, H., & Wang, A., et al., (2017). 20.8% industrial PERC solar cell: ALD Al2O3 rear surface passivation, efficiency loss mechanisms analysis and roadmap to 24%. Solar Energy Materials and Solar Cells, 161, 14–30. https://doi.org/10.1016/j.solmat.2016.11.018.

26. Jasim, K. E. (2015). Quantum dots solar cells. Solar Cells–New Approaches Reviews, 3, 303–331. https://doi.org/10.5772/59159.

27. Jean, J., Brown, P. R., Jaffe, R. L., Buonassisi, T., & Bulovi⊠, V. (2015). Pathways for solar photovoltaics. Energy & Environmental Science, 8(4), 1200–1219. https://doi.org/10.1039/C4EE04073B.

28. Jia, G., Plentz, J., Dellith, J., Dellith, A., Wahyuono, R. A., & Andrä, G. (2019). Large area graphene deposition on hydrophobic surfaces, flexible textiles, glass fibers and 3D structures. *Coatings, 9*, 183. https://doi.org/10.3390/coatings9030183.

29. Kant, N., & Singh, P. (2022). Review of next generation photovoltaic solar cell technology and comparative materialistic development. *Materials Today: Proceedings, 56*, 3460–3470. https://doi.org/10.1016/j.matpr.2021.11.116.

30. Keis, K., Magnusson, E., Lindström, H., Lindquist, S. E., & Hagfeldt, A. (2002). A 5% efficient photoelectrochemical solar cell based on nanostructured ZnO electrodes. Solar Energy Materials and Solar Cells, 73(1), 51–58. https://doi.org/10.1016/S0927-0248(01)00110-6.

31. Kim, H. S., Lee, C. R., Im, J. H., Lee, K. B., Moehl, T., Marchioro, A., Moon, S.-J., Humphry-Baker, R., Yum, J.-H., Moser, J. E., et al., (2012). Lead iodide perovskite sensitized all-solid-state submicron thin film mesoscopic solar cell with efficiency exceeding 9%. Scientific Reports, 2(1), 591. https://doi.org/10.1038/srep00591.

32. Kim, S., Kim, D., Hong, J., Elmughrabi, A., Melis, A., Yeom, J.-Y., Park, C., & Cho, S. (2022). Performance comparison of CdTe, CdTe, and CdTe single crystals for solar cell applications. Materials, 15(4), 1408. https://doi.org/10.3390/ma15041408.

33. Kowalski, M., Partyka, J., Węgierek, P., Żukowski, P., Komarov, F. F., Jurchenko, A. V., & Freik, D. (2005). Frequency-dependent annealing characteristics of the implant-isolated GaAs layers. Vacuum, 78(3), 311–317. https://doi.org/10.1016/j.vacuum.2005.01.112.

34. Kranz, L., Buecheler, S., & Tiwari, A. N. (2013). Technological status of CdTe photovoltaics. Solar Energy Materials and Solar Cells, 119, 278–280. https://doi.org/10.1016/j.solmat.2013.08.028.

35. Krügener, J., Osten, H. J., Kiefer, F., Haase, F., & Peibst, R. (2016). Ion implantation for photovoltaic applications: Review and outlook for n-type silicon solar cells. In *Proceedings of the 2016 21st International Conference on Ion Implantation Technology (IIT)*. Tainan, China.

36. Kuczyńska-Łażewska, A., Klugmann-Radziemska, E., & Witkowska, A. (2021). Recovery of valuable materials and methods for their management when recycling thin-film CdTe photovoltaic modules. Materials, 14(24), 7836. https://doi.org/10.3390/ma14247836.

37. Law, M., Greene, L. E., Johnson, J. C., Saykally, R., & Yang, P. (2005). Nanowire dye-sensitized solar cells. Nature Materials, 4(6), 455–459. https://doi.org/10.1038/nmat1387.

38. Lee, M. M., Teuscher, J., Miyasaka, T., Murakami, T. N., & Snaith, H. J. (2012). Efficient hybrid solar cells based on meso-super structured organometal halide perovskites. Science, 338(6107), 643–647. https://doi.org/10.1126/science.1228604.

39. Li, X., Zhu, H., Wang, K., Cao, A., Wei, J., Li, C., Jia, Y., Li, Z., Li, X., & Wu, D. (2010). Graphene-on-silicon Schottky junction solar cells. Advanced Materials, 22, 2743–2748. https://doi.org/10.1002/adma.200904383.

40. Lim, D. H., Ha, J. W., Choi, H., Yoon, S. C., Lee, B. R., & Ko, S. J. (2021). Recent progress of ultra-narrow-bandgap polymer donors for NIR-absorbing organic solar cells. Nanoscale Advances, 3(16), 4306–4320. https://doi.org/10.1039/D1NA00245G.

41. Liu, Y., & Chen, Y. (2020). Integrated perovskite/bulk-heterojunction organic solar cells. Advanced Materials, 32(30), 1805843. https://doi.org/10.1002/adma.201805843.

42. López, E., Martí, A., Antolín, E., & Luque, A. (2020). On the potential of silicon intermediate band solar cells. Energies, 13, 3044. https://doi.org/10.3390/en13123044.

43. Luque, A., & Hegedus, S. (Eds.). (2011). *Handbook of Photovoltaic Science and Engineering* (2nd ed., pp. 4–36). John Wiley & Sons.

44. Mahmoudi, T., Wang, Y., & Hahn, Y. B. (2018). Graphene and its derivatives for solar cells application. Nano Energy, 47, 51–65. https://doi.org/10.1016/j.nanoen.2018.02.047.

45. Marques Lameirinhas, R. A., Torres, J. P. N., & de Melo Cunha, J. P. (2022). A photovoltaic technology review: History, fundamentals and applications. *Energies, 15*, 1823. https://doi.org/10.3390/en15051823.

46. Metz, A., Adler, D., Bagus, S., Blanke, H., Bothar, M., Brouwer, E., Dauwe, S., Dressler, K., Droessler, R., & Droste, T., et al., (2014). Industrial high performance crystalline silicon solar cells and modules based on rear surface passivation technology. Solar Energy Materials and Solar Cells, 120, 417–425. https://doi.org/10.1016/j.solmat.2013.06.025.

47. Mozaffari, S., Nateghi, M. R., & Zarandi, M. B. (2017). An overview of the challenges in the commercialization of dye-sensitized solar cells. Renewable and Sustainable Energy Reviews, 71, 675–686. https://doi.org/10.1016/j.rser.2016.12.096.

48. Muhammad, J. Y. U., Waziri, A. B., Shitu, A. M., Ahmad, U. M., Muhammad, M. H., Alhaji, Y., Olaniyi, A. T., & Bala, A. A. (2019). Recent progressive status of materials for solar photovoltaic cell: A comprehensive review. *Science Journal of Energy Engineering,* 7, 77–89. https://doi.org/10.11648/j.sjee.20190704.14.

49. Nakamura, M., Yamaguchi, K., Kimoto, Y., Yasaki, Y., Kato, T., & Sugimoto, H. (2019). Cd-free Cu (In, Ga) (Se, S)2 thin-film solar cell with record efficiency of 23.35%. IEEE Journal of Photovoltaics, 9(6), 1863–1867. https://doi.org/10.1109/JPHOTOV.2019.2937218.

50. Nayeripour, M., Mansouri, M., Orooji, F., & Waffenschmidt, E. (Eds.). (2020). Solar Cells. IntechOpen Limited.

51. Ojo, A. A., Cranton, W. M., & Dharmadasa, I. M. (2019). Next generation multilayer graded bandgap solar cells. In *Springer*, 17–40.

52. Otte, K., Makhova, L., Braun, A., & Konovalov, I. (2006). Flexible Cu (In,Ga)Se2 thin-film solar cells for space application. Thin Solid Films, 511, 613–622. https://doi.org/10.1016/j.tsf.2005.11.068.

53. Parida, B., Iniyan, S., & Goic, R. (2011). A review of solar photovoltaic technologies.

Renewable and Sustainable Energy Reviews, 15(3), 1625–1636. https://doi.org/10.1016/j.rser.2010.11.032.

54. Pérez, E., Duenas, S., Castán, H., García, H., Bailón, L., Montero, D., García-Hernansanz, R., Hemme, E. G., Olea, J., & González-Díaz, G. (2015). A detailed analysis of the energy levels configuration existing in the band gap of supersaturated silicon with titanium for photovoltaic applications. *Journal of Applied Physics, 118*, 245704. https://doi.org/10.1063/1.4939198.

55. Petrova-Koch, V., Hezel, R., & Goetzberger, A. (2009). High-Efficient Low-Cost Photovoltaics. Springer International Publishing.

56. Peumans, P., Yakimov, A., & Forrest, S. R. (2003). Small molecular weight organic thin-film photodetectors and solar cells. Journal of Applied Physics, 93(7), 3693–3723. https://doi.org/10.1063/1.1534621.

57. Rajan, G., Karki, S., Collins, R. W., Podraza, N. J., & Marsillac, S. (2020). Real-time optimization of anti-reflective coatings for CIGS solar cells. Materials, 13(19), 4259. https://doi.org/10.3390/ma13194259.

58. Ren, F., Yao, M., Li, M., & Wang, H. (2021). Tailoring the structural and electronic properties of graphene through ion implantation. *Materials, 14*(5080). https://doi.org/10.3390/ma14175080.

59. Richter, A., Hermle, M., & Glunz, S. W. (2013). Reassessment of the limiting efficiency for crystalline silicon solar cells. *IEEE Journal of Photovoltaics, 3*, 1184–1191. https://doi.org/10.1109/JPHOTOV.2013.2270351.

60. Roy, S., Baruah, M. S., Sahu, S., & Nayak, B. B. (2021). Computational analysis on the thermal and mechanical properties of thin film solar cells. Materials Today: Proceedings, 44, 1207–1213. https://doi.org/10.1016/j.matpr.2020.11.241.

61. Saga, T. (2010). Advances in crystalline silicon solar cell technology for industrial mass production. NPG Asia Materials, 2(2), 96–102. https://doi.org/10.1038/asiamat.2010.82.

62. Salhi, B. (2022). The photovoltaic cell based on CIGS: Principles and technologies. Materials, 15(5), 1908. https://doi.org/10.3390/ma15051908.

63. Sharma, D., Mehra, R., & Raj, B. (2021). Comparative analysis of photovoltaic technologies for high efficiency solar cell design. *Superlattices and Microstructures, 153*, 106861. https://doi.org/10.1016/j.spmi.2021.106861.

64. Sharma, P., & Goyal, P. (2020). Evolution of PV technology from conventional to nano-materials. *Materials Today: Proceedings, 28*, 1593–1597. https://doi.org/10.1016/j.matpr.2020.04.846.

65. Singh, B. P., Goyal, S. K., & Kumar, P. (2021). Solar PV cell materials and technologies: Analyzing the recent developments. *Materials Today: Proceedings, 43*, 2843–2849. https://doi.org/10.1016/j.matpr.2021.01.003.

66. Stamford, L., & Azapagic, A. (2019). Environmental impacts of copper indium gallium-selenide (CIGS) photovoltaics and the elimination of cadmium through atomic layer deposition. Science of the Total Environment, 688, 1092–1101. https://doi.org/10.1016/j.scitotenv.2019.06.343.

67. Taguchi, M., Kawamoto, K., Tsuge, S., Baba, T., Sakata, H., Morizane, M., Uchihashi,

K., Nakamura, N., Kiyama, S., & Oota, O. (2000). HITTM cells—High-efficiency crystalline Si cells with novel structure. Progress in Photovoltaics: Research and Applications, 8(5), 503–513. https://doi.org/10.1002/1099-159X(200009/10)8:5<503:AID-PIP347>3.0.CO;2-G.

68. Taguchi, M., Yano, A., Tohoda, S., Matsuyama, K., Nakamura, Y., Nishiwaki, T., Fujita, K., & Maruyama, E. (2013). 24.7% record efficiency HIT solar cell on thin silicon wafer. IEEE Journal of Photovoltaics, 4(1), 96–99. https://doi.org/10.1109/JPHOTOV.2013.2282737.

69. Tian, J., & Cao, G. (2013). Semiconductor quantum dot-sensitized solar cells. Nano Reviews, 4, 22578. https://doi.org/10.3402/nano.v4i0.22578.

70. Tsakalakos, L. (2010). *Nanotechnology for Photovoltaics* (1st ed., pp. 1–48). CRC Press.

71. Ushasree, P. M., & Bora, B. (Eds.). (2019). Solar Energy Capture Materials. London, UK: The Royal Society of Chemistry.

72. Van Deelen, J., Tezsevin, Y., & Barink, M. (2016). Multi-material front contact for 19% thin film solar cells. Materials, 9(2), 96. https://doi.org/10.3390/ma9020096.

73. Wang, W., Winkler, M. T., Gunawan, O., Gokmen, T., Todorov, T. K., Zhu, Y., & Mitzi, D. B. (2014). Device characteristics of CZTSSe thin-film solar cells with 12.6% efficiency. Advanced Energy Materials, 4(7), 1301465. https://doi.org/10.1002/aenm.201301465.

74. Wang, Y., Chen, X., Zhong, Y., Zhu, F., & Loh, K. P. (2009). Large area, continuous, few-layered graphene as anodes in organic photovoltaic devices. *Applied Physics Letters, 95*, 209. https://doi.org/10.1063/1.3204698.

75. Węgierek, P., & Billewicz, P. (2011). Jump mechanism of electric conduction in n-type silicon implanted with Ne++ neon ions. *Acta Physica Polonica A, 120*, 122–124. https://doi.org/10.12693/APhysPolA.120.122.

76. Węgierek, P., & Billewicz, P. (2013). Research on mechanisms of electric conduction in the p-type silicon implanted with Ne+ ions. *Acta Physica Polonica A, 123*, 948–951. https://doi.org/10.12693/APhysPolA.123.948.

77. Węgierek, P., & Pastuszak, J. (2021). Application of neon ion implantation to generate intermediate energy levels in the band gap of boron-doped silicon as a material for photovoltaic cells. *Materials, 14*, 6950. https://doi.org/10.3390/ma14226950.

78. Węgierek, P., & Pietraszek, J. (2019). Analysis of the influence of annealing temperature on mechanisms of charge carrier transfer in GaAs in the aspect of possible applications in photovoltaics. *Acta Physica Polonica A, 136*, 299–302. https://doi.org/10.12693/APhysPolA.136.299.

79. Węgierek, P., & Pietraszek, J. (2019). Application of poly-energy implantation with H+ ions for additional energy levels formation in GaAs dedicated to photovoltaic cells. *Archives of Electrical Engineering, 68*, 925–931. https://doi.org/10.24425/aee.2019.130692.

80. Węgierek, P., Pastuszak, J., & Dziadosz, K. (2020). Influence of substrate type and dose of implanted ions on the electrical parameters of silicon in terms of improving

the efficiency of photovoltaic cells. *Energies, 13*, 6708. https://doi.org/10.3390/en13246708.

81. Wilkins, M. M., Dumitrescu, E. C., & Krich, J. J. (2020). Material quality requirements for intermediate band solar cells. IEEE Journal of Photovoltaics, 10(2), 467–474. https://doi.org/10.1109/JPHOTOV.2019.2959934.
82. Wilson, G. M., Al-Jassim, M., Metzger, W. K., Glunz, S. W., Verlinden, P., Xiong, G., Mansfield, L. M., Stanbery, B. J., Zhu, K., & Yan, Y. (2020). The 2020 photovoltaic technologies roadmap. *Journal of Physics D: Applied Physics, 53*, 493001. https://doi.org/10.1088/1361-6463/ab9c6a.
83. Wu, C., Wang, K., Batmunkh, M., Bati, A. S., Yang, D., Jiang, Y., Hou, Y., Shapter, J. G., & Priya, S. (2020). Multifunctional nanostructured materials for next generation photovoltaics. Nano Energy, 70, 104480. https://doi.org/10.1016/j.nanoen.2020.104480.
84. Wu, X. (2004). High-efficiency polycrystalline CdTe thin-film solar cells. Solar Energy, 77(6), 803–814. https://doi.org/10.1016/j.solener.2004.06.006.
85. Yamaguchi, M., Takamoto, T., Araki, K., & Ekins-Daukes, N. (2005). Multi-junction III–V solar cells: Current status and future potential. Solar Energy, 79, 78–85. https://doi.org/10.1016/j.solener.2004.09.018.
86. Yuan, J., Hazarika, A., Zhao, Q., Ling, X., Moot, T., Ma, W., & Luther, J. M. (2020). Metal halide perovskites in quantum dot solar cells: Progress and prospects. Joule, 4, 1160–1185. https://doi.org/10.1016/j.joule.2020.04.006.
87. Zandi, S., Seresht, M. J., Khan, A., & Gorji, N. E. (2022). Simulation of heat loss in Cu2ZnSn4SxSe4−x thin film solar cells: A coupled optical-electrical-thermal modeling. Renewable Energy, 181, 320–328. https://doi.org/10.1016/j.renene.2021.09.035.

CHAPTER

5

Metal–Semiconductor and Semiconductor Heterojunctions

LEARNING OBJECTIVES

After studying this chapter, you will be able to:

- Learn about the schottky barrier diode
- Discuss the metal–semiconductor ohmic contacts
- Understand about heterojunctions

KEY TERMS FROM THIS CHAPTER

Anisotype junction
Fast ion conductor
Homojunction
Isotype junction
Richardson constant
Schottky effect
Electron affinity rule
Heterojunction
Image force–induced lowering
Ohmic contact
Schottky barrier height
Semiconductors

5.1. INTRODUCTION

An electrical junction known as a metal-semiconductor (M-S) junction occurs when a metal and a semiconductor material are in close proximity to one another. It is one of the earliest semiconductor devices still in use today. There are two types of M-S junctions: rectifying and non-rectifying. A Schottky diode is a device that has a rectifying metal-semiconductor junction that forms a Schottky barrier; an ohmic contact is a non-rectifying junction.

All semiconductor devices depend on metal-semiconductor junctions to function. In order to facilitate easy electrical charge conductivity between a transistor's active region and external circuitry, an ohmic contact is typically preferred. But occasionally, a Schottky barrier is helpful, as in the case of metal-semiconductor field effect transistors, Schottky diodes, and Schottky transistors.

The metal-semiconductor junction and the semiconductor heterojunction, in which the materials on either side of the junction differ, are covered in this chapter. Diodes can also be produced by these junctions.

Integrated circuits, sometimes known as semiconductor devices, need to communicate with the outside world. Ohmic contacts, also known as non-rectifying metal-semiconductor junctions, are used to create this contact. A low-resistance junction that allows current to flow in both directions is known as an ohmic contact. In this chapter, we study the circumstances that lead to metal-semiconductor ohmic contacts.

5.2. THE SCHOTTKY BARRIER DIODE

Potential energy barriers for electrons created at a metal-semiconductor junction are known as Schottky barriers. Schottky barriers can be used as diodes because of their rectifying properties. The Schottky barrier height, represented by Φ_B, is one of the main features of a Schottky barrier (see Figure 5.1). The combination of semiconductor and metal determines the value of Φ_B (Tung, Raymond (2014)).

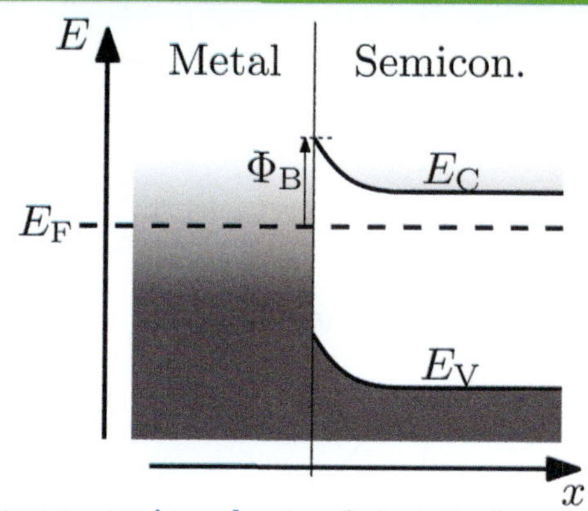

Figure 5.1. Band diagram for n-type semiconductor Schottky barrier at zero bias (equilibrium) with graphical definition of the Schottky barrier height, Φ_B, as the difference between the interfacial conduction band edge E_C and Fermi level E_F.

Source: https://upload.wikimedia.org/wikipedia/commons/thumb/c/cc/Schottky_barrier_zero_bias.svg/800px-Schottky_barrier_zero_bias.svg.png.

The metal-semiconductor diode was among the first useful semiconductor devices used in the early 1900s. The process of creating this diode, also known as a point contact diode, involved touching an exposed semiconductor surface with a metallic whisker. In the 1950s, the pn junction took the place of these metal-semiconductor diodes since they were more mechanically reliable and easier to duplicate. But today, metal-semiconductor contacts that are dependable and repeatable are created using semiconductor and vacuum technologies. The metal-semiconductor rectifying contact, also known as the Schottky barrier diode, is discussed in this section. We focus on n-type semiconductors since they are the ones on which the rectifying contacts are typically made.

5.2.1. Qualitative Characteristics

Figure 5.2a displays the optimal energy-band diagram for a specific metal and n-type semiconductor prior to making contact. As a point of reference, the vacuum level is employed. The metal work function (φm) is expressed in volts, the semiconductor work function (φs) is expressed in volts, and the electron affinity (χ) is expressed as a parameter. Table 5.1 lists the work functions of different metals, and Table 5.2 lists the electron affinities of various semiconductors. It is assumed in Figure 5.2a that $\varphi_m > \varphi_s$. Figure 5.2b depicts the optimal thermal-equilibrium metal-semiconductor energy-band diagram in this case. The semiconductor's Fermi level was higher than the metal's prior to contact. Electrons from the semiconductor flow into the lower energy states in the metal in order for the Fermi level to become a constant throughout the system in thermal equilibrium. A space charge region is formed in the semiconductor by positively charged donor atoms that stay there.

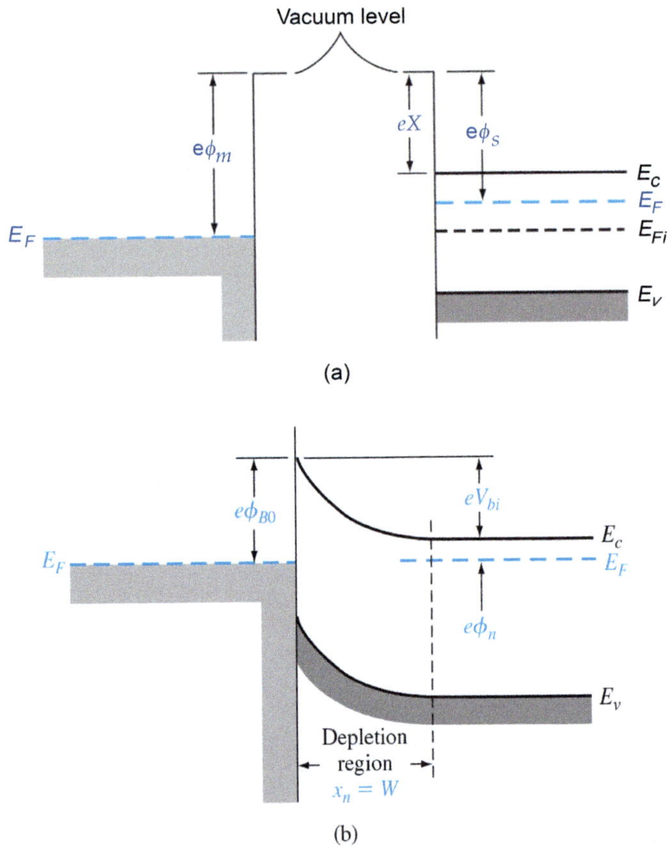

Figure 5.2. (a) Energy-band diagram of a metal and semiconductor before contact; (b) ideal energy-band diagram of a metal–n-semiconductor junction for $\varphi_m > \varphi_s$.

Source: Donald A. Neamen. Semiconductor Physics and Devices: Basic Principles, Fourth Edition. McGraw-Hill. ISBN 978-0-07-352958-5.

Table 5.1. Work Functions of Some Elements

Element	Work function, ϕ_m
Ag, silver	4.26
Al, aluminum	4.28
Au, gold	5.1
Cr, chromium	4.5
Mo, molybdenum	4.6
Ni, nickel	5.15
Pd, palladium	5.12
Pt, platinum	5.65
Ti, titanium	4.33
W, tungsten	4.55

Source: Donald A. Neamen. Semiconductor physics and devices: basic principles, Fourth edition. mcGraw-hill. ISBN 978-0-07-352958-5.

Table 5.2. Electron Affinity of Some Semiconductors

Element	Electron affinity, χ
Ge, germanium	4.13
Si, silicon	4.01
GaAs, gallium arsenide	4.07
AlAs, aluminum arsenide	3.5

Source: Donald A. Neamen. Semiconductor physics and devices: basic principles, fourth edition. McGraw-Hill. ISBN 978-0-07-352958-5.

The ideal barrier height of the semiconductor contact, or the potential barrier that electrons in the metal encounter when attempting to enter the semiconductor, is represented by the parameter Φ_{B0} This barrier, which goes by the name of Schottky barrier, is ideally provided by.

$$\phi_{B0} = (\phi_m - \chi) \tag{5.1}$$

The built-in potential barrier, or V_{bi}, is present in semiconductors. When electrons in the conduction band attempt to enter the metal, they encounter this barrier, which is comparable to the situation of the pn junction. The inherent possible barrier is provided by

$$V_{bi} = \phi_{B0} - \phi_n \tag{5.2}$$

The idealized case of a semiconductor-to-metal barrier height increasing with an applied positive voltage to the semiconductor relative to the metal results in a constant Φ_{B0}. It's the reverse bias that we have here. With respect to the semiconductor, a positive voltage applied to the metal reduces the semiconductor-to-metal barrier V_{bi}, but Φ_{B0} again stays relatively constant. Because the barrier has been lowered in this instance, electrons can move from the semiconductor more readily into the metal. The forward bias is this particular bias condition. Figures 5.3a,b showcase the energy-band diagrams for both the forward and reverse bias, where V_a represents the forward-biased voltage and V_R represents the reverse-biased voltage.

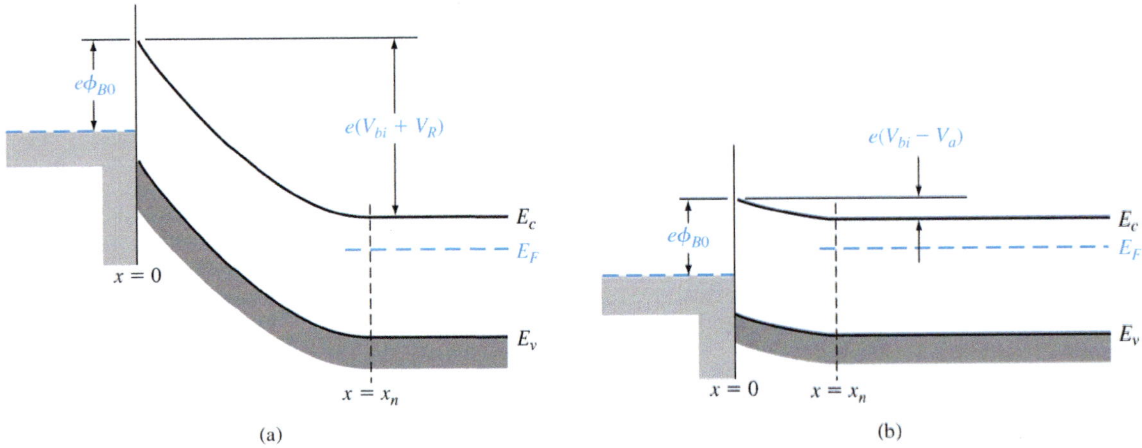

Figure 5.3. Ideal energy-band diagram of a metal–semiconductor junction (a) under reverse bias and (b) under forward bias.

Source: Donald A. Neamen. Semiconductor Physics and Devices: Basic Principles, Fourth Edition. McGraw-Hill. ISBN 978–0-07–352958–5.

The metal-semiconductor junction energy-band diagrams versus voltage displayed in Figure 5.3 are strikingly similar to the pn junction diagrams provided in the preceding chapter. We anticipate that the current-voltage characteristics of the Schottky barrier junction will resemble the pn junction diode's exponential behavior due to this similarity. However, the majority carrier electron flow is the current mechanism at play here. Forward bias facilitates the majority carrier electrons' easier passage from the semiconductor into the metal by lowering the barrier that the semiconductor's electrons see. As an exponential function of the forward-bias voltage V_a, the forward-bias current flows from metal to semiconductor.

5.2.2. Ideal Junction Properties

The same methods we use to ascertain the pn junction's electrostatic characteristics can be applied to this one as well. Poisson's equation provides information about the electric field in the space charge area. We possess that

$$\frac{dE}{dx} = \frac{\rho(x)}{\epsilon_s} \tag{5.3}$$

where ε_s is the semiconductor's permittivity and $\rho(x)$ is the space charge volume density. Assuming uniform semiconductor doping, we can obtain by integrating Equation (5.3).

$$E = \int \frac{eN_d}{\epsilon_s} dx = \frac{eN_d x}{\epsilon_s} + C_1 \tag{5.4}$$

where the integration constant C_1 is present. Since there is no electric field at the semiconductor's space charge edge, the integration constant can be found as

$$C_1 = -\frac{eN_d x_n}{\epsilon_s} \tag{5.5}$$

The electric field can then be written as

$$E = -\frac{eN_d}{\epsilon_s}(x_n - x) \tag{5.6}$$

which is a linear function of distance, for the uniformly doped semiconductor, and reaches a peak value at the metal–semiconductor interface. Since the E-field is zero inside the metal, a negative surface charge must exist in the metal at the metal–semiconductor junction.

The space charge region width, W, may be calculated as we do for the pn junction. The result is identical to that of a one-sided p^+n junction. For the uniformly doped semiconductor, we have

$$W = x_n = \left[\frac{2\epsilon_s(V_{bi} + V_R)}{eN_d}\right]^{1/2} \tag{5.7}$$

where VR is the magnitude of the applied reverse-biased voltage. We are again assuming an abrupt junction approximation.

A junction capacitance can also be determined in the same way as we do for the pn junction. We have that

$$C' = eN_d \frac{dx_n}{dV_R} = \left[\frac{e\epsilon_s N_d}{2(V_{bi} + V_R)}\right]^{1/2} \tag{5.8}$$

where C' is the capacitance per unit area. If we square the reciprocal of Equation (5.8), we obtain

$$\left(\frac{1}{C'}\right)^2 = \frac{2(V_{bi} + V_R)}{e\epsilon_s N_d} \tag{5.9}$$

Equation (5.9) can be used to determine the built-in potential barrier V_{bi}, as well as the semiconductor doping N_d, to a first approximation. The slope of the curve obtained from Equation (5.9) can also be used. Equation (5.2) allows us to compute the potential

φ_n and, from there, the Schottky barrier Φ_{B0}. It is evident that the silicon diode's intrinsic potential barrier is smaller than the gallium arsenide Schottky diode's. This experimental outcome is typically seen with all kinds of metal connections.

5.2.3. Nonideal Effects on the Barrier Height

5.2.3.1. Schottky Barrier Lowering

The actual Schottky barrier height differs from the theoretical value provided by Equation (5.1) due to a number of effects. The Schottky effect, or image-force-induced lowering of the potential barrier, is the first such effect that we examine.

An electric field is produced by an electron in a dielectric material that is x distances away from the metal. By adding an image charge, +e, inside the metal at the same distance, |x|, from the interface, the electric field can be calculated. Figure 5.4a illustrates this image effect. Observe that, as predicted, the E-field lines are perpendicular to the metal surface. The force on the electron, due to the coulomb attraction with the image force, is

$$F = \frac{-e^2}{4\pi\epsilon_s(2x)^2} = -e\mathrm{E} \tag{5.10}$$

The potential can then be found as

$$-\phi(x) = +\int_x^\infty \mathrm{E}\,dx' = +\int_x^\infty \frac{e}{4\pi\epsilon_s \cdot 4(x')^2}\,dx' = \frac{-e}{16\pi\epsilon_s x} \tag{5.11}$$

where x' is the integration variable and where we have assumed that the potential is zero at x = ∞.

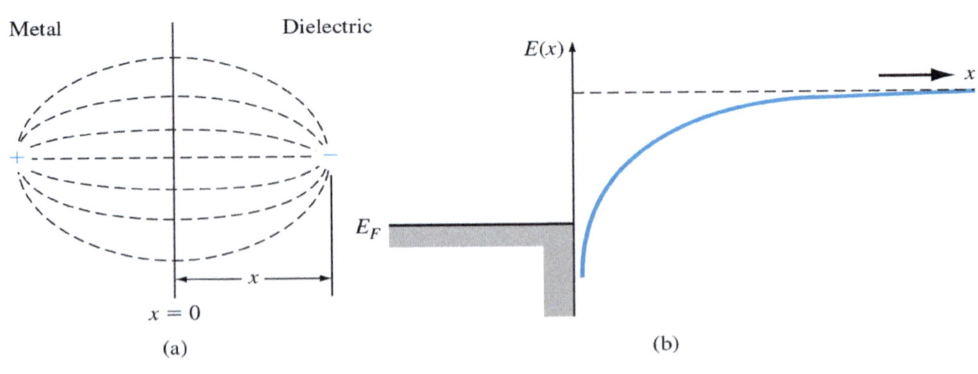

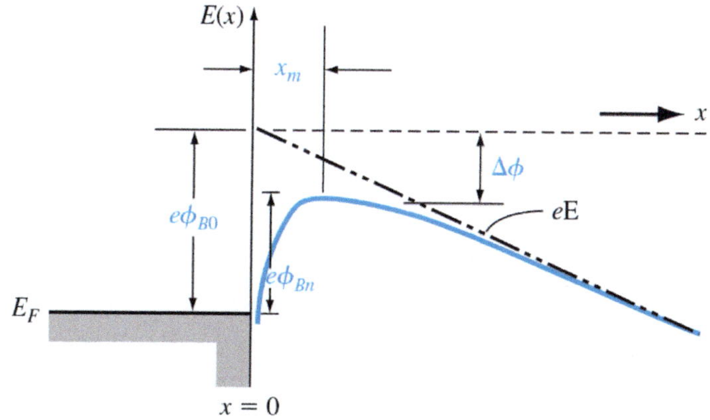

(c)

Figure 5.4. (a) Image charge and electric field lines at a metal–dielectric interface. (b) Distortion of the potential barrier due to image forces with zero electric field and (c) with a constant electric field.

Source: Donald A. Neamen. Semiconductor physics and devices: basic princ

iples, fourth edition. McGraw-Hill. ISBN 978–0-07–352958–5.

The electron has a potential energy of $-e\varphi(x)$. The potential energy is plotted in Figure 5.4b, assuming that there are no external electric fields. The potential changes when there is an electric field in the dielectric and can be expressed as

$$-\phi(x) = \frac{-e}{16\pi\epsilon_s x} - \mathrm{E}x \tag{5.12}$$

Figure 5.4c shows a plot of the electron's potential energy that takes into account the impact of a steady electric field. There is currently a lower peak potential barrier. This is known as the Schottky effect, or image force-induced lowering, which lowers the potential barrier.

We can find the Schottky barrier lowering, $\Delta\varphi$, and the position of the maximum barrier, x_m, from the condition that

$$\frac{d[e\phi(x)]}{dx} = 0 \tag{5.13}$$

We find that

$$x_m = \sqrt{\frac{e}{16\pi\epsilon_s \mathrm{E}}} \tag{5.14}$$

And

$$\Delta\phi = \sqrt{\frac{eE}{4\pi\epsilon_s}} \quad (5.15)$$

5.2.3.2. Interface States

Gallium arsenide and silicon Schottky diode measured barrier heights as a function of metal work functions are plotted in Figure 5.5. Although there is a monotonic relationship between the metal work function and the measured barrier height, the curves do not fit the straightforward relationship found in Equation (5.1). Both the semiconductor surface or interface states and the metal work function influence the metal–semiconductor junction barrier height.

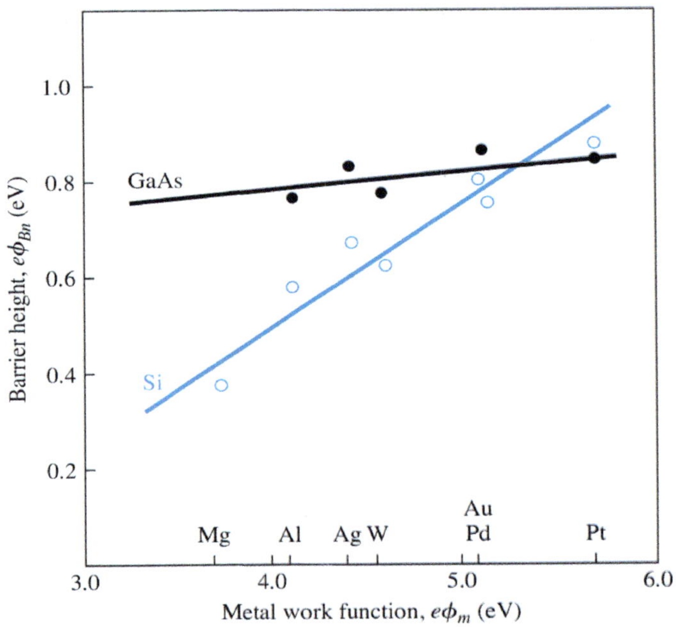

Figure 5.5. Experimental barrier heights as a function of metal work functions for GaAs and Si.

Source: Donald A. Neamen. Semiconductor Physics and Devices: Basic Principles, Fourth Edition. McGraw-Hill. ISBN 978–0-07–352958–5.

Figure 5.6 displays a more thorough energy-band diagram of a metal to n-type semiconductor contact in thermal equilibrium. We consider that there is a thin insulator interfacial layer between the semiconductor and the metal. The interfacial layer will be transparent to the electron flow between the semiconductor and metal, but it can support a potential difference. At the metal-semiconductor interface, the semiconductor additionally displays a distribution of surface states. All states below the surface potential φ0 are assumed to be donor states; in the event that an electron is present, the state will be positively charged; otherwise, it will be neutral. Furthermore, we assume that all states above φ_0 are acceptor states, which are negatively charged if they contain an electron and neutral otherwise.

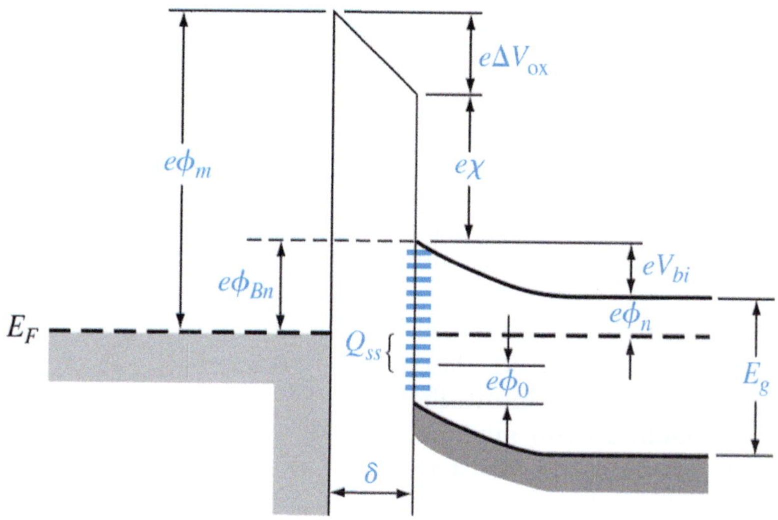

Figure 5.6. Energy-band diagram of a metal–semiconductor junction with an interfacial layer and interface states.

Source: Donald A. Neamen. Semiconductor Physics and Devices: Basic Principles, Fourth Edition. McGraw-Hill. ISBN 978–0-07–352958–5.

Some acceptor states are shown above φ_0 and below E_F in the diagram in Figure 5.6. Usually negatively charged, these states have electrons. Since the surface state density is equal to D_{it} states/cm^2-eV, we can assume that it is constant. The relationship between surface potential, surface state density, and additional semiconductor characteristics is discovered to be

$$(E_g - e\phi_0 - e\phi_{Bn}) = \frac{1}{eD_{it}}\sqrt{2e\epsilon_s N_d(\phi_{Bn} - \phi_n)} - \frac{\epsilon_i}{eD_{it}\delta}[\phi_m - (\chi + \phi_{Bn})] \tag{5.16}$$

We consider two limiting cases.

Case 1

Let $D_{it} \rightarrow \infty$. In this case, the right side of Equation (5.16) goes to zero. We then have

$$\phi_{Bn} = \frac{1}{e}(E_g - e\phi_0) \tag{5.17}$$

The bandgap energy and potential φ_0 now set the barrier height. Both the semiconductor electron affinity and the metal work function have no effect whatsoever on the barrier height. At the surface, specifically the surface potential φ_0, the Fermi level becomes "pinned."

Case 2

Let $D_{it}\ \delta \rightarrow 0$. Equation (5.16) reduces to

$$\phi_{Bn} = (\phi_m - \chi)$$

which is the original ideal expression.

Through the barrier lowering effect, the Schottky barrier height depends on the electric field inside the semiconductor. The semiconductor's surface states also influence the barrier height. Thus, the ideal theoretical value of the barrier height is altered. The barrier height needs to be an experimentally determined parameter because the surface state density is not predictable to any significant extent.

5.2.4. Current–Voltage Relationship

Unlike minority carriers in a pn junction, majority carriers in a metal-semiconductor junction are primarily responsible for current transport. The thermionic emission theory describes the fundamental mechanism of electron transport across the potential barrier in the rectifying contact with an n-type semiconductor.

By assuming that the barrier height is significantly greater than kT, the Maxwell-Boltzmann approximation is applied and thermal equilibrium is unaffected, which leads to the derivation of the thermionic emission characteristics. The one-dimensional barrier with two electron current density components and an applied forward-bias voltage V_a is depicted in Figure 5.7. The electron current density resulting from electrons flowing from the semiconductor into the metal is represented by the current Js → m, while the electron current density resulting from electrons flowing from the metal into the semiconductor is represented by the current $_{Jm \to s}$. The direction of electron flow is indicated by the currents' subscripts. The direction of electron flow is opposite to that of conventional current flow.

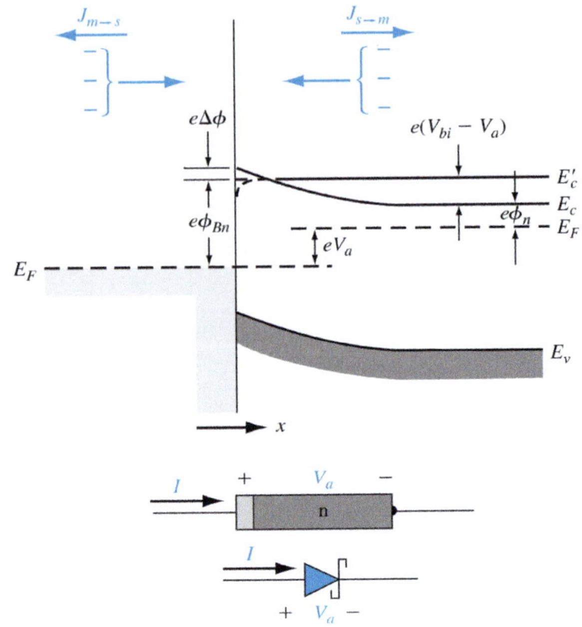

Figure 5.7. Energy-band diagram of a forward-biased metal– semiconductor junction including the image lowering effect.

Source: Donald A. Neamen. Semiconductor Physics and Devices: Basic Principles, Fourth Edition. McGraw-Hill. ISBN 978–0-07–352958–5.

The concentration of electrons with x-directed velocities that are sufficient to cross the barrier determines the current density $J_{s\to m}$. We may write

$$J_{s\to m} = e\int_{E'_c}^{\infty} v_x\,dn \tag{5.18}$$

where E'_c is the minimum energy required for thermionic emission into the metal, v_x is the carrier velocity in the direction of transport, and e is the magnitude of the electronic charge. The incremental electron concentration is given by

$$dn = g_c(E)\,f_F(E)\,dE \tag{5.19}$$

where gc(E) is the density of states in the conduction band and fF(E) is the

Fermi– Dirac probability function. Assuming that the Maxwell–Boltzmann approximation applies, we may write

$$dn = \frac{4\pi(2m_n^*)^{3/2}}{h^3}\sqrt{E-E_c}\,\exp\left[\frac{-(E-E_F)}{kT}\right]dE \tag{5.20}$$

If all of the electron energy above E_c is assumed to be kinetic energy, then we have

$$\frac{1}{2}m_n^*\,v^2 = E - E_c \tag{5.21}$$

The net current density in the metal-to-semiconductor junction can be written as

$$J = J_{s\rightarrow m} - J_{m\rightarrow s} \tag{5.22}$$

which is defined to be positive in the direction from the metal to the semiconductor. We find that

$$J = \left[A^*\,T^2\exp\left(\frac{-e\phi_{Bn}}{kT}\right)\right]\left[\exp\left(\frac{eV_a}{kT}\right)-1\right] \tag{5.23}$$

Where

$$A^* \equiv \frac{4\pi e m_n^* k^2}{h^3} \tag{5.24}$$

The parameter A* is called the effective Richardson constant for thermionic emission.

Equation (5.23) can be written in the usual diode form as

$$J = J_{sT}\left[\exp\left(\frac{eV_a}{kT}\right)-1\right] \tag{5.25}$$

where J_{sT} is the reverse-saturation current density and is given by

$$J_{sT} = A^*T^2\exp\left(\frac{-e\phi_{Bn}}{kT}\right) \tag{5.26}$$

We may recall that the Schottky barrier height φ_{Bn} changes because of the image-force lowering. We have that $\varphi_{Bn} = \varphi_{B0} - \Delta\varphi$. Then we can write Equation (5.26) as

$$J_{sT} = A^*T^2\exp\left(\frac{-e\phi_{B0}}{kT}\right)\exp\left(\frac{e\Delta\phi}{kT}\right) \tag{5.27}$$

When the applied reverse-biased voltage or the electric field both rise, the change in barrier height, Δφ, also rises accordingly. A reverse-biased current-voltage characteristic of a Schottky barrier diode is depicted in Figure 5.8. The barrier lowering effect causes the reverse-biased current to increase with reverse-biased voltage. The breakdown of the Schottky barrier diode is also depicted in this figure.

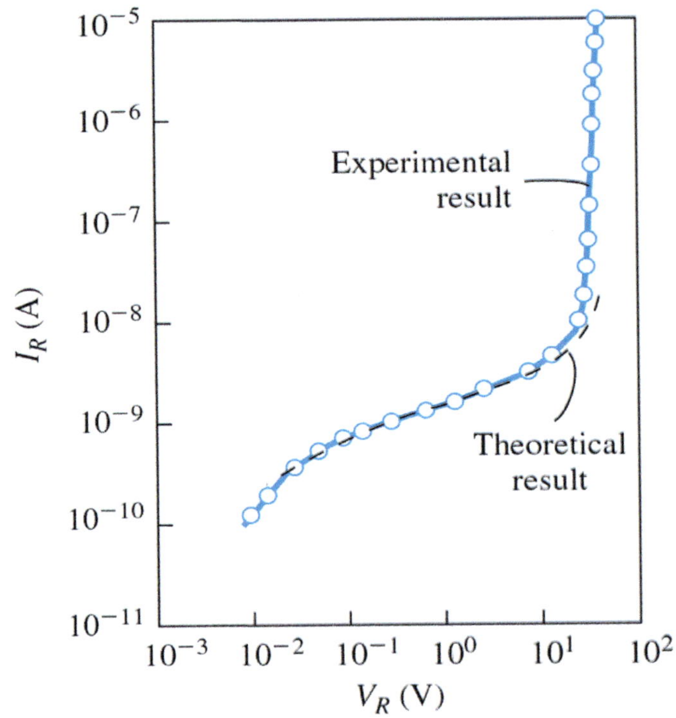

Figure 5.8. Experimental and theoretical reverse-biased currents in a PtSi–Si diode.

Source: Donald A. Neamen. Semiconductor Physics and Devices: Basic Principles, Fourth Edition. McGraw-Hill. ISBN 978-0-07-352958-5.

We can observe that there is a roughly two-order-of-magnitude difference in the reverse saturation current densities of the tungsten-silicon and tungsten-gallium arsenide diodes in Figure 5.9.

If the barrier heights in the two diodes are nearly the same, then the effective Richardson constant will exhibit a two-order-of-magnitude difference. Equation (5.24) defines the effective Richardson constant and includes the electron effective mass, which varies significantly between gallium arsenide and silicon.

Using the effective density of states function in the thermionic emission theory directly leads to the effective mass being in the expression for the Richardson constant. As a result, there will be significant differences in A^* and J_{sT} between silicon and gallium arsenide.

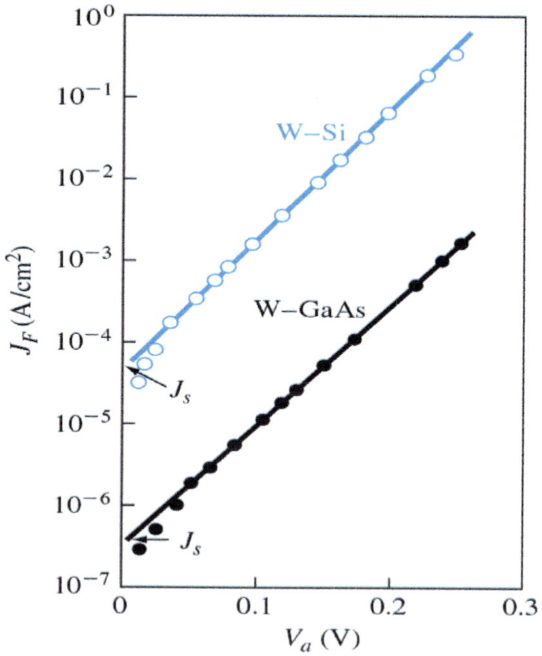

Figure 5.9. Forward-bias current density JF versus Va for Wi–Si and W–GaAs diodes.

Source: Donald A. Neamen. Semiconductor Physics and Devices: Basic Principles, Fourth Edition. McGraw-Hill. ISBN 978-0-07-352958-5.

5.2.5. Comparison of the Schottky Barrier Diode and the PN Junction Diode

There are two significant distinctions between a Schottky diode and a pn junction diode, despite the fact that their ideal current–voltage relationship, as given by Equation (5.25), is of the same form. The first is in the magnitudes of the reverse-saturation current densities, and the second is in the switching characteristics.

The reverse-saturation current density of the Schottky barrier diode was given by Equation (5.26) and is

$$J_{sT} = A^* T^2 \exp\left(\frac{-e\phi_{Bn}}{kT}\right)$$

The ideal reverse-saturation current density of the PN junction diode can be written as

$$J_s = \frac{eD_n n_{po}}{L_n} + \frac{eD_p p_{no}}{L_p} \tag{5.28}$$

The current mechanisms in the two devices differ, as does the form of the two equations. While the thermionic emission of majority carriers over a potential barrier determines the current in a Schottky barrier diode, the diffusion of minority carriers determines the current in a pn junction.

Remember that the generation current dominates the reverse-biased current in a silicon pn junction diode. The reverse-saturation current density of the Schottky barrier diode is two to three orders of magnitude greater than the typical generation current density of about 10^{-7} A/cm^2. The reverse-biased Schottky barrier diode also has a generation current, but it is very small in comparison to the J_{sT} value.

Given that $J_{sT} \gg J_s$, the forward-bias properties of the two varieties of diodes will also differ. A Schottky barrier diode and a pn junction diode's typical I–V characteristics are displayed in Figure 5.10. The Schottky diode has a lower effective turn-on voltage than the pn junction diode.

The metal-semiconductor contact barrier height and the pn junction doping concentrations will determine the actual difference in turn-on voltages, but a comparatively large difference will always be observed. We'll look at one use for the variation in turn-on voltage—the Schottky clamped transistor—in this application.

The frequency response, or switching characteristics, is the second main distinction between a pn junction diode and a Schottky barrier diode. We have discussed how the injection of majority carriers across a potential barrier is what causes the current in a Schottky diode. For instance, the energy-band diagram in Figure 5.2 demonstrates that electrons in the metal may be present in close proximity to semiconductor empty states. The effect of an electron flowing into the metal from the semiconductor's valence band would be the same as injecting holes into the semiconductor. The n region would have an excess of minority carrier holes as a result of this hole injection. Measurements and computations, however, have demonstrated that, for the most part, the minority carrier hole current to total current ratio is incredibly low.

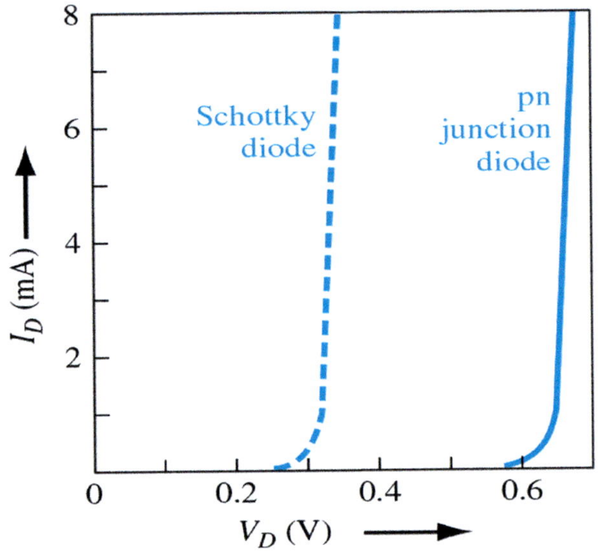

Figure 5.10. Comparison of forwardbias I–V characteristics between a Schottky diode and a pn junction diode.

Source: Donald A. Neamen. Semiconductor physics and devices: basic principles, fourth edition. McGraw-Hill. ISBN 978-0-07-352958-5.

Therefore, a majority carrier device is the Schottky barrier diode. This indicates that a forward-biased Schottky diode has no diffusion capacitance connected to it. The Schottky diode operates at a higher frequency than the pn junction diode due to the removal of the diffusion capacitance. Furthermore, unlike a pn junction diode, there is no minority carrier stored charge to remove when a Schottky diode is switched from forward to reverse bias. Schottky diodes have no minority carrier storage time, which makes them suitable for fast-switching applications. A pn junction's switching time is typically in the nanosecond range, whereas a Schottky diode's switching time is typically in the picosecond range.

5.3. METAL–SEMICONDUCTOR OHMIC CONTACTS

Any semiconductor device, also known as an integrated circuit, needs to be connected to the outside world. Ohmic contacts are used to make these connections. Although they are metal-to-semiconductor connections, ohmic contacts are not rectifying contacts in this context. An ohmic contact is a low-resistance junction that allows for bidirectional conduction between the semiconductor and the metal. Ideally, the applied voltage should be very small and the current through the ohmic contact should be a linear function of it. There are two general kinds of ohmic contacts that can occur: the ideal non-rectifying barrier and the tunneling barrier. In this section, we define a particular contact resistance that is used to describe ohmic contacts.

5.3.1. Ideal Nonrectifying Barrier

Under the scenario where $\varphi_m > \varphi_s$, we have examined an optimal metal-n-type semiconductor contact in Figure 5.2. In the scenario where $\varphi_m < \varphi_s$, Figure 5.11 illustrates the identical optimal contact. The energy levels for thermal equilibrium are shown in Figure 5.11a prior to contact and in Figure 5.11b following contact. As a result of electrons moving from the metal into the semiconductor's lower energy states in order to reach thermal equilibrium in this junction, the semiconductor's surface becomes more n-type. A surface charge density is essentially the form that the excess electron charge takes in n-type semiconductors. There is nothing stopping electrons from moving from the semiconductor into the metal when a positive voltage is applied to it. A moderately to heavily doped semiconductor will have an effective barrier height of about $\varphi_{Bn} = \varphi_n$ for electrons flowing from the metal into the semiconductor if a positive voltage is applied to it. This is a relatively small barrier height. Electrons can move from the metal into the semiconductor with ease under this bias condition.

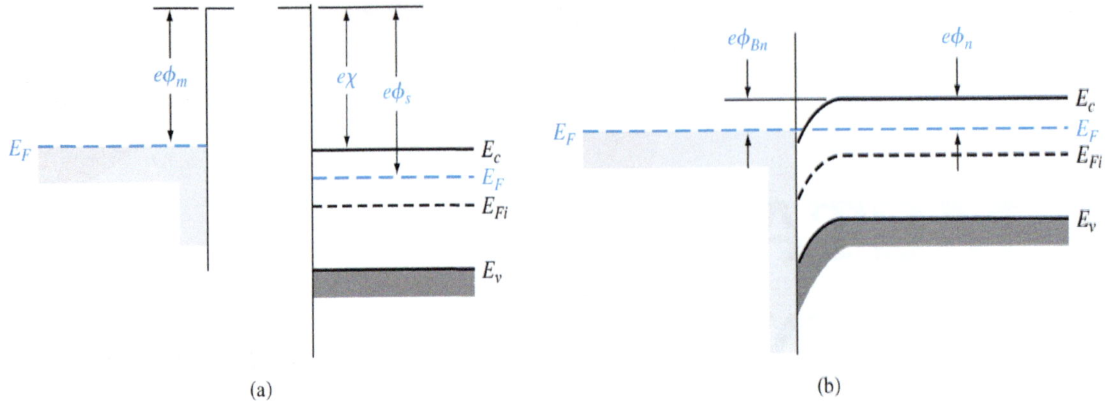

Figure 5.11. Ideal energy-band diagram (a) before contact and (b) after contact for a metal-n-type semiconductor junction for $\varphi_m < \varphi_s$.

Source: Donald A. Neamen. Semiconductor Physics and Devices: Basic Principles, Fourth Edition. McGraw-Hill. ISBN 978–0-07–352958–5.

When a positive voltage is applied to the metal in relation to the semiconductor, Figure 5.12a displays the energy-band diagram. From the semiconductor into the metal, electrons can move "downhill" with ease. When a positive voltage is applied to the semiconductor in relation to the metal, Figure 5.12b depicts the situation. The barrier preventing electrons from moving from the metal into the semiconductor is easily crossed. Therefore, this junction is an ohmic contact.

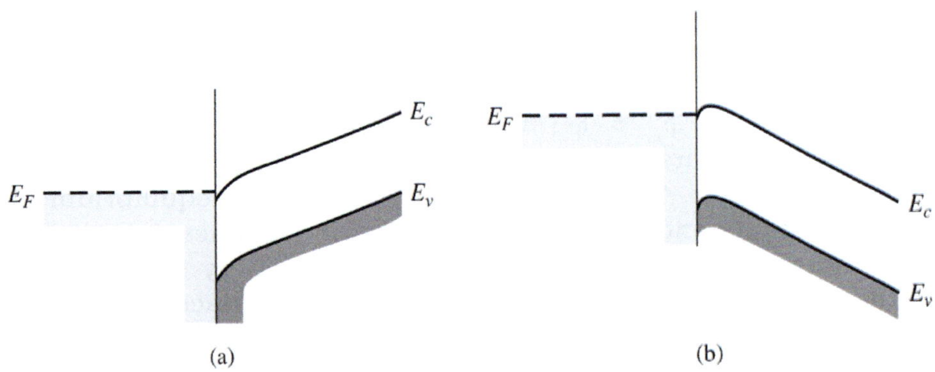

Figure 5.12. Ideal energy-band diagram of a metal-n-type semiconductor ohmic contact (a) with a positive voltage applied to the metal and (b) with a positive voltage applied to the semiconductor.

Source: Donald A. Neamen. Semiconductor physics and devices: basic principles, fourth edition. McGraw-Hill. ISBN 978–0-07–352958–5.

An ideal non-rectifying contact between a metal and a p-type semiconductor is depicted in Figure 5.13. The energy levels prior to contact for the scenario where $\varphi_m > \varphi_s$ are displayed in Figure 5.13a. In order to reach thermal equilibrium, electrons from the semiconductor move into the metal upon contact, leaving behind more empty states, or holes. The semiconductor's surface is more p-type due to the excess concentration of holes at the surface. The semiconductor's empty states are easily reached by electrons from the metal. Holes moving into the metal from the semiconductor are reflected by this charge movement. Additionally, we can see the metal's holes flowing into the semiconductor. This junction is also an ohmic contact.

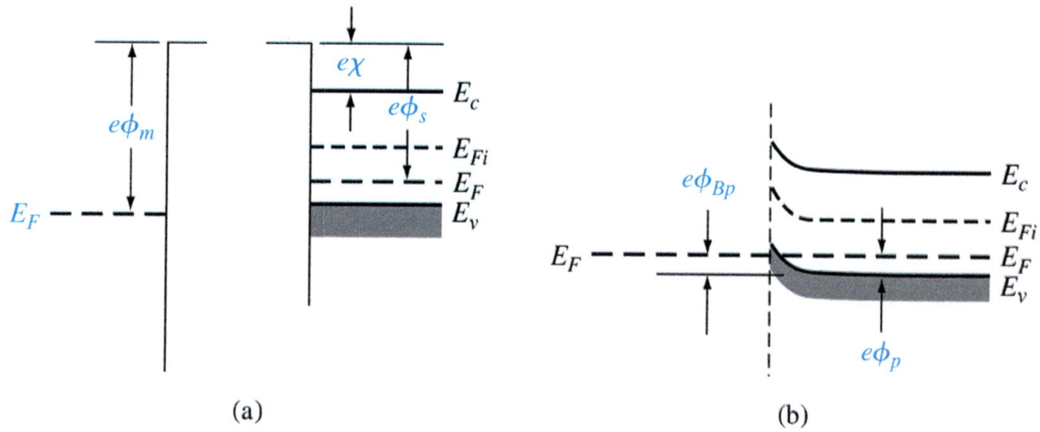

Figure 5.13. Ideal energy-band diagram (a) before contact and (b) after contact for a metal–p-type semiconductor junction for $\varphi_m < \varphi_s$

Source: Donald A. Neamen. Semiconductor Physics and Devices: Basic Principles, Fourth Edition. McGraw-Hill. ISBN 978-0-07-352958-5.

The influence of surface states is not taken into consideration in the ideal energy bands displayed in Figures 5.11 and 5.13. Since all acceptor states in the scenario depicted in Figure 5.11b are below E_F, it can be assumed that acceptor surface states exist in the upper half of the semiconductor bandgap. As a result, these surface states will be negatively charged and will change the energy-band diagram. Similarly, for the case depicted in Figure 5.13b, all donor states will be positively charged if we assume that they exist in the lower half of the bandgap; the positively charged surface states will also change this energy-band diagram. Therefore, if $\varphi_m < \varphi_s$ for the metal-n-type semiconductor contact, and if $\varphi_m > \varphi_s$ for the metal-p-type semiconductor contact, we may not necessarily form a good ohmic contact.

5.3.2. Tunneling Barrier

In a rectifying metal-semiconductor contact, the square root of the semiconductor doping has an inverse relationship with the space charge width. The probability of tunneling through the barrier increases as the doping concentration in the semiconductor increases because the width of the depletion region decreases with increasing doping concentration. A junction where the metal and heavily doped n-type epitaxial layer are shown in Figure 5.14.

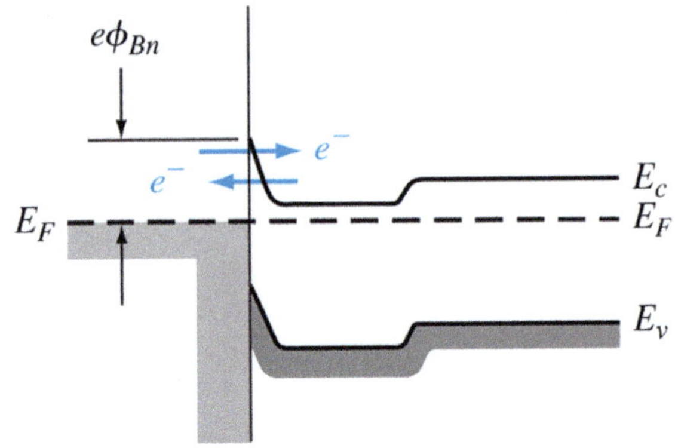

Figure 5.14. Energy-band diagram of a heavily doped n-semiconductor-to-metal junction.

Source: Donald A. Neamen. Semiconductor Physics and Devices: Basic Principles, Fourth Edition. McGraw-Hill. ISBN 978–0-07–352958–5.

The tunneling current has the form

$$J_t \propto \exp\left(\frac{-e\phi_{Bn}}{E_{oo}}\right) \tag{5.29}$$

where

$$E_{oo} = \frac{e\hbar}{2}\sqrt{\frac{N_d}{\epsilon_s m_n^*}} \tag{5.30}$$

The tunneling current increases exponentially with doping concentration.

5.3.3. Specific Contact Resistance

The particular contact resistance, or R_c, is a figure of merit for ohmic contacts. The definition of this parameter is the reciprocal of the current density derivative evaluated at zero bias with respect to voltage. We may write

$$R_c = \left(\frac{\partial J}{\partial V}\right)^{-1}\Bigg|_{V=0} \qquad \Omega\text{-cm}^2 \tag{5.31}$$

We want R_c to be as small as possible for an ohmic contact.

For a rectifying contact with a low to moderate semiconductor doping concentration, the current–voltage relation is given by Equation (5.23) as

$$J_n = A^*T^2 \exp\left(\frac{-e\phi_{Bn}}{kT}\right)\left[\exp\left(\frac{eV}{kT}\right) - 1\right]$$

The thermionic emission current is dominant in this junction. The specific contact resistance for this case is then

$$R_c = \frac{\left(\frac{kT}{e}\right)\exp\left(\frac{+e\phi_{Bn}}{kT}\right)}{A^*T^2} \tag{5.32}$$

The specific contact resistance decreases rapidly as the barrier height decreases.

For a metal–semiconductor junction with a high impurity doping concentration, the tunneling process will dominate. From Equations (5.29) and (5.30), the specific contact resistance is found to be

$$R_c \propto \exp\left(\frac{+2\sqrt{\epsilon_s m_n^*}}{\hbar} \cdot \frac{\phi_{Bn}}{\sqrt{N_d}}\right) \tag{5.33}$$

which shows that the specific contact resistance is a very strong function of semiconductor doping.

A plot of the theoretical values of R_c versus semiconductor doping is presented in Figure 5.15. R_c exhibits an exponential dependence on N_d, and the tunneling process takes over at doping concentrations higher than roughly 10^{19} cm^{-3}. The R_c values depend on the barrier heights at lower doping concentrations and become nearly independent of the doping. Experiments on aluminum-silicon and platinum silicide-silicon junctions are also displayed in the figure.

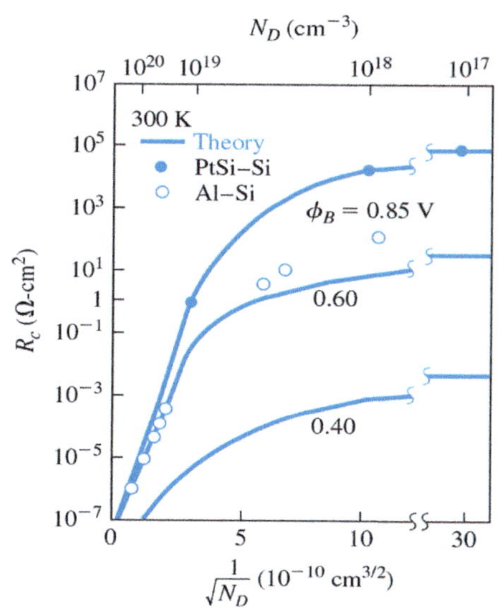

Figure 5.15. Theoretical and experimental specific contact resistance as a function of doping.

Source: Donald A. Neamen. Semiconductor Physics and Devices: Basic Principles, Fourth Edition. McGraw-Hill. ISBN 978–0-07–352958–5.

The specific contact resistance of the tunneling junction, represented by equation (5.33), is the same as the metal-n^+ contact depicted in Figure 5.14. However, because there is a barrier connected to the n^+n junction, it also has a particular contact resistance. In a region with relatively low doping, the contact resistance could potentially outweigh the junction's overall resistance.

The formation of ohmic contacts has a simple theory. We must use a highly doped semiconductor at the surface and establish a low barrier in order to form a good ohmic contact. But in reality, it is more difficult to create dependable, high-quality ohmic contacts than it is in theory. Additionally, creating strong ohmic contacts on materials with large bandgaps is more challenging. These materials typically do not allow for low barriers, so a heavily doped semiconductor at the surface is required to create a tunneling contact. It takes either diffusion, ion implantation, or possibly epitaxial growth for a tunneling junction to form. The semiconductor's surface doping concentration could be restricted by the impurity solubility, which for n-type GaAs is roughly 5×10^{19} cm^{-3}. It is also possible that non-uniformities in the surface doping concentration will keep the specific contact resistance from reaching its theoretical limit. In actuality, obtaining a good ohmic contact typically requires a significant amount of empirical processing.

5.4. HETEROJUNCTIONS

An interface separating two layers or areas of different semiconductor types is called a heterojunction. Rather than a homojunction, these semiconducting materials have unequal band gaps. Engineering the electronic energy bands is often beneficial in a variety of solid-state device applications, such as transistors, solar cells, and semiconductor lasers. Although the terms are frequently used interchangeably, a heterostructure is the combination of multiple heterojunctions in a device. The condition that all materials be semiconductors with uneven band gaps is not strictly enforced, particularly at small length scales where spatial characteristics influence electronic characteristics. The interface between any two solid-state materials, including crystalline and amorphous structures of metallic, insulating, fast-ion conducting, and semiconducting materials, is a more contemporary definition of heterojunction.

When we talked about pn junctions, we assumed that the semiconductor material was the same all the way around the structure. We refer to this kind of junction as a homojunction. A semiconductor heterojunction is created when two distinct semiconductor materials are combined to form the junction.

As with many other topics in this text, our aim is to present the fundamental ideas related to the heterojunction. A thorough examination of heterojunction structures necessitates intricate computations and quantum mechanics, which are outside the purview of this book. Therefore, the discussion of heterojunctions will be restricted to the introduction of a few fundamental ideas.

5.4.1. Heterojunction Materials

The energy band will have a discontinuity at the junction interface because the two materials that form a heterojunction will have different energy bandgaps. We could have an abrupt junction, where the semiconductor suddenly transitions from a material with a narrow bandgap to one with a wide bandgap. In contrast, a graded heterojunction could form in a GaAs– $Al_xGa_{1-x}As$ system, for instance, if the value of x varies continuously over a few nanometers. We can engineer, or design, the bandgap energy by varying the value of x in the $Al_xGa_{1-x}As$ system.

The lattice constants of the two materials must match well for there to be a useful heterojunction. Any lattice mismatch has the potential to introduce dislocations that lead to interface states, which is why the lattice match is crucial. The lattice constants of germanium and gallium arsenide, for instance, are within 0.13% of each other. Many studies have been conducted on germanium–gallium arsenide heterojunctions. Since there is only a 0.14% difference in the lattice constants of GaAs and the AlGaAs system, there has been a lot of recent research done on gallium arsenide–aluminum gallium arsenide (GaAs–AlGaAs) junctions.

5.4.2. Energy-Band Diagrams

The bandgap energies must align in order for a heterojunction made of two materials with different bandgaps to form. This will determine the properties of the heterojunction. Three potential scenarios are depicted in Figure 5.16. The situation where the bandgap of the narrow-gap material completely overlaps the forbidden bandgap of the wide-gap material is depicted in Figure 5.16a. Straddling refers to this situation, which most heterojunctions fall under. This case is the only one we look at. The alternate options, referred to as staggered and broken gaps, are displayed in Figure 5.16b, c.

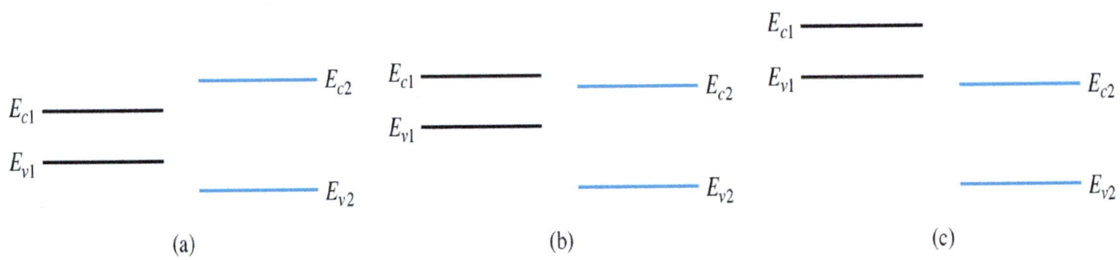

Figure 5.16. Relation between narrow-bandgap and wide-bandgap energies: (a) straddling, (b) staggered, and (c) broken gap.

Source: Donald A. Neamen. Semiconductor physics and devices: basic principles, fourth edition. McGraw-Hill. ISBN 978–0-07–352958–5.

There are four fundamental heterojunction types. Anisotype refers to those in which the dopant type varies at the junction. We are able to create nP or Np junctions, where the larger-bandgap material is indicated by the capital letter. Isotype heterojunctions are heterojunctions having the same type of dopant on both sides of the junction. nN and pP isotype heterojunctions can be formed.

With the vacuum level serving as a reference, the energy-band diagrams of isolated n-type and p-type materials are displayed in Figure 5.17. Relative to the narrow-bandgap material, the wide-bandgap material has a lower electron affinity. Whereas ΔE_v represents the difference between the two valence band energies, ΔE_c represents the difference between the two conduction band energies. From Figure 5.17, we can see that

$$\Delta E_c = e(\chi_n - \chi_P) \qquad (5.34s)$$

And

$$\Delta E_c + \Delta E_v = E_{gP} - E_{gn} = \Delta E_g \tag{5.34b}$$

The vacuum level in an ideal abrupt heterojunction made of nondegenerately doped semiconductors is parallel to the valence and conduction bands. The heterojunction interface will exhibit the same ΔE_c and ΔE_v discontinuities if the vacuum level is continuous. The electron affinity rule refers to this ideal state. Although the applicability of this rule is still somewhat unclear, it offers a useful starting point for the discussion of heterojunctions.

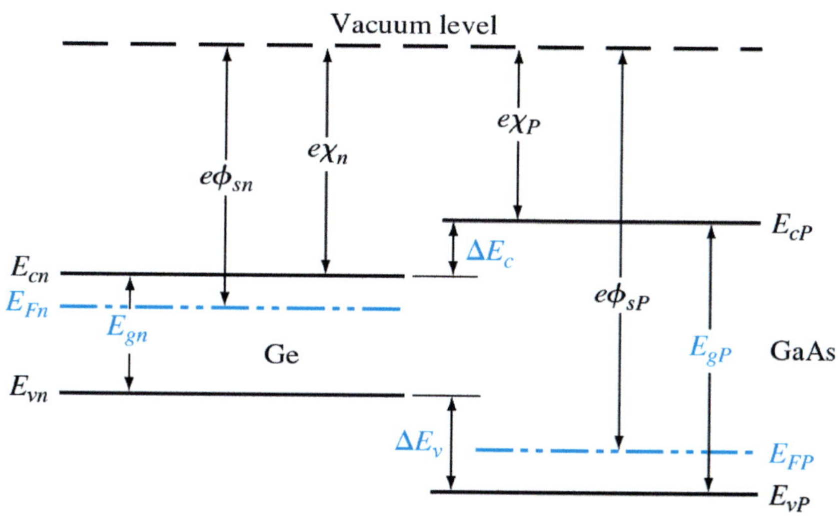

Figure 5.17. Energy-band diagrams of a narrow-bandgap and a wide-bandgap material before contact.

Source: Donald A. Neamen. Semiconductor physics and devices: basic principles, fourth edition. McGraw-Hill. ISBN 978–0-07–352958–5.

A typical ideal nP heterojunction in thermal equilibrium is depicted in Figure 5.18. Electrons from the narrow-gap n region and holes from the wide-gap P region must flow across the junction for the Fermi levels in the two materials to align. This charge flow produces a space charge region close to the metallurgical junction, just like it does in the case of a homojunction. The symbols x_n and x_p represent the space charge width into the n-type region and the P-type region, respectively. The figure displays the shift in the vacuum level as well as the discontinuities in the valence and conduction bands.

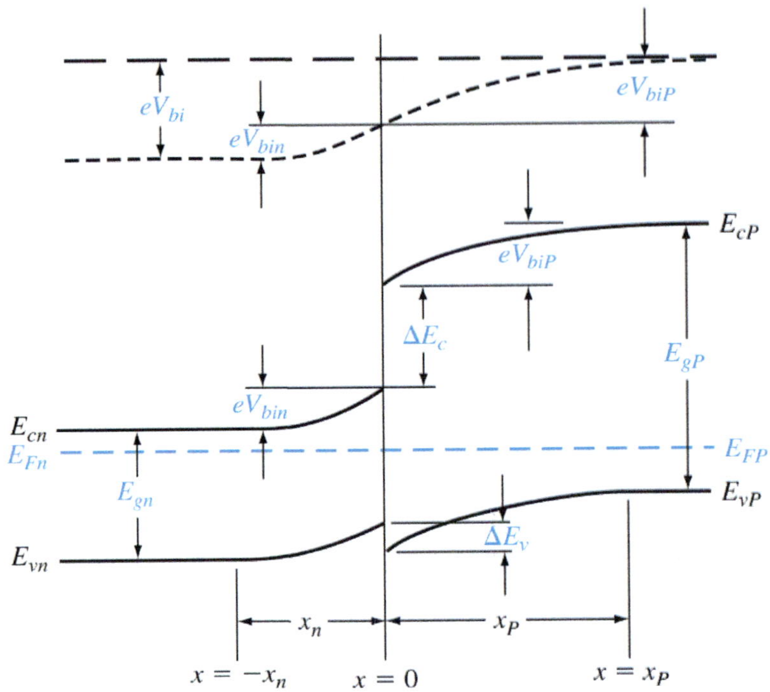

Figure 5.18. Ideal energy-band diagram of an nP heterojunction in thermal equilibrium.

Source: Donald A. Neamen. Semiconductor Physics and Devices: Basic Principles, Fourth Edition. McGraw-Hill. ISBN 978–0-07–352958–5.

5.4.3. Two-Dimensional Electron Gas

We will talk about a special property of an isotype junction before moving on to the heterojunction's electrostatics. The energy-band diagram of a thermally balanced nN GaAs–AlGaAs heterojunction is displayed in Figure 5.19. GaAs can be more lightly doped or even intrinsic, whereas AlGaAs can be moderately to heavily doped n type. As previously stated, electrons from the wide-bandgap AlGaAs flow into the GaAs to reach thermal equilibrium, forming an electron accumulation layer in the potential well next to the interface. The energy of an electron contained in a potential well is quantized, which is a fundamental outcome of quantum mechanics that we have previously discovered. When electrons are free to travel in two spatial directions but have quantized energy levels in one (perpendicular to the interface), this state is referred to as a "two-dimensional electron gas."

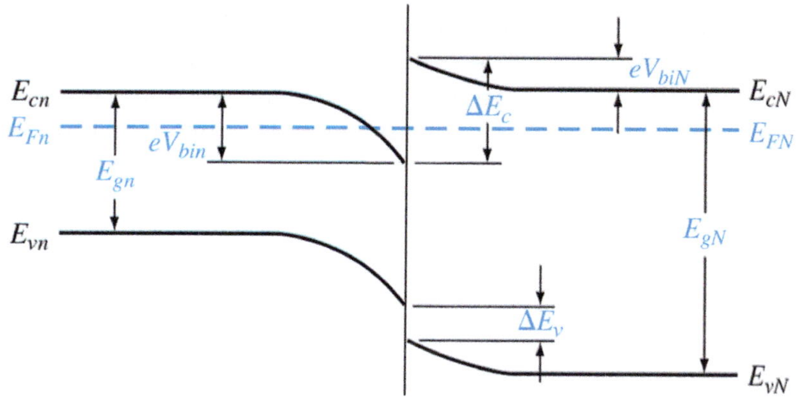

Figure 5.19. Ideal energy-band diagram of an nN heterojunction in thermal equilibrium.

Source: Donald A. Neamen. Semiconductor physics and devices: basic principles, fourth edition. McGraw-Hill. ISBN 978–0-07–352958–5.

A triangular potential well can be used to approximate the potential function close to the interface. The approximate triangular potential well is shown in Figure 5.20b, while Figure 5.20a depicts the conduction band edges close to the abrupt junction interface. We can write

$$V(x) = e\mathrm{E}z \qquad z > 0 \tag{5.35a}$$

$$V(z) = \infty \qquad z < 0 \tag{5.35b}$$

This potential function can be used to solve Schrodinger's wave equation. In Figure 5.20b, the quantized energy levels are displayed. Elevated energy levels are typically disregarded.

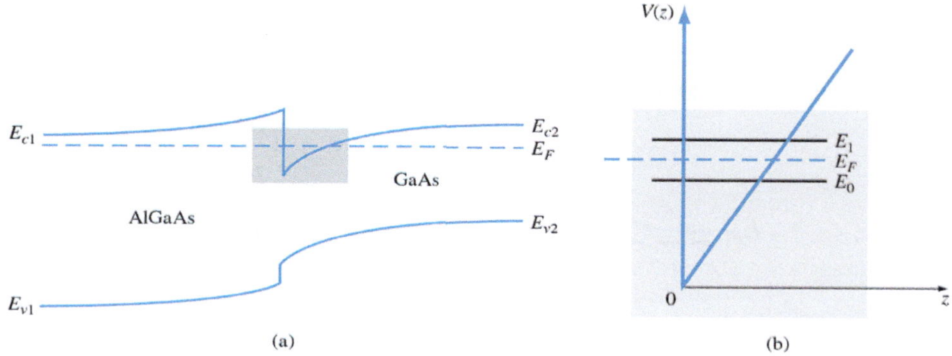

Figure 5.20. (a) Conduction-band edge at N-AlGaAs, n-GaAs heterojunction; (b) triangular well approximation with discrete electron energies.

Source: Donald A. Neamen. Semiconductor physics and devices: basic principles, fourth edition. McGraw-Hill. ISBN 978–0-07–352958–5.

Figure 5.21 displays the electrons' qualitative distribution in the potential well. A current flowing parallel to the interface will depend on both the electron mobility and this concentration. The two-dimensional electron gas is in a region of low impurity doping, which minimizes the effects of impurity scattering, because the GaAs can be intrinsic or lightly doped. Compared to the case when the electrons were in the same area as the ionized donors, the electron mobility will be substantially greater.

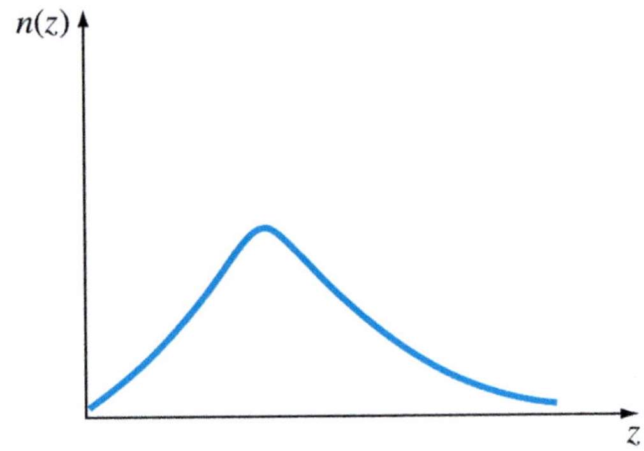

Figure 5.21. Electron density in triangular potential well.

Source: Donald A. Neamen. Semiconductor physics and devices: basic principles, fourth edition. McGraw-Hill. ISBN 978-0-07-352958-5.

The Coulomb attraction of the ionized impurities in the AlGaAs will continue to affect the electrons' motion parallel to the interface. By employing a graded AlGaAs–GaAs heterojunction, the impact of these forces can be further diminished. $Al_xGa_{1-x}As$ is the graded layer, where the mole fraction x changes with distance. In this instance, the N-type AlGaAs and the intrinsic GaAs can be positioned between intrinsic layers of graded AlGaAs. The conduction-band edges across a graded AlGaAs–GaAs heterojunction in thermal equilibrium are depicted in Figure 5.22. To raise the electron mobility above that of an abrupt heterojunction, the electrons in the potential well are further isolated from the ionized impurities.

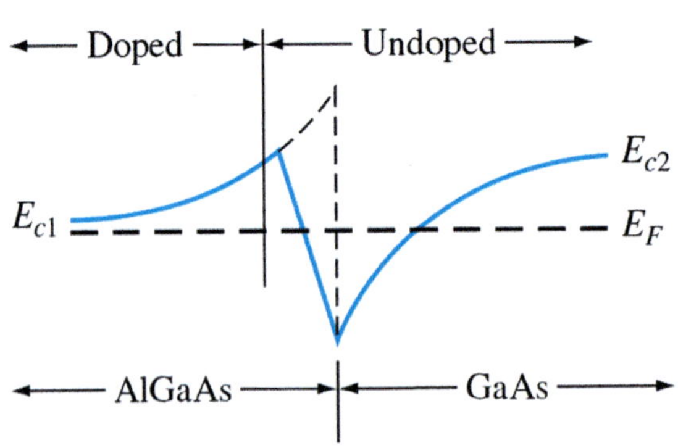

Figure 5.22. Conduction-band edge at a graded heterojunction.

Source: Donald A. Neamen. Semiconductor physics and devices: basic principles, fourth edition. McGraw-Hill. ISBN 978-0-07-352958-5.

5.4.4. Equilibrium Electrostatics

The electrostatics of the nP heterojunction depicted in Figure 5.18 will now be discussed. Similar to the homojunction, there are possible variations between the n and P regions' space charge regions. The inherent potential barriers on either side of the intersection line up with these potential differences. The potential difference across the vacuum level is the defined built-in potential barrier for this ideal case, as illustrated in Figure 5.18. The total of the potential differences in each of the space charge regions is the built-in potential barrier. However, unlike what we defined for the homojunction, the heterojunction built-in potential barrier is not equal to the difference between the valence and conduction bands across the junction.

Ideally, the total built-in potential barrier V_{bi} can be found as the difference between the work functions, or

$$V_{bi} = \phi_{sP} - \phi_{sn} \tag{5.36}$$

Equation (5.36), from Figure 5.17, can be written as

$$eV_{bi} = [e\chi_P + E_{gP} - (E_{FP} - E_{vP})] - [e\chi_n + E_{gn} - (E_{Fn} - E_{vn})] \tag{5.37a}$$

Or

$$eV_{bi} = e(\chi_P - \chi_n) + (E_{gP} - E_{gn}) + (E_{Fn} - E_{vn}) - (E_{FP} - E_{vP}) \tag{5.37b}$$

which can be expressed as

$$eV_{bi} = -\Delta E_c + \Delta E_g + kT\ln\left(\frac{N_{vn}}{p_{no}}\right) - kT\ln\left(\frac{N_{vP}}{p_{po}}\right) \tag{5.38}$$

Finally, we can write Equation (5.38) as

$$eV_{bi} = \Delta E_v + kT\ln\left(\frac{p_{po}}{p_{no}}\cdot\frac{N_{vn}}{N_{vP}}\right) \tag{5.39}$$

where p_{po} and p_{no} are the hole concentrations in the P and n materials, respectively, and N_{vn} and N_{vP} are the effective density of states functions in the n and P materials, respectively. We can also obtain an expression for the built-in potential barrier in terms of the conduction band shift as

$$eV_{bi} = -\Delta E_c + kT\ln\left(\frac{n_{no}}{n_{po}}\cdot\frac{N_{cP}}{N_{cn}}\right) \tag{5.40}$$

The electric field and potential in the junction can be found using Poisson's equation in the same manner as the homojunction. For homogeneous doping on each side of the junction, we have in the n region

$$\mathrm{E}_n = \frac{eN_{dn}}{\epsilon_n}(x_n + x) \qquad (-x_n \leq x < 0) \tag{5.41a}$$

and in the P region

$$\mathrm{E}_P = \frac{eN_{aP}}{\epsilon_P}(x_P - x) \qquad (0 < x \leq x_P) \tag{5.41b}$$

where ε_n and ε_P are the permittivities of the n and P materials, respectively. We may note that $E_n = 0$ at $x = -x_n$ and $E_P = 0$ at $x = x_P$. The electric flux density D is continuous across the junction, so

$$\epsilon_n \mathrm{E}_n(x = 0) = \epsilon_P \mathrm{E}_P(x = 0) \tag{5.42a}$$

which gives

$$N_{dn} x_n = N_{aP} x_P \tag{5.42b}$$

The net positive charge in the n region is equal to the net negative charge in the P region, which is the same situation we had in a pn homojunction, according to equation (5.42b). Any possible interface states at the heterojunction are disregarded.

The potential difference across each region can be found by integrating the electric field through the space charge region in order to find the electric potential. We find that

$$V_{bin} = \frac{eN_{dn} x_n^2}{2\epsilon_n} \tag{5.43a}$$

And

$$V_{biP} = \frac{eN_{aP} x_P^2}{2\epsilon_P} \tag{5.43b}$$

Equation (5.42b) can be rewritten as

$$\frac{x_n}{x_P} = \frac{N_{aP}}{N_{dn}} \tag{5.44}$$

The ratio of the built-in potential barriers can then be determined as

$$\frac{V_{bin}}{V_{biP}} = \frac{\epsilon_P}{\epsilon_n} \cdot \frac{N_{dn}}{N_{aP}} \cdot \frac{x_n^2}{x_P^2} = \frac{\epsilon_P N_{aP}}{\epsilon_n N_{dn}} \tag{5.45}$$

In the event that ε_n and ε_P have similar magnitudes, the lower-doped region has a greater potential difference.

The total built-in potential barrier is

$$V_{bi} = V_{bin} + V_{biP} = \frac{eN_{dn}x_n^2}{2\epsilon_n} + \frac{eN_{aP}x_P^2}{2\epsilon_P} \tag{5.46}$$

If we solve for x_P, for example, from Equation (5.42b) and substitute into Equation (5.46), we can solve for x_n as

$$x_n = \left[\frac{2\epsilon_n \epsilon_P N_{aP} V_{bi}}{eN_{dn}(\epsilon_n N_{dn} + \epsilon_P N_{aP})}\right]^{1/2} \tag{5.47a}$$

We can also find

$$x_P = \left[\frac{2\epsilon_n \epsilon_P N_{dn} V_{bi}}{eN_{aP}(\epsilon_n N_{dn} + \epsilon_P N_{aP})}\right]^{1/2} \tag{5.47b}$$

The total depletion width is found to be

$$W = x_n + x_P = \left[\frac{2\epsilon_n \epsilon_P (N_{dn} + N_{aP})^2 V_{bi}}{eN_{dn}N_{aP}(\epsilon_n N_{dn} + \epsilon_P N_{aP})}\right]^{1/2} \tag{5.48}$$

If a reverse-biased voltage is applied across the heterojunction, the same equations apply if V_{bi} is replaced by $V_{bi} + V_R$. Similarly, if a forward bias is applied, the same equations also apply if V_{bi} is replaced by $V_{bi} - V_a$. As explained earlier, V_R is the magnitude of the reverse-biased voltage and V_a is the magnitude of the forward-bias voltage.

As in the case of a homojunction, a change in depletion width with a change in junction voltage yields a junction capacitance. We can find for the nP junction

$$C'_j = \left[\frac{eN_{dn}N_{aP}\epsilon_n\epsilon_P}{2(\epsilon_n N_{dn} + \epsilon_P N_{aP})(V_{bi} + V_R)}\right]^{1/2} \quad (\text{F/cm}^2) \tag{5.49}$$

A plot of $(1/C'_j)^2$ versus V_R again yields a straight line. The extrapolation of this plot of $(1/C'_j)^2 = 0$ is used to find the built-in potential barrier, V_{bi}.

The optimal energy-band diagram for the nP abrupt heterojunction is displayed in Figure 5.18. It is possible for the ideal values of ΔE_c and ΔE_v, as established by the electron affinity rule, to deviate from the values obtained through experimentation. Given that most heterojunctions have interface states, this discrepancy could have several causes. Because of the surface charge trapped in the interface states, the electric flux density at the heterojunction will be discontinuous if the electrostatic potential is assumed to be continuous through the junction. Then, in the same way that they altered the energy-band diagram of the metal-semiconductor junction, the interface states will alter the semiconductor heterojunction's energy-band diagram. The electron orbitals of the two materials interact as they are brought together to form the heterojunction, leading to a transition region of a few angstroms at the interface. This is another explanation for the deviation from the ideal. Through this transition region, the energy bandgap remains continuous and is not specific to either material. However, we still have the relation

$$\Delta E_c + \Delta E_v = \Delta E_g \tag{5.50}$$

for the straddling type of heterojunction, although the ΔE_c and ΔE_v values may differ from those determined from the electron affinity rule.

We can examine the overall features of the energy-band diagrams pertaining to the other varieties of heterojunctions. An Np heterojunction's energy-band diagram is displayed in Figure 5.23. In the nP and Np junctions, for example, the general shape of the conduction band is different, but the ΔE_c and ΔE_v discontinuities are the same. The I–V characteristics of the two junctions will be impacted by this difference in energy bands.

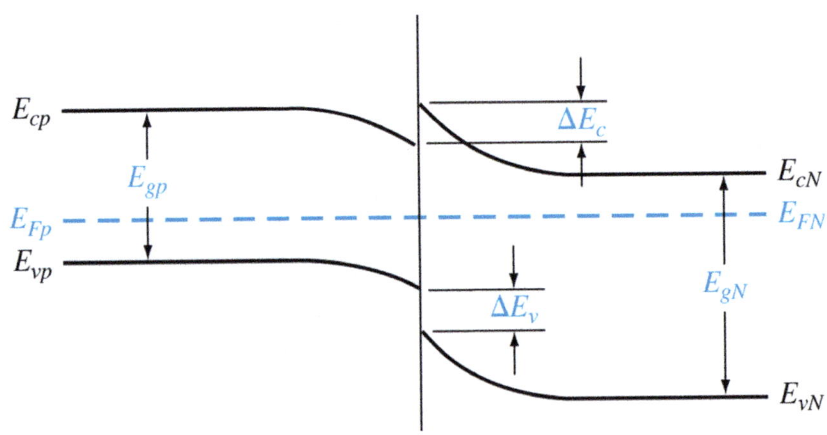

Figure 5.23. Ideal energy-band diagram of an Np heterojunction in thermal equilibrium.

Source: Donald A. Neamen. Semiconductor physics and devices: basic principles, fourth edition. McGraw-Hill. ISBN 978-0-07-352958-5.

The nN and pP isotype junctions are the other two types of heterojunctions. Figure 5.19 displays the nN junction's energy-band diagram. Electrons from the wide-bandgap material will move into the narrow-bandgap material in order to reach thermal equilibrium.

The narrow-bandgap material now has an electron accumulation layer at the interface, while the wide-bandgap material has a positive space charge region. The space charge width x_n and the built-in potential barrier V_{bin} are expected to be small in the narrow-gap material because of the large number of allowed energy states in the conduction band. Figure 5.24 displays the pP heterojunction's energy-band diagram when it is in thermal equilibrium. An accumulation layer of holes forms in the narrow-bandgap material at the interface as a result of holes from the wide-bandgap material flowing into the narrow-bandgap material to reach thermal equilibrium. It is evident that a homojunction cannot support these kinds of isotype heterojunctions.

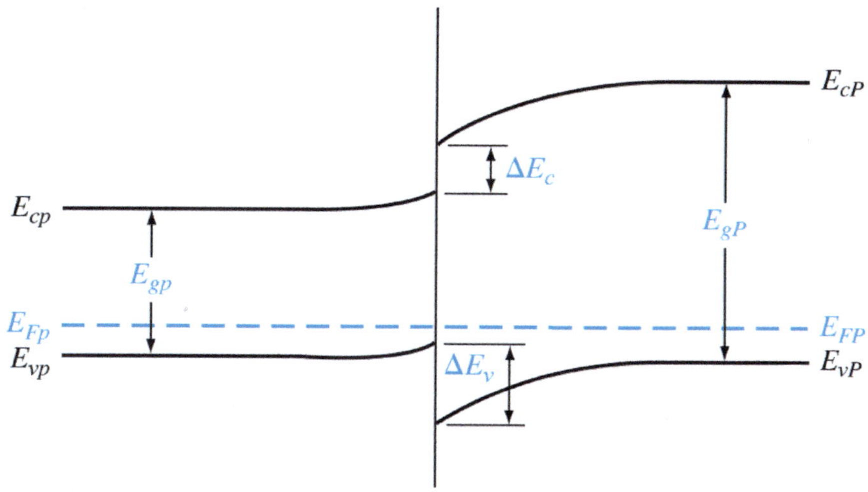

Figure 5.24. Ideal energy-band diagram of a pP heterojunction in thermal equilibrium.

Source: Donald A. Neamen. Semiconductor physics and devices: basic principles, fourth edition. McGraw-Hill. ISBN 978–0-07–352958–5.

5.4.5. Current–Voltage Characteristics

A relationship, usually shown as a chart or graph, between the electric current flowing through a material, device, or circuit and the corresponding voltage, or potential difference, across it is known as the current-voltage characteristic, or I-V curve (current–voltage curve).

We would anticipate that the I-V characteristics of the two junctions would vary because the energy-band diagram of a heterojunction is more intricate than that of a homojunction.

The barrier heights observed by the electrons and holes are one obvious distinction between a homojunction and a heterojunction. The relative doping levels control the relative magnitude of the electron and hole currents because the intrinsic potential barrier for electrons and holes in a homojunction is the same. The barrier heights that electrons and holes perceive in a heterojunction are different. The energy-band diagrams shown in Figures 5.18 and 5.23 show how different the barrier heights can be between electrons and holes in a heterojunction. Since the electron barrier height in Figure 5.18 is greater than the hole barrier height, we would anticipate that the electron current would be negligible in comparison to the hole current. Assuming all other parameters remain constant, the electron current will be about 10^4 times smaller than the hole current if the electron barrier height is 0.2 eV greater than the hole barrier height. With regard to the band diagram displayed in Figure 5.23, the situation is the opposite.

A rectifying metal-semiconductor contact's conduction-band edge and valence-band edge are comparable in certain ways in Figures 5.23 and 5.18. As with metal-semiconductor junctions, we determine the current-voltage characteristics of heterojunctions generally based on the thermionic emission of carriers over the barrier. We can then write

$$J = A^* T^2 \exp\left(\frac{-E_w}{kT}\right) \tag{5.51}$$

where E_w is an effective barrier height. As in the case of a pn homojunction or a Schottky barrier junction, the barrier height can be altered by applying a potential across the junction. However, it might be necessary to alter the heterojunction I-V characteristics to account for diffusion and tunneling effects. The effective mass of a carrier varies from one side of the junction to the other, which adds to the complexity. The overall form of the I-V equation is still comparable to that of a Schottky barrier diode and is typically dominated by one type of carrier, despite the heterojunction's actual I-V relationship having a complex derivation.

5.4.6. Heterojunction Manufacture and Applications

Molecular beam epitaxy (MBE) is typically needed for heterojunction manufacturing (Smith, C. G. (1996)) or chemical vapor deposition (CVD) techniques to accurately regulate the thickness of the deposition and produce an abrupt interface that is neatly lattice-matched. The mechanical stacking of layered materials into van der Waals heterostructures is a recent alternative that is being studied (Geim, A. K.; Grigorieva, I. V. (2013)).

Heterojunctions, despite their high cost, have found use in many specialized applications where their special qualities are essential:

- Solar cells: In certain solar cell architectures, heterojunctions are created at the interface between an amorphous silicon thin film (band gap 1.7 eV) and a crystalline silicon substrate (band gap 1.1 eV) (Leu, Sylvère; Sontag, Detlef (2020)). Like a p–n junction, the heterojunction is used to divide charge carriers. First developed in 1983 (Okuda, Koji et al., 1983), the Heterojunction with Intrinsic Thin-Layer (HIT) solar cell structure was later commercialized by Sanyo/Panasonic. With a conversion efficiency of 26.7%, HIT solar cells are currently the most efficient single-junction silicon solar cells (Yamamoto, Kenji et al, (2018)).
- Lasers: It was initially suggested to use heterojunctions in lasers in 1963 (Kroemer, H. (1963)), when eminent researcher Herbert Kroemer proposed that heterostructures could significantly improve population inversion. Carriers can be confined to enable room temperature lasing with low threshold currents by

sandwiching a smaller direct band gap material, such as GaAs, between two larger band gap layers, such as AlAs. Although it took many years for the material science of heterostructure fabrication to advance beyond Kroemer's concepts, they are now the accepted norm in the field. Later on, it was found that by utilizing the quantum size effects in quantum well heterostructures, the band gap could be manipulated. Another important benefit of using heterostructures in semiconductor lasers is that they can be utilized as waveguides for the index step that happens at the interface. Alternating layers of different III-V and II-VI compound semiconductors are used in the manufacturing of semiconductor diode lasers, which are used in CD and DVD players as well as fiber optic transceivers, to create lasing heterostructures.

- Bipolar transistors: A bipolar junction transistor with a heterojunction as its base-emitter junction has a very high forward gain and a low reverse gain. This results in low leakage currents and very good high-frequency operation (values in the tens to hundreds of GHz range). A heterojunction bipolar transistor (HBT) is the name given to this device.
- Field-effect transistors: High electron mobility transistors (HEMTs), which can function at much higher frequencies (above 500 GHz), use heterojunctions. High electron mobilities are produced by a two-dimensional electron gas inside a dopant-free region with minimal scattering when the doping profile and band alignment are right.

CASE STUDY

WALTER H. SCHOTTKY

Walter Hermann Schottky was a German physicist who invented the screen-grid vacuum tube in 1915 and the tetrode in 1919 while working at Siemens. In 1938, Schottky formulated a theory predicting the Schottky effect, now used in Schottky diodes.

Source: https://engineersblog.net/biography-of-famous-scientist-walter-schottky/

Early Life and Education

Walter Hans Schottky was born on July 23, 1886, in Zurich, Switzerland. He came from a family with a strong academic background. Schottky studied physics at the University of Berlin, where he earned his doctorate in 1912 under the supervision of Max Planck. During his early academic years, he developed a keen interest in experimental physics and electronic phenomena.

World War I Service

With the outbreak of World War I, Schottky served in the German military. After the war, he resumed his scientific pursuits, focusing on the emerging field of solid-state physics.

Contributions to Semiconductor Physics

Schottky made significant contributions to the understanding of the behavior of electrons in solids, particularly in semiconductors. In 1919, he proposed the existence of a barrier at the metal-semiconductor interface, which came to be known as the Schottky barrier. This concept played a crucial role in the development of semiconductor devices.

Academic Positions and Collaborations

Throughout his career, Schottky held various academic positions. He worked at the Siemens Research Laboratory and the Physikalisch-Technische Reichsanstalt (PTR) in Berlin. He also collaborated with other notable physicists of his time, including Werner Heisenberg and Max von Laue.

Research on Electron Emission

In the 1920s, Schottky conducted extensive research on electron emission phenomena, contributing to the understanding of thermionic emission and the development of vacuum tubes. His work in this area laid the foundation for advancements in electronic devices and communication technology.

Later Career and World War II

During the 1930s, Schottky continued his research in semiconductor physics. However, due to political reasons and the rise of the Nazi regime, he faced challenges in his career. Schottky left Germany in 1934 and moved to Switzerland, where he continued his research.

Post-World War II Period

After World War II, Schottky's contributions to semiconductor physics gained broader recognition. He returned to Germany and held academic positions at the Technical University of Munich. His work laid the groundwork for the development of semiconductor devices crucial to the electronics industry.

Awards and Honors

In recognition of his significant contributions, Walter Schottky received several awards and honors. Notably, he was awarded the Max Planck Medal in 1956 for his achievements in theoretical physics.

Legacy

Walter Schottky's work had a profound impact on the field of semiconductor physics and electronics. The Schottky barrier and the Schottky diode are named after him. His insights into electron behavior in semiconductors paved the way for the development of modern electronic devices and solid-state technology.

Walter Hans Schottky passed away on March 4, 1976, leaving behind a lasting legacy in the world of physics and technology. His pioneering contributions continue to influence the advancement of semiconductor research and applications in the contemporary era.

CLASS ACTIVITY

Student calculate the Schottky barrier lowering and the position of the maximum barrier height. Consider a gallium arsenide metal–semiconductor contact in which the electric field in the semiconductor is assumed to be $E = 6.8 \times 10^4$ V/cm.

SUMMARY

- A metal-semiconductor (M-S) junction is a type of electrical junction in which a metal comes in close contact with a semiconductor material. It is the oldest practical semiconductor device. M-S junctions can either be rectifying or non-rectifying.
- A metal on a lightly doped semiconductor can produce a rectifying contact that is known as a Schottky barrier diode. The ideal barrier height between the metal and semiconductor is the difference between the metal work function and the semiconductor electron affinity.
- When a positive voltage is applied to an n-type semiconductor with respect to the metal (reverse bias), the barrier between the semiconductor and metal increases so that there is essentially no flow of charged carriers. When a positive voltage is applied to the metal with respect to an n-type semiconductor (forward bias), the barrier between the semiconductor and metal is lowered so that electrons can easily flow from the semiconductor into the metal by a process called thermionic emission.
- The ideal current-voltage relationship of the Schottky barrier diode is the same as that of the pn junction diode. However, since the current mechanism is different from that of the pn junction diode, the switching speed of the Schottky diode is faster. In addition, the reverse saturation current of the Schottky diode is larger than that of the pn junction diode, so a Schottky diode requires less forward bias voltage to achieve a given current compared to a pn junction diode.
- Metal-semiconductor junctions can also form ohmic contacts, which are low-resistance junctions providing conduction in both directions with very little voltage drop across the junction.
- Semiconductor heterojunctions are formed between two semiconductor materials with different bandgap energies. One useful property of a heterojunction is the creation of a potential well at the interface. Electrons are confined to the potential well in the direction perpendicular to the interface, but are free to move in the other two directions.

REVIEW QUESTIONS

1. What is the ideal Schottky barrier height? Indicate the Schottky barrier height on an energy-band diagram.
2. Using an energy-band diagram, indicate the effect of the Schottky barrier lowering.
3. What is the mechanism of charge flow in a forward-biased Schottky barrier diode?
4. Compare the forward-biased current–voltage characteristic of a Schottky barrier diode to that of pn junction diode.

5. Explain the difference in switching characteristics between a Schottky diode and a pn junction diode. Discuss charge storage effects.
6. Sketch the energy-band diagram of a tunneling junction. Why is this an ohmic contact?
7. What is a heterojunction?
8. What is a 2-D electron gas?

REFERENCES

1. Anderson, R. L. (1962). Experiments on Ge–GaAs heterojunctions. *Solid-State Electronics, 5*(5), 341–351.
2. Crowley, A. M., & Sze, S. M. (1965). Surface states and barrier height of metal–semiconductor systems. *Journal of Applied Physics, 36*, 3212. https://doi.org/10.1063/1.1714294.
3. Geim, A. K., & Grigorieva, I. V. (2013). Van der Waals heterostructures. *Nature, 499*(7459), 419–425. https://doi.org/10.1038/nature12385.
4. Hu, C. C. (2010). *Modern Semiconductor Devices for Integrated Circuits*. Pearson Prentice Hall.
5. Kroemer, H. (1963). A proposed class of hetero-junction injection lasers. *Proceedings of the IEEE, 51*(12), 1782–1783. https://doi.org/10.1109/PROC.1963.2706.
6. Leu, S., & Sontag, D. (2020). Crystalline silicon solar cells: Heterojunction cells. In A. Shah (Ed.), *Solar Cells and Modules* (Vol. 301, pp. 163–195). Springer International Publishing. https://doi.org/10.1007/978-3-030-46487-5_7.
7. MacMillan, H. F., Hamaker, H. C., Virshup, G. F., & Werthen, J. G. (1988). Multijunction III-V solar cells: Recent and projected results. In *Twentieth IEEE Photovoltaic Specialists Conference* (pp. 48–54).
8. Michaelson, H. B. (1978). Relation between an atomic electronegativity scale and the work function. *IBM Journal of Research and Development, 22*(1), 72–80. https://doi.org/10.1147/rd.221.0072.
9. Okuda, K., Okamoto, H., & Hamakawa, Y. (1983). Amorphous Si/Polycrystalline Si stacked solar cell having more than 12% conversion efficiency. *Japanese Journal of Applied Physics, 22*(9), L605–L607. https://doi.org/10.1143/JJAP.22.L605.
10. Rideout, V. L. (1978). A review of the theory, technology and applications of metal–semiconductor rectifiers. *Thin Solid Films, 48*(3), 261–291. https://doi.org/10.1016/0040-6090(78)90205-4.
11. Smith, C. G. (1996). Low-dimensional quantum devices. *Reports on Progress in Physics, 59*(3), 235–282. https://doi.org/10.1088/0034-4885/59/3/001.
12. Streetman, B. G., & Banerjee, S. K. (2006). *Solid State Electronic Devices* (6th ed.). Pearson Prentice Hall.
13. Sze, S. M., & Ng, K. K. (2007). *Physics of Semiconductor Devices* (3rd ed.). John Wiley and Sons.

14. Tung, R. T. (2014). The physics and chemistry of the Schottky barrier height. *Applied Physics Reviews, 1*(1), 011304. https://doi.org/10.1063/1.4858400.
15. Yamamoto, K., Yoshikawa, K., Uzu, H., & Adachi, D. (2018). High-efficiency heterojunction crystalline Si solar cells. *Japanese Journal of Applied Physics,* 57(8S3), 08RB20. https://doi.org/10.7567/JJAP.57.08RB20.
16. Yang, E. S. (1988). *Microelectronic Devices.* McGraw-Hill.
17. Yuan, J. S. (1999). *SiGe, GaAs, and InP Heterojunction Bipolar Transistors.* John Wiley and Sons.

CHAPTER

6 Fundamentals of the Metal-Oxide-Semiconductor Field-Effect Transistor

LEARNING OBJECTIVES

After studying this chapter, you will be able to:

- Discuss the two-terminal MOS structure
- Describe the capacitance-voltage characteristics
- Discuss on MOSFET operation
- Focus on frequency limitations
- Define the CMOS technology

KEY TERMS FROM THIS CHAPTER

Accumulation layer charge
Channel conductance modulation
Conduction parameter
Depletion mode MOSFET
Field-effect
Interface states
Channel conductance
CMOS
Cutoff frequency
Enhancement mode MOSFET
Flat-band voltage
Inversion layer charge

6.1. INTRODUCTION

Electronic switching circuits can be created by utilizing the single-junction semiconductor devices we have examined, such as the pn homojunction diode, to produce rectifying current-voltage characteristics. Transistors are multijunction semiconductor devices that can provide voltage, current, and signal power gains when combined with other circuit components. The fundamental function of a transistor is to control current at one terminal by applying voltage across the other two terminals. One of the two main categories of transistors is the Metal-Oxide-Semiconductor Field-Effect Transistor (MOSFET). This chapter develops the basic physics of the MOSFET. Because of its small size, the MOSFET is widely used in digital circuit applications, allowing millions of devices to be fabricated in a single integrated circuit. It is possible to create the n-channel MOSFET and the p-channel MOSFET, two complementary MOS transistor configurations. When the two kinds of devices are combined into one circuit, electronic circuit design becomes incredibly flexible. Complementary MOS (CMOS) circuits are the name given to these circuits.

6.2. THE TWO-TERMINAL MOS STRUCTURE

The MOS capacitor, which is depicted in Figure 6.1, is the MOSFET's heart. Even though the metal is frequently actually high-conductivity polycrystalline silicon that has been deposited on the oxide, the term "metal" is still frequently used. The metal may be aluminum or another kind of metal. In this figure, the oxide's thickness (t_{ox}) and permittivity (εox) are represented as parameters.

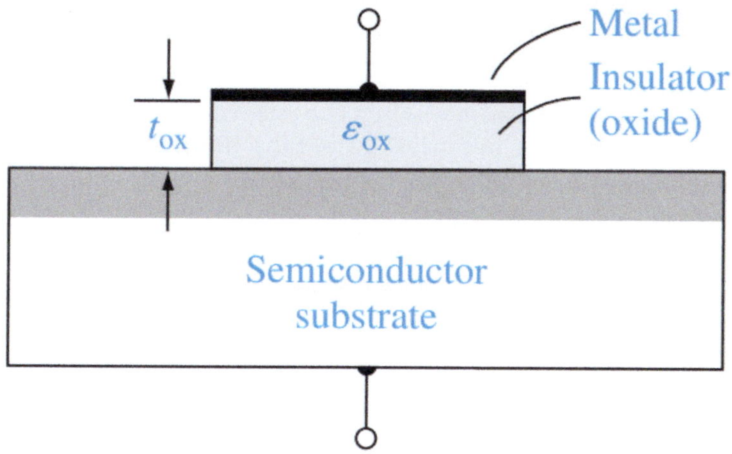

Figure 6.1. The basic MOS capacitor structure.

Source: https://www.optima.ufam.edu.br/SemPhys/Downloads/Neamen.pdf.

6.2.1. Energy-Band Diagrams

The use of a basic parallel-plate capacitor helps to clarify the physics of the MOS structure. A parallel-plate capacitor with the top plate at a negative voltage in relation to the bottom plate is depicted in Figure 6.2a. The two plates are separated by an insulating substance. This bias causes an electric field to be induced between the two plates as indicated, with a positive charge on the bottom plate and a negative charge on the top plate. The capacitance per unit area for this geometry is

$$C' = \frac{\epsilon}{d} \qquad (1)$$

where ∈is the permittivity of the insulator and d is the distance between the two plates. The magnitude of the charge per unit area on either plate is

$$Q' = C'V \qquad (2)$$

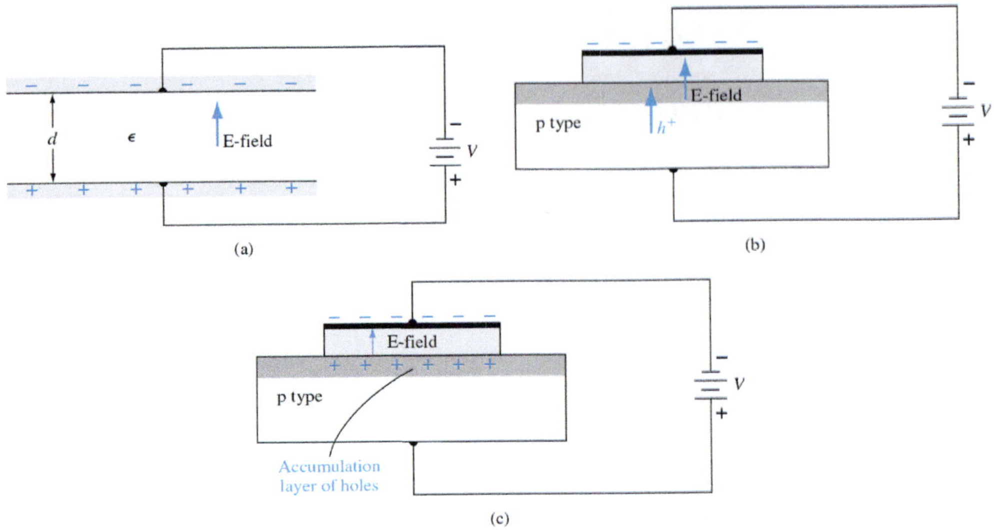

Figure 6.2. (a) A parallel-plate capacitor showing the electric field and conductor charges. (b) A corresponding MOS capacitor with a negative gate bias showing the electric field and charge flow. (c) The MOS capacitor with an accumulation layer of holes.

Source: https://www.optima.ufam.edu.br/SemPhys/Downloads/Neamen.pdf.

where the prime indicates charge or capacitance per unit area. The magnitude of the electric field is

$$E = \frac{V}{d} \qquad (3)$$

A MOS capacitor with a p-type semiconductor substrate is depicted in Figure 6.2b. The voltage across the top metal gate and the semiconductor substrate is negative. We can observe from the parallel-plate capacitor example that an electric field will be induced in the direction depicted in the figure, and a negative charge will exist on the top metal plate. The majority of carrier holes would feel a force toward the oxide-semiconductor interface if the electric field were to enter the semiconductor. The MOS capacitor's equilibrium charge distribution at this specific applied voltage is displayed in Figure 6.2c. At the oxide-semiconductor junction, an accumulation layer of holes reflects the positive charge on the MOS capacitor's bottom "plate."

The same MOS capacitor with the applied voltage's polarity reversed is depicted in Figure 6.3a. The top metal plate is now positively charged, and as seen, the induced electric field is pointing in the opposite direction. In this scenario, the majority of carrier holes will feel a force that pushes them away from the oxide–semiconductor interface if the electric field enters the semiconductor. The fixed ionized acceptor atoms cause a negative space charge region to form as the holes are pushed away from the interface. The negative charge on the MOS capacitor's bottom "plate" and the negative charge in the induced depletion region match. The MOS capacitor's equilibrium charge distribution at this applied voltage is depicted in Figure 6.3b.

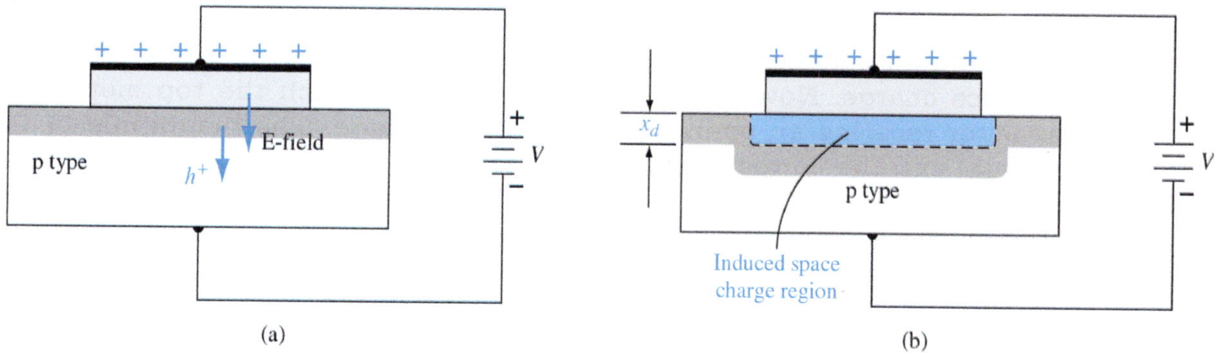

Figure 6.3. The MOS capacitor with a moderate positive gate bias, showing (a) the electric field and charge flow and (b) the induced space charge region.

Source: https://www.optima.ufam.edu.br/SemPhys/Downloads/Neamen.pdf.

The MOS capacitor with a p-type substrate energy-band diagrams for different gate biases are displayed in Figure 6.4. When zero bias is applied across the MOS device, Figure 6.4a illustrates the ideal scenario. The semiconductor's flat energy bands signify the absence of any net charge within the material.

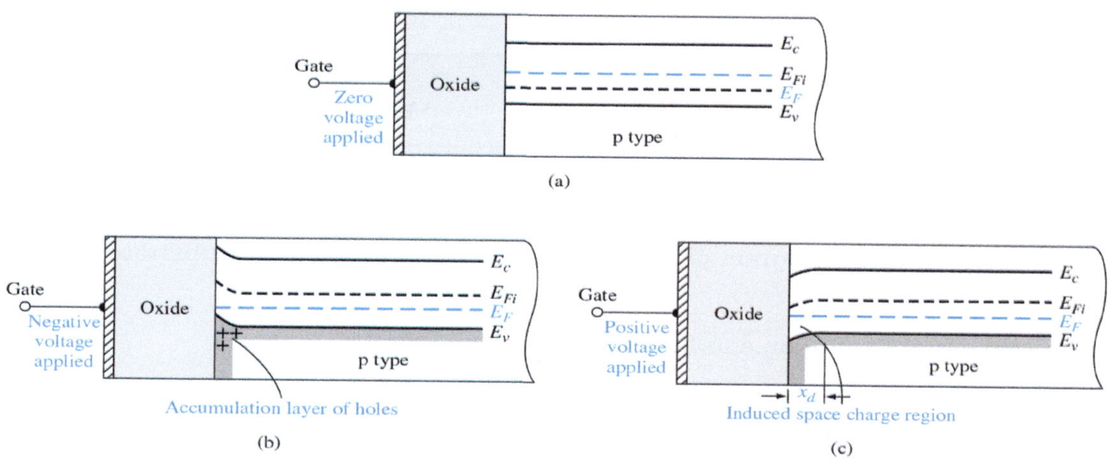

Figure 6.4. The energy-band diagram of a MOS capacitor with a p-type substrate for (a) a zero applied gate bias showing the ideal case, (b) a negative gate bias, and (c) a moderate positive gate bias.

Source: https://www.optima.ufam.edu.br/SemPhys/Downloads/Neamen.pdf.

The energy-band diagram for the scenario where a negative bias is applied to the gate is displayed in Figure 6.4b. At the oxide-semiconductor interface, the valence-band edge is nearer the Fermi level than it is in the bulk material, suggesting the presence of a hole accumulation. Compared to the bulk material, the semiconductor surface seems to be more p-type. Since there is no current flowing through the oxide and the MOS system is in thermal equilibrium, the Fermi level in the semiconductor is constant. The energy-band diagram of the MOS system with a positive voltage applied to the gate is displayed in Figure 6.4c. The figure illustrates how the conduction- and valence-band edges bend, revealing a space charge region akin to that found in a pn junction. The intrinsic Fermi levels and conduction band approach the Fermi level. xd is the width of the induced space charge. Now consider the scenario in which the top metal gate of the MOS capacitor receives an even higher positive voltage. The magnitude of the induced electric field and the corresponding increases in positive and negative charges on the MOS capacitor are what we anticipate. More band bending and a larger induced space charge region are implied by a higher negative charge in the MOS capacitor. Such a condition is depicted in Figure 6.5. At the surface, the intrinsic Fermi level is now lower than the Fermi level. While the valence band in the bulk semiconductor is approaching the Fermi level, the conduction band at the surface is currently not far from it. This finding suggests that the semiconductor surface that is next to the oxide-semiconductor interface is n-type. We have inverted the semiconductor's surface from a p-type to an n-type semiconductor by applying a sufficiently high positive gate voltage. An electron inversion layer at the oxide-semiconductor interface has been produced.

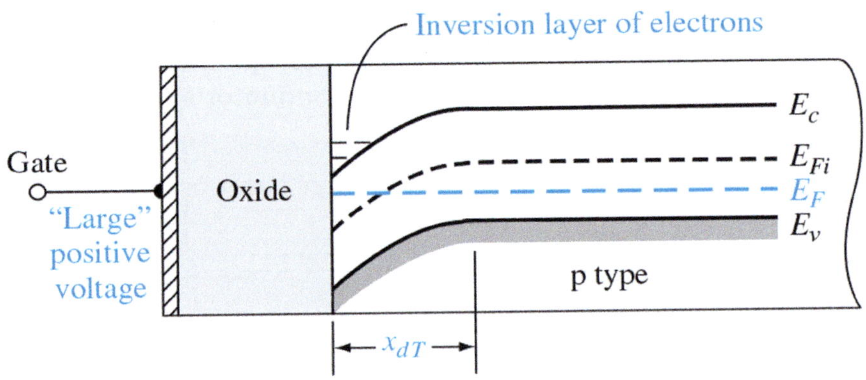

Figure 6.5. The energy-band diagram of the MOS capacitor with a p-type substrate for a "large" positive gate bias.

Source: https://www.optima.ufam.edu.br/SemPhys/Downloads/Neamen.pdf.

In the MOS capacitor structure, we just looked at, a p-type semiconductor substrate was taken into consideration. When constructing an energy-band diagram for a MOS capacitor with an n-type semiconductor substrate, the same approach can be used. The MOS capacitor structure is depicted in Figure 6.6a with a positive voltage applied to the upper gate terminal. The top gate is positively charged, and the direction of the induced electric field is depicted in the figure. In the n-type substrate, an electron accumulation layer will be created. Figure 6.6b illustrates the scenario in which a negative voltage is applied to the top gate. In this case, the n-type semiconductor experiences the induction of a positive space charge region.

This MOS capacitor with the n-type substrate's energy-band diagrams is displayed in Figure 6.7. The scenario where an accumulation layer of electrons forms and a positive voltage is applied to the gate is depicted in Figure 6.7a. When a negative voltage is applied to the gate, the energy bands are displayed in Figure 6.7b. Now that a space charge region has been induced in the n-type substrate, the conduction and valence bands bend upward.

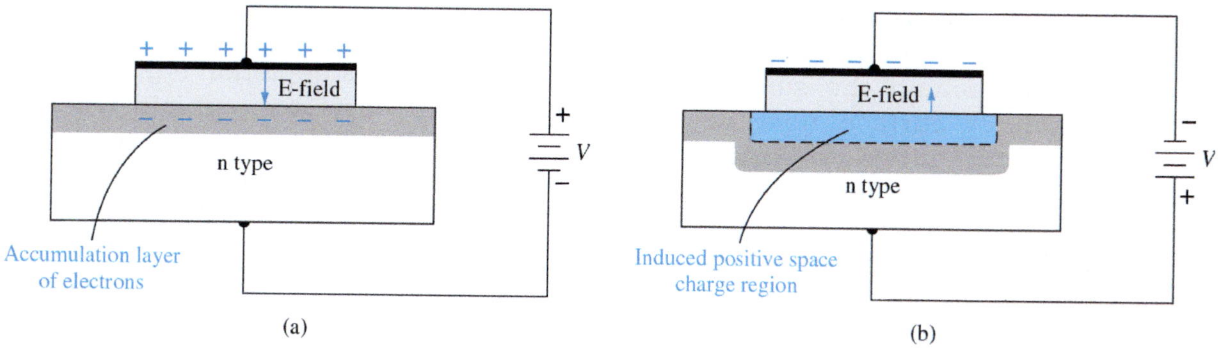

Figure 6.6. The MOS capacitor with an n-type substrate for (a) a positive gate bias and (b) a moderate negative gate bias.

Source: https://www.optima.ufam.edu.br/SemPhys/Downloads/Neamen.pdf.

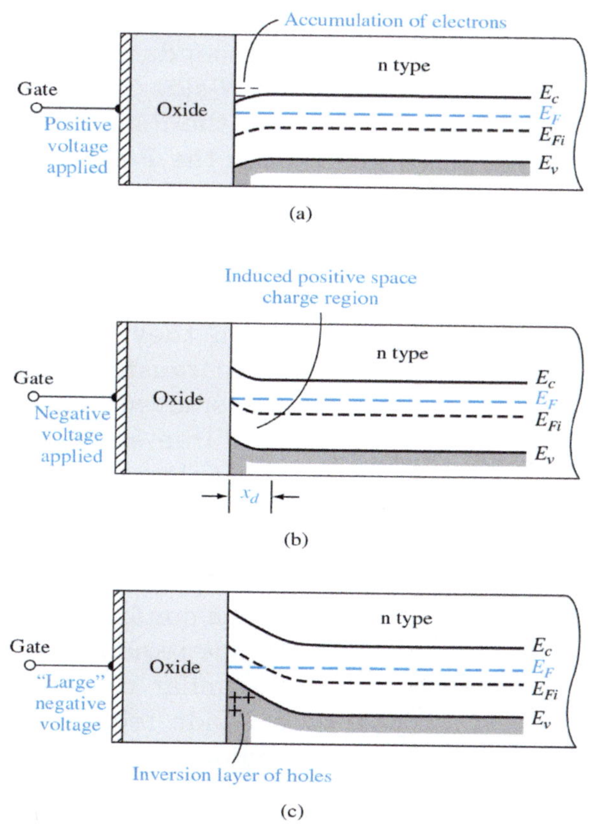

Figure 6.7. The energy-band diagram of the MOS capacitor with an n-type substrate for (a) a positive gate bias, (b) a moderate negative bias, and (c) a "large" negative gate bias.

Source: https://www.optima.ufam.edu.br/SemPhys/Downloads/Neamen.pdf.

The energy bands are displayed in Figure 6.7c when a greater negative voltage is applied to the gate. In addition to the intrinsic Fermi level rising above the Fermi level, the conduction and valence bands are bent even further. In the bulk semiconductor, the conduction band is near the Fermi level, while the valence band at the surface is currently approaching the same level. The semiconductor surface next to the oxide–semiconductor interface appears to be p-type, based on this result. The MOS capacitor's semiconductor surface has been inverted from n-type to p-type by applying a sufficiently high negative voltage to the gate. At the oxide–semiconductor interface, an inversion layer of holes has been produced.

6.2.2. Depletion Layer Thickness

The depletion layer thickness, often referred to as the depletion region or space-charge region, is a crucial parameter in semiconductor physics, particularly in the context of p-n junctions. This region forms at the interface between p-type and n-type semiconductor materials, where mobile charge carriers (holes from the p-side and electrons from the n-side) diffuse across the junction and recombine, leaving behind fixed ionized donor and acceptor atoms. As a result, an electric field is established, creating a barrier that prevents further diffusion of carriers. The thickness of the depletion layer depends on several factors, including the doping concentrations of the p-type and n-type materials, the applied voltage across the junction, and the temperature. In equilibrium, the depletion layer grows until the built-in potential balances the diffusion forces, reaching a steady-state thickness. This thickness is significant because it influences the junction's capacitance, the electric field distribution, and the overall behavior of the semiconductor device, affecting its performance in applications such as diodes, transistors, and photovoltaic cells. Understanding and controlling the depletion layer thickness is essential for optimizing the design and function of these semiconductor devices.

6.2.2.1. Formation in a p–n Junction

At a p-n junction, a depletion region appears instantly. It is easiest to explain when the junction is in thermal equilibrium or a steady state, as these situations result in dynamic equilibrium, which is characterized by time-invariant system properties. In the same way that ink diffuses into water until it is evenly distributed, electrons and holes diffuse into areas where their concentrations are lower. The N-type semiconductor is defined as having more free electrons (in the conduction band) than the P-type semiconductor, and the P-type semiconductor has more holes (in the valence band) than the N-type semiconductor. Thus, free electrons from the N-side conduction band migrate (diffuse) into the P-side conduction band, and holes from the P-side valence band migrate into the N-side valence band when N-doped and P-doped semiconductors are stacked together to form a junction. The diffused electrons are eliminated by recombination in the P-side when they come into contact with holes after transfer. Similarly, the diffused holes are removed from the N-side by recombining with free electrons. The diffused electrons and holes are eliminated as a whole. Due to two processes, (1) electron diffusion to the P-side and (2) electron recombination to holes diffused from the P-side, free electrons in the conduction band are eliminated in an N-side region close to the junction interface. A similar reason also disappears holes in a P-side region close to the interface. The region surrounding the junction interface is referred to as the depletion region or depletion zone because it is the result of the majority charge carriers—free electrons for N-type semiconductors and holes for P-type semiconductors—being depleted there. The depletion region has a positive charge on its N-side and a negative charge on its P-side as a result of the majority charge carrier diffusion mentioned earlier. By doing this, an electric field is produced, providing a force that prevents the diffusion of charges. The depletion region reaches equilibrium when the electric field is high enough to prevent additional hole and electron diffusion. The built-in voltage, also referred to as the junction voltage, barrier voltage, or contact potential, is calculated by integrating the electric field across the depletion region.

Physically speaking, charge transfer in semiconductor devices is from (1) the charge carrier drift by the electric field and (2) the charge carrier diffusion due to the spatially varying carrier concentration. In the P-side of the depletion region, where holes drift by the electric field with the electrical conductivity σ and diffuse with the diffusion constant D, the net current density is given by

$$\mathbf{J} = \sigma\mathbf{E} - eD\nabla p,$$

where E is the electric field, e is the elementary charge (1.6×10^{-19} coulomb), and p is the hole density (number per unit volume). Due to the electric field's ability to cause holes to drift in the direction of the field and diffusion holes to move in the direction of decreasing concentration, a positive density gradient is produced by holes with a negative current. (In certain situations, both electrons and holes must be included.) If the carriers are electrons, the hole density p is substituted with the electron density n with a negative sign. The current is zero because of the Einstein relation, which relates D to σ, when the two current components balance, as in the p–n junction depletion region at dynamic equilibrium.

6.2.3. Surface Charge Density

Surface charge density is a fundamental concept in electromagnetism, representing the quantity of electric charge per unit area on a surface. It is typically denoted by the Greek letter sigma (σ) and measured in coulombs per square meter (C/m^2). This parameter is crucial in understanding the behavior of electric fields at material boundaries and plays a vital role in various physical phenomena and technological applications.

Surface charge density arises due to the presence of excess charges on the surface of a conductor or a dielectric material. In conductors, charges reside on the surface because they are free to move and will redistribute themselves to minimize the system's energy, resulting in a uniform surface charge density in equilibrium. In contrast, in dielectric materials, surface charge density can result from polarization effects, where bound charges align under an external electric field, creating surface charges without free charge movement.

The concept of surface charge density is essential in electrostatics, where it is used to determine the electric field near surfaces. According to Gauss's law, the electric field just outside a charged surface is proportional to the surface charge density and inversely proportional to the permittivity of the surrounding medium. Mathematically, the electric field (E) near a surface with charge density (σ) is given by $E = \sigma / \varepsilon_0$ for a surface in a vacuum, where ε_0 is the permittivity of free space. This relationship helps in calculating the forces and potential distributions around charged surfaces.

Surface charge density is also pivotal in capacitance calculations, particularly in parallel plate capacitors, where the charge stored on the plates is directly proportional to the surface charge density. For a capacitor with plates of area A and surface charge density σ, the total charge (Q) on each plate is Q = σA. The capacitance (C) of the capacitor can then be related to the surface charge density through the voltage (V) between the plates: C = Q / V.

In semiconductor physics, surface charge density affects the operation of devices like metal-oxide-semiconductor field-effect transistors (MOSFETs). The control of surface charges at the semiconductor-oxide interface is crucial for modulating the channel conductivity, directly influencing the device's performance.

Moreover, surface charge density is significant in environmental science and engineering, such as in electrostatic precipitation, where charged particles are removed from gases using electric fields. The efficiency of such processes depends on the surface charge density of the particles and the collecting plates.

In biological systems, surface charge density influences cell membrane properties and interactions, affecting processes like cell adhesion, signaling, and ion transport. For instance, the distribution of charges on cell membranes can dictate the behavior of ions and molecules near the surface, impacting physiological functions.

Overall, surface charge density is a critical parameter across various disciplines, providing insights into the behavior of electric fields, the design and function of electronic devices, environmental applications, and biological systems. Understanding and manipulating surface charge density enables advancements in technology and a deeper comprehension of natural phenomena.

6.2.4. Work Function Differences

The work function is a crucial material property that refers to the minimum energy required to extract an electron from the surface of a material to a point just outside its influence. This value varies greatly among different materials due to differences in their electronic structure and surface properties. In metals, the work function typically falls between 2 to 5 electron volts (eV), with alkali metals like cesium having lower work functions around 2.1 eV due to their loosely bound valence electrons, while noble metals like gold and platinum have higher work functions around 5.1 eV and 5.6 eV respectively, due to their tightly bound electrons and stable surface electron configurations. Semiconductors have work functions that are influenced by doping levels and types; for example, n-type silicon has a work function of about 4.05 eV, while p-type silicon has a higher work function around 5.15 eV due to changes in electron density and the Fermi level resulting from doping. Insulators, with their large band gaps, have more complex work functions, but surface states and defects can create localized energy levels that impact the effective work function, as seen in materials like diamond which has a high intrinsic work function around 5.5 eV, but can have lower values when surface states are altered, such as with hydrogen termination. Understanding these variations is crucial for customizing materials for specific applications: in electron emission technologies, materials with lower work functions like cesium are preferred for efficient electron emission at lower temperatures; in photovoltaic devices, the work function influences charge separation and carrier collection efficiency; in catalysis, particularly heterogeneous catalysis involving metals, the work function influences the adsorption and activation of reactants on the catalyst surface; and in electronic devices like field-effect transistors and diodes, it determines the Schottky barrier height, which impacts carrier injection and device performance. Therefore, the work function is a key parameter in a wide range of applications in electronics, photonics, and catalysis, with its variability playing a crucial role in material selection and device optimization.

6.2.5. Flat-Band Voltage

The flat-band voltage (V_{FB}) is a critical parameter in the study of metal-oxide-semiconductor (MOS) structures, influencing the performance of devices such as MOS capacitors and field-effect transistors (FETs). It is defined as the voltage at which the energy bands of the semiconductor become flat, meaning there is no band bending at the semiconductor-oxide interface. This condition implies that the electric field within the semiconductor is zero, and the surface potential is equal to the bulk potential.

The flat-band voltage is influenced by several factors, including the work function difference between the metal gate and the semiconductor, the fixed charge within the oxide layer, and the interface trap charges at the semiconductor-oxide boundary. Mathematically, the flat-band voltage can be expressed as:

$$V_{FB} = \Phi_{MS} - \frac{Q_{ox}}{C_{ox}}$$

where Φ_{MS} is the work function difference between the metal and the semiconductor, Q_{ox} is the charge density in the oxide layer, and C_{ox} is the oxide capacitance per unit area.

The work function difference Φ_{MS} is determined by the inherent properties of the materials used. For instance, if a high work function metal like platinum is used as the gate material on an n-type silicon substrate, the flat-band voltage will be positive. Conversely, using a low work function metal like aluminum on a p-type silicon substrate will result in a negative flat-band voltage.

The fixed charges in the oxide layer (Q_{ox}) can arise from various sources, such as impurities, defects, and trapped charges introduced during the oxide growth process. These charges can significantly affect the flat-band voltage by shifting it from its ideal value. For example, positive charges in the oxide will increase the flat-band voltage, while negative charges will decrease it.

Interface trap charges (Q_{it}) at the semiconductor-oxide interface also play a crucial role. These traps can capture and release carriers, altering the effective charge distribution and thus influencing the flat-band voltage. The presence of these traps is often due to imperfections at the interface, such as dangling bonds and other structural defects.

Understanding and controlling the flat-band voltage is essential for the design and optimization of MOS devices. It affects the threshold voltage (V_{th}) of MOSFETs, which is the minimum gate voltage required to create a conductive channel between the source and drain terminals. Precise control over V_{FB} ensures that devices operate within the desired electrical characteristics, improving performance and reliability.

The flat-band voltage is a fundamental parameter in MOS structures, determined by the work function difference between the gate and semiconductor, the fixed oxide charges, and the interface trap charges. Accurate control and understanding of V_{FB} are essential for optimizing the performance of MOS devices, affecting key characteristics such as the threshold voltage and overall device behavior.

6.2.6. Threshold Voltage

The threshold voltage of a field-effect transistor (FET) is the lowest gate-to-source voltage (V_{GS}) required to establish a conducting path between the source and drain terminals. It is often shortened as V_{th} or $V_{GS(th)}$. It is a crucial scaling component for preserving power efficiency. When referring to a junction field-effect transistor (JFET), the threshold voltage is often called pinch-off voltage instead (Marco Delaurenti 1999). This is a little confusing because, even though the current is always on, pinch off in the context of an insulated-gate field-effect transistor (IGFET) refers to the channel pinching that causes current saturation behavior under high source-drain bias. In contrast to pinch off, threshold voltage is a clear term that describes the same idea in any field-effect transistor. The transistor in n-channel enhancement-mode devices does not naturally have a conductive channel. Dopant ions added to the FET's body without a V_{GS} create a depletion region, which is a region devoid of mobile carriers. Free-floating electrons within the body are drawn toward the gate by a positive V_{GS}. To counter the dopant ions and create a conductive channel, however, a sufficient number of electrons must be drawn in close proximity to the gate. We refer to this procedure as inversion. From source to drain, the conductive channel is connected at the threshold voltage of the FET. At higher V_{GS}, more electrons are drawn to the gate and the channel widens.

The p-channel "enhancement-mode" MOS transistor is off when VGS = 0, with the channel open and non-conducting. Applying a negative gate voltage turns the transistor on by enhancing channel conductivity. In contrast, n-channel depletion-mode devices have a conductive channel already present. The threshold voltage in this case refers to when the channel is wide enough for electrons to flow easily. This concept also applies to p-channel depletion-mode devices, where a negative gate voltage creates a depletion layer, exposing a carrier-free region of immobile, negatively charged acceptor ions. For the n-channel depletion MOS transistor, a sufficient negative V_{GS} depletes the channel of free electrons, turning the transistor off. Similarly, for a p-channel depletion-mode MOS transistor, a sufficient positive gate-source voltage depletes the channel of free holes, also turning it off. The threshold voltage of wide planar transistors remains constant regardless of the drain-source voltage (V_{DS}), making it a clearly defined characteristic. However, in modern nanometer-sized MOSFETs, the threshold voltage is less predictable due to drain-induced barrier lowering. In figures 6.8, the source (left side) and drain (right side) are labeled n^+ to indicate heavily doped (blue) n-regions. The depletion layer dopant is labeled N_A- to indicate that the ions in the (pink) depletion layer are negatively charged and there are very few holes. In the (red) bulk, the number of holes $p = N_A$ making the bulk charge neutral.

If the gate voltage is lower than the threshold voltage as shown in the left figure, the "enhancement-mode" transistor will be deactivated and ideally, there will be no current flowing from the drain to the source of the transistor. However, there will still be a small amount of current present, known as subthreshold leakage, which increases exponentially with gate bias even below the threshold. Datasheets will typically list the threshold voltage in relation to a specific measurable current amount, commonly 250 µA or 1 mA.

Because there are many electrons in the channel at the oxide-silicon interface, creating a low-resistance channel where charge can flow from drain to source, the enhancement-mode transistor is activated if the gate voltage is higher than the threshold voltage (right figure). This state is known as strong inversion when the voltage is much higher than the threshold. The channel is tapered when $V_D > 0$ because the voltage drop due to the current in the resistive channel reduces the oxide field supporting the channel as the drain is approached.

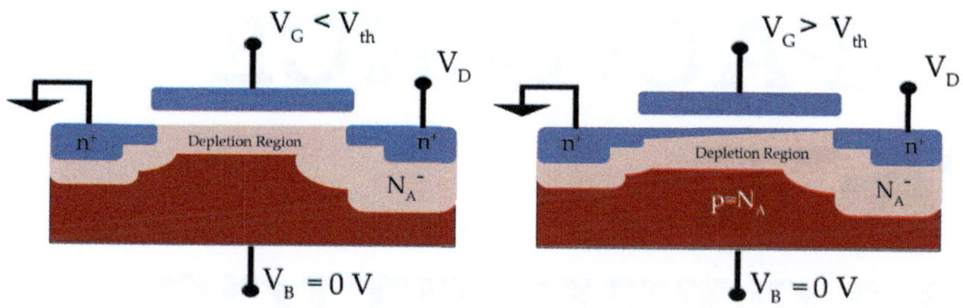

Figure 6.8. MOSFET cross section above threshold.

Source: https://en.wikipedia.org/wiki/Threshold_voltage#/media/File:MOSFET_above_threshold.png.

6.3. CAPACITANCE-VOLTAGE CHARACTERISTICS

The core component of the MOSFET is the MOS capacitor structure. Valuable insights about the MOS device and the oxide-semiconductor interface can be gleaned from analyzing the capacitance versus voltage (C-V) characteristics of the device. The capacitance of a device is defined as

$$C = \frac{dQ}{dV}$$

where dQ is the magnitude of the differential change in charge on one plate as a function of the differential change in voltage dV across the capacitor. Capacitance is typically measured as a small signal or AC parameter by adding a small AC voltage to a DC gate voltage. This allows for the measurement of capacitance as a function of the applied DC gate voltage.

6.3.1. Application of Capacitance Voltage

Capacitance-voltage (C-V) testing is a widely used technique in semiconductor research, especially for MOSCAP and MOSFET structures, to determine semiconductor parameters. However, other semiconductor device and technology types, such as bipolar junction transistors, JFETs, III-V compound devices, photovoltaic cells, MEMS devices, organic thin-film transistor (TFT) displays, photodiodes, and carbon nanotubes (CNTs), are also extensively characterized by C-V measurements.

Because of their fundamental nature, these measurements can be used in a variety of research tasks and disciplines. In the labs of semiconductor manufacturers and universities, for instance, researchers use them to assess novel materials, devices, circuits, and processes. Product and yield enhancement engineers, who are in charge of enhancing procedures and device performance, find great value in these measurements. These measurements are also used by reliability engineers to evaluate failure mechanisms, monitor process parameters, and qualify suppliers of the materials they use. When the right techniques, tools, and software are used, C-V measurements can yield a wide range of semiconductor device and material parameters. This data is evaluated for epitaxially grown crystals, including characteristics like average doping concentration, doping profiles, and carrier lifetimes. It is then used throughout the semiconductor production chain.

In wafer processes, C-V measurements can disclose the interface trap density, mobile ion contamination, oxide thickness, and oxide charges. A C-V profile for bulk MOSFETs with varying oxide thicknesses produced using nanoHUB. It is evident that the blue curve depicts the high-frequency C-V profile, while the red curve denotes low frequency. Observe closely how the threshold voltage changes with varying oxide thicknesses. Following the completion of additional process steps like lithography, etching, cleaning, polysilicon and dielectric depositions, and metallization, among others, these measurements remain crucial. C-V profiling is frequently used to model device performance and characterize threshold voltages and other parameters during basic and reliability device testing after devices have been fully fabricated. Electronic instrumentation capacitance-voltage meters are used for C-V measurements. By using the obtained C-V graphs, they are used to analyze the doping profiles of semiconductor devices.

6.3.2. C–V Characteristics of Metal–Oxide–Semiconductor Structure

The metal-oxide-semiconductor structure is an essential component of a MOSFET, regulating the level of the potential barrier in the channel through the gate oxide. An n-channel MOSFET's operation can be divided into three regions:

6.3.2.1. Depletion

The valence band edge is driven far from the Fermi level and the holes from the body are driven away from the gate when a small positive bias voltage is applied to the metal. This results in a low carrier density, which lowers the capacitance.

6.3.2.2. Inversion

An inversion layer or n-channel at the interface between the semiconductor and the oxide is created on the semiconductor surface when the gate bias is increased even further, bringing the conduction band edge closer to the Fermi level.

6.3.2.3. Accumulation

An analogous p-channel to the n-channel case, but with opposite polarity of charges and voltages, is formed at the surface of the n region by applying a negative gate-source voltage (positive source-gate). The capacitance increases in proportion to the increase in hole density.

6.3.3. Frequency Effects

The MOS capacitor is biased in the inversion condition and has a p-type substrate. As we have argued, in the ideal scenario, a differential change in the capacitor voltage results in a differential change in the charge density of the inversion layer. But we also need to take into account the source of electrons that alters the inversion charge density. The inversion layer's charge density can be altered by two different electron sources. Diffusion of minority carrier electrons from the p-type substrate across the space charge region is the first source. The ideal reverse saturation current is produced by the same diffusion process found in a reverse-biased pn junction.

Thermal generation of electron-hole pairs within the space charge region is the second source of electrons. The reverse-biased generation current is generated by the same process that also occurs in a reverse-biased pn junction. These two procedures produce electrons at different rates. Therefore, it is not possible for the electron concentration in the inversion layer to change instantly. The inversion layer charge change won't be able to react if the ac voltage across the MOS capacitor changes quickly. The frequency of the ac signal used to measure the capacitance will then determine the C-V characteristics. The inversion layer charge will not react to a differential change in capacitor voltage up to a very high frequency.

The charge distribution in the MOS capacitor with a p-type substrate. At a high signal frequency, the differential change in charge occurs at the metal and in the space charge width in the semiconductor. The capacitance of the MOS capacitor is then C'_{min}, which we discussed earlier. The high-frequency and low-frequency limits of the C-V characteristics. In general, high frequency corresponds to a value on the order of 1 MHz and low frequency corresponds to values in the range of 5 to 100 Hz. Typically, the high-frequency characteristics of the MOS capacitor are measured.

6.3.4. Fixed Oxide and Interface Charge Effects

Metal-oxide-semiconductor (MOS) devices are affected by fixed oxide and interface charges, which modify their electrical properties and change the way they behave in terms of capacitance-voltage (C-V) and other aspects. During oxide growth, impurities, defects, or stress can result in fixed oxide charges, which are immobile and found within the oxide layer near the semiconductor interface. By causing a horizontal shift in the C-V curve, these charges modify the device's threshold voltage (V_{th}) and flat-band voltage (V_{FB}). The shift is positive for positive fixed oxide charges and negative for negative fixed oxide charges. Interface charges have the ability to both capture and release carriers. They are also known as interface trap charges (Q_{it}), and they are found at the semiconductor-oxide interface. Dangling bonds, contaminants at the interface, or structural flaws can all cause these traps. They provide evidence of the interface states' response to the AC signal and their impact on the measured capacitance by contributing to capacitance-voltage stretch-out. The stretched-out state is a result of the higher density of states present at the interface, which can trap carriers and lead to unstable device functioning. By introducing variations in threshold voltage, increased leakage currents, and decreased carrier mobility, both types of charges impair MOS devices' ideal behavior. The significance of superior oxide growth and interface passivation methods in reducing fixed oxide and interface charges and guaranteeing dependable and effective MOS device functioning is underscored by these phenomena.

6.4. MOSFET OPERATION

The flow of charge in the channel region or inversion layer next to the oxide-semiconductor interface is what drives current in a MOSFET. The formation of the inversion layer charge in enhancement-type MOS capacitors has been covered. Additionally, we might have depletion-type devices, where a channel is present even at zero gate voltage.

6.4.1. MOSFET Structures

There are four primary types of MOSFET devices. Shown in Figure 6.9 is an n-channel enhancement mode MOSFET. In enhancement mode, the semiconductor substrate does not have an inverted area directly under the oxide when the gate voltage is zero. Applying a positive gate voltage creates an electron inversion layer that connects the n-type source and drain regions. Carriers flow from the source terminal through the channel to the drain terminal. This n-channel device allows electrons to flow from the source to the drain, causing conventional current to enter the drain and exit the source. The figure also displays the conventional circuit symbol for this n-channel enhancement mode device. In Figure 6.10, an n-channel depletion mode MOSFET is depicted, with a region under the oxide containing 0 V at the gate. It has been demonstrated that the threshold voltage of a MOS device with a p-type substrate can be negative, resulting in an existing electron inversion layer even with zero gate voltage applied. This type of device is referred to as a depletion mode device. The n-channel in the figure may represent an electron inversion layer or a purposely doped n-region. The conventional circuit symbol for the n-channel depletion mode MOSFET is also illustrated in the figure.

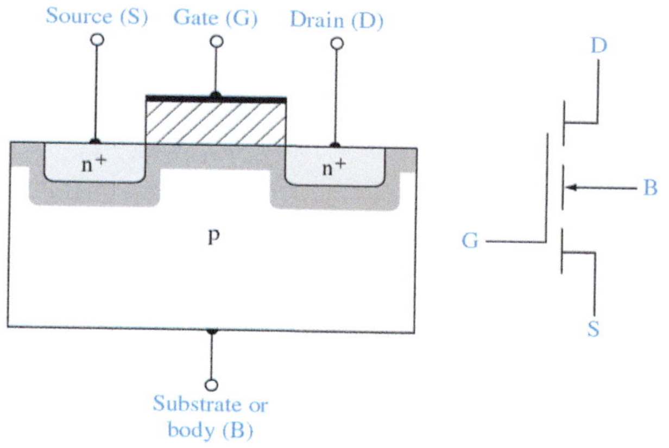

Figure 6.9. Cross section and circuit symbol for an n-channel enhancement mode MOSFET.

Source: https://www.optima.ufam.edu.br/SemPhys/Downloads/Neamen.pdf.

A p-channel enhancement mode MOSFET and a p-channel depletion mode MOSFET are depicted in Figures 6.11a, b. To "connect" the p-type source and drain regions of the p-channel enhancement mode device, an inversion layer of holes must be created via the application of a negative gate voltage. Conventional current enters the source and exits the drain because holes flow from the source to the drain. The depletion mode device has a p-channel region even when the gate voltage is zero. The figure displays the standard circuit symbols.

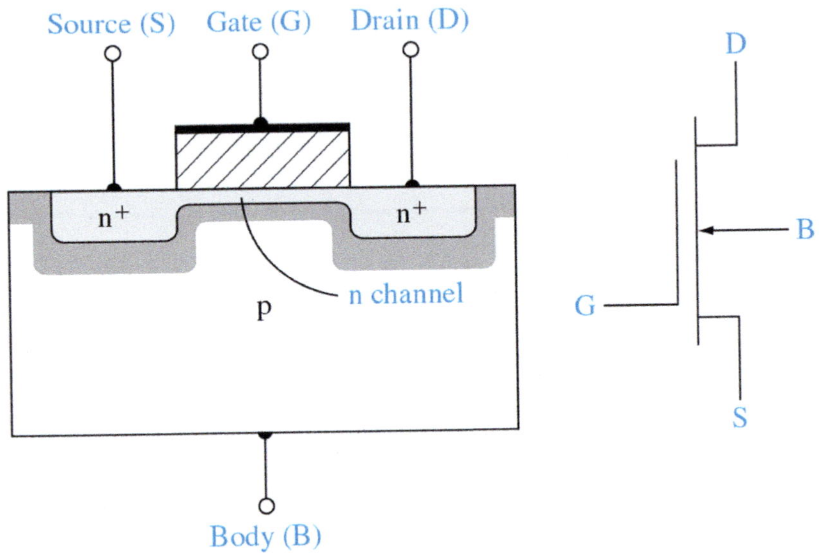

Figure 6.10. Cross section and circuit symbol for an n-channel depletion mode MOSFET.

Source: https://www.optima.ufam.edu.br/SemPhys/Downloads/Neamen.pdf.

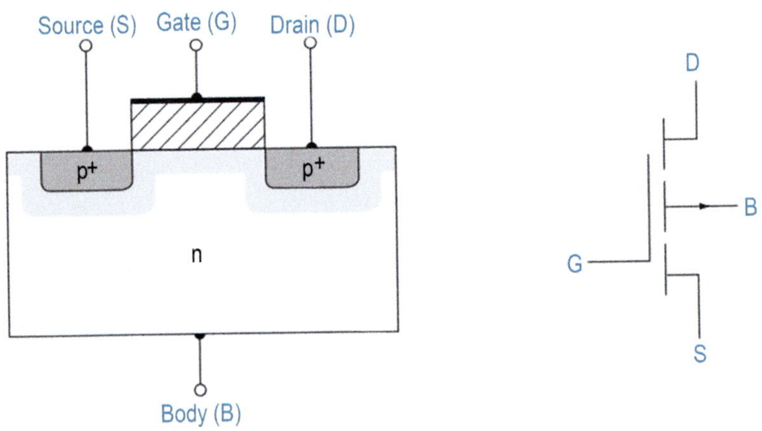

(a)

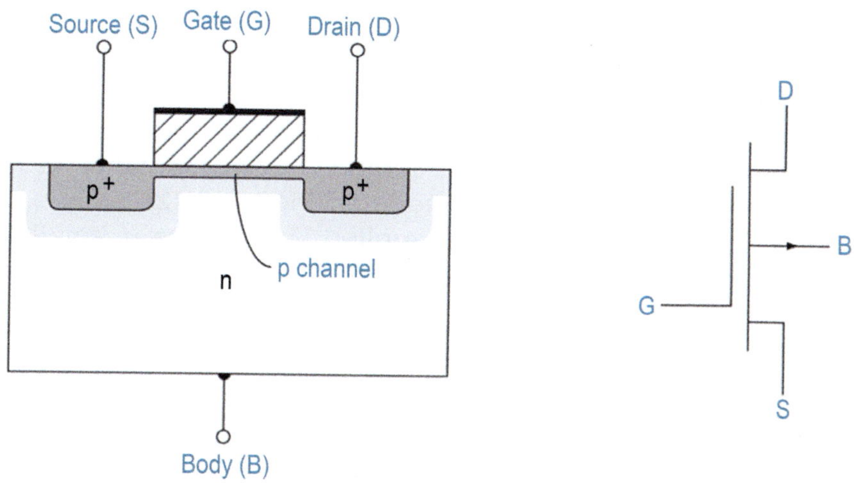

(b)

Figure 6.11. Cross section and circuit symbol for (a) a p-channel enhance ment mode MOSFET and (b) a p-channel depletion mode MOSFET.

Source: https://www.optima.ufam.edu.br/SemPhys/Downloads/Neamen.pdf.

6.4.2. Current–Voltage Relationship

An n-channel enhancement mode MOSFET is being used with a gate-to-source voltage lower than the threshold voltage and a minimal drain-to-source voltage. The source and substrate terminals are both grounded. As a result of this configuration, there is no electron inversion layer, the drain-to-substrate pn junction is reverse biased, and the drain current is zero. Alternatively, if the gate voltage is applied such that $V_{GS} > V_T$, an electron inversion layer is formed. When a small drain voltage is then applied, electrons within the inversion layer will flow from the source to the positive drain terminal. Conventional current flows into the drain terminal and exits the source terminal. Ideally, there should be no current passing through the oxide to the gate terminal. For small V_{DS} values, when the MOSFET is in the linear or triode region, the channel region has the characteristics of a resistor, so we can write

$$I_D = g_d V_{DS}$$

where gd is defined as the channel conductance in the limit as $V_{DS} \to 0$. The channel conductance is given by

$$g_d = \frac{W}{L} \cdot \mu_n |Q_n'|$$

where μ_n is the mobility of the electrons in the inversion layer and $|Q_n'|$ is the magnitude of the inversion layer charge per unit area. Since the inversion layer charge depends on the gate voltage, the fundamental function of a MOS transistor is the gate voltage's modulation of the channel conductance. The drain current is then determined by the channel conductance. Initially, we're going to assume that there is constant mobility.

The I_D versus V_{DS} characteristics, for small values of V_{DS}. When $V_{GS} < V_T$, the drain current is zero. As V_{GS} becomes larger than V_T, channel inversion charge density increases, which increases the channel conductance. A larger value of gd produces a larger initial slope of the I_D versus V_{DS} characteristic. Figure 6.12a shows the basic MOS structure for the case when $V_{GS} < V_T$ and the applied V_{DS} voltage is small. The thickness of the inversion channel layer in the figure qualitatively indicates the relative charge density, which is essentially constant along the entire channel length for this case. The corresponding I_D versus V_{DS} curve is shown in the figure.

Figure 6.12b illustrates the scenario in which the V_{DS} value rises. The voltage drop across the oxide near the drain terminal decreases with increasing drain voltage, which implies a decrease in the induced inversion charge density near the drain. When the channel's incremental conductance at the drain decreases, the slope of the I_D versus V_{DS} curve also decreases. The I_D versus V_{DS} curve in the figure illustrates this effect. At the drain terminal, the induced inversion charge density is zero when V_{DS} rises to the point where the potential drop across the oxide is equal to V_T. An illustration of this effect can be found in Figure 6.12c. Since there is now zero incremental conductance at the drain, the slope of the I_D versus V_{DS} curve is also zero. We can write

$$V_{GS} - V_{DS}(\text{sat}) = V_T$$

Or

$$V_{DS}(\text{sat}) = V_{GS} - V_T$$

where V_{DS} (sat) is the drain-to-source voltage producing zero inversion charge density at the drain terminal.

The channel's point where the inversion charge is exactly zero advances in the direction of the source terminal when V_{DS} exceeds the V_{DS} (sat) value. In this instance, electrons start at the source of the channel, move through it in the direction of the drain, and are injected into the space charge region at the point where the charge is zero. From there, they are swept by the E-field to the drain contact. If we assume that the change in channel length L is small compared to the original length L, then the drain current will be constant for $V_{DS} > V_{DS}$ (sat). The region of the I_D versus V_{DS} characteristic is referred to as the saturation region. Figure 6.12d shows this region of operation.

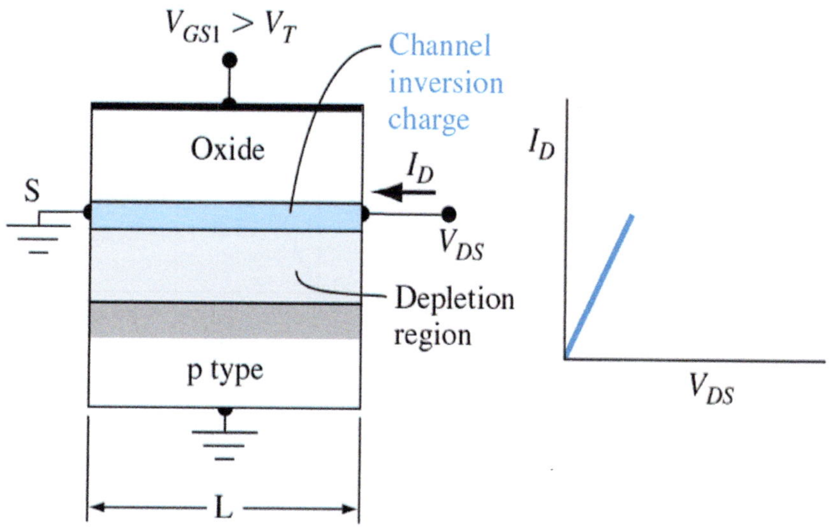
$V_{GS1} > V_T$
Channel inversion charge
Oxide
S
I_D
V_{DS}
Depletion region
p type
L
I_D
V_{DS}

(a)

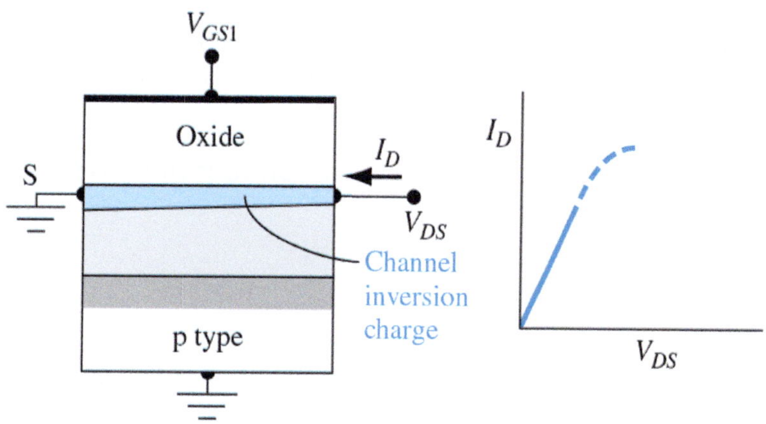
V_{GS1}
Oxide
S
I_D
V_{DS}
Channel inversion charge
p type
I_D
V_{DS}

(b)

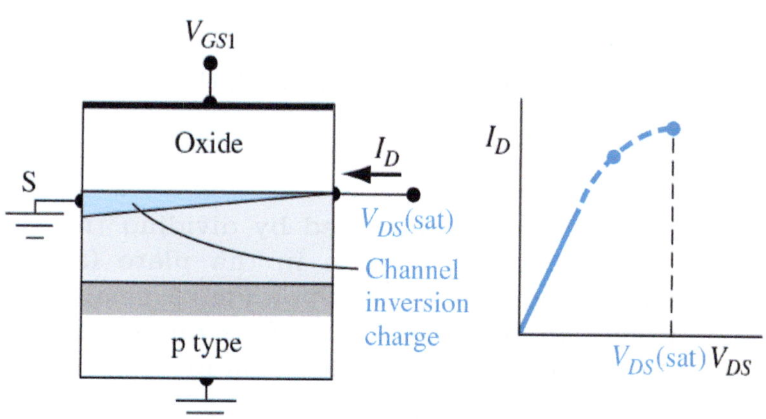
V_{GS1}
Oxide
S
I_D
V_{DS}(sat)
Channel inversion charge
p type
I_D
V_{DS}(sat) V_{DS}

(c)

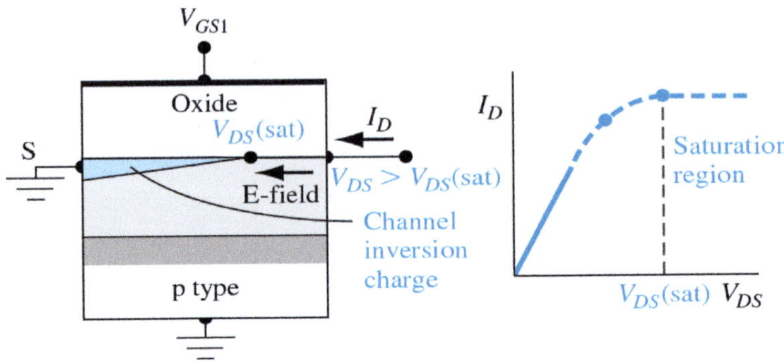

(d)

Figure 6.12. Cross section and I_D versus V_{DS} curve when $V_{GS} < V_T$ for (a) a small V_{DS} value, (b) a larger V_{DS} value, (c) a value of $V_{DS} = V_{DS}$ (sat), and (d) a value of $V_{DS} > V_{DS}$(sat).

Source: https://www.optima.ufam.edu.br/SemPhys/Downloads/Neamen.pdf.

6.4.3. Transconductance

The electrical property that connects a device's output current to its input voltage is known as transconductance (or transfer conductance), sometimes referred to as mutual conductance.

Transconductance is very often denoted as a conductance, g_m, with a subscript, m, for mutual. It is defined as follows:

$$g_m = \frac{\Delta I_{\text{out}}}{\Delta V_{\text{in}}}$$

For small signal alternating current, the definition is simpler:

$$g_m = \frac{i_{\text{out}}}{v_{\text{in}}}$$

The SI unit for transconductance is the siemens, with the symbol S, as in conductance.

6.4.3.1. Vacuum Tubes

Transconductance for vacuum tubes is calculated by dividing the change in the grid/cathode voltage by the corresponding change in the plate (anode) current, while maintaining a constant plate (anode) to cathode voltage. A small-signal vacuum tube's typical g_m values range from 1 to 10 millisiemens. It is one of the three characteristic constants of a vacuum tube, along with the plate resistance (r_p or r_a) and gain (μ) (mu).

The Van der Bijl equation defines their relation as follows:

$$g_m = \frac{\mu}{r_p}$$

6.4.3.2. Field-effect transistors

Comparably, transconductance in field-effect transistors, and MOSFETs specifically, is calculated by dividing the change in drain current by the slight variation in gate-source voltage while maintaining a constant drain-source voltage. A small-signal field-effect transistor's typical g_m values range from 1 to 30 millisiemens.

Using the Shichman–Hodges model, the transconductance for the MOSFET can be expressed as

$$g_m = \frac{2I_D}{V_{OV}},$$

where I_D is the DC drain current at the bias point, and V_{OV} is the overdrive voltage, which is the difference between the bias point gate–source voltage and the threshold voltage ((i.e., $V_{OV} \equiv V_{GS} - V_{th}$). The overdrive voltage (sometimes known as the effective voltage) is customarily chosen at about 70–200 mV for the 65 nm technology node ($I_D \approx 1.13$ mA/µm of width) for a g_m of 11–32 mS/µm.

Additionally, the transconductance for the junction FET is given by

$$g_m = \frac{2I_{DSS}}{|V_P|}\left(1 - \frac{V_{GS}}{V_P}\right),$$

where V_P is the pinchoff voltage, and I_{DSS} is the maximum drain current.

6.4.3.3. Bipolar transistors

Bipolar small-signal transistors have a highly variable g_m that is proportional to the collector current. Its usual range is between 1 and 400 millisiemens. The output is a change in collector current flowing between the collector and emitter with a constant collector/emitter voltage. The input voltage change is applied between the base and emitter.

The transconductance for the bipolar transistor can be expressed as

$$g_m = \frac{I_C}{V_T}$$

where I_C = DC collector current at the Q-point, and V_T = thermal voltage, typically about 26 mV at room temperature. For a typical current of 10 mA, $g_m \approx 385$ mS. The input impedance is the current gain (β) divided by the transconductance.

The output (collector) conductance is determined by the Early voltage and is proportional to the collector current. For most transistors in linear operation, it is well below 100 μS.

6.4.4. Substrate Bias Effects

Substrate bias effects, prevalent in semiconductor devices, play a critical role in their performance and functionality. The substrate, or the material upon which a semiconductor is built, influences various parameters such as threshold voltage, leakage current, and overall device characteristics. One significant aspect of substrate bias effects is their impact on threshold voltage. When a substrate bias is applied, it alters the electric field within the device, thereby modifying the threshold voltage required to initiate conduction. This phenomenon is particularly crucial in MOSFETs (Metal-Oxide-Semiconductor Field-Effect Transistors), where a change in threshold voltage can significantly affect switching speed and power consumption. Additionally, substrate bias effects influence leakage current, which refers to the unwanted flow of current in a device when it is in the off-state. By manipulating substrate bias, engineers can mitigate leakage current, improving device efficiency and reducing power consumption. Furthermore, substrate bias effects play a role in device reliability and radiation tolerance. Properly biased substrates can enhance a device's resilience to radiation-induced faults, crucial in applications such as aerospace and nuclear environments. However, excessive substrate bias can also lead to detrimental effects such as hot carrier injection and substrate-induced noise, impacting device lifespan and performance. In conclusion, substrate bias effects are fundamental considerations in semiconductor device design and operation, influencing parameters ranging from threshold voltage and leakage current to reliability and radiation tolerance. Understanding and appropriately managing these effects are paramount in optimizing device performance and ensuring reliability across various applications.

6.5. FREQUENCY LIMITATIONS

In many applications, the MOSFET is used in a linear amplifier circuit. A small-signal equivalent circuit for the MOSFET is needed in order to mathematically analyze the electronic circuit. The equivalent circuit contains capacitances and resistances that introduce frequency effects.

Frequency limitations are inherent constraints that affect the performance of electronic circuits and systems, particularly in high-speed applications such as communication and signal processing. These limitations arise due to various factors including device characteristics, circuit topology, and external factors such as parasitic elements and environmental conditions.

One significant factor contributing to frequency limitations is the intrinsic properties of electronic components. For instance, in active devices like transistors, there is a limit to how quickly they can switch between on and off states, known as the transit frequency or fT. This limit is determined by the device's physical structure and material properties. Similarly, passive components such as resistors, capacitors, and inductors also exhibit frequency-dependent behavior due to factors like parasitic capacitance and inductance, which can attenuate or distort high-frequency signals.

Circuit topology and design also impose constraints on frequency response. Complex circuits with multiple stages or feedback loops may introduce phase shifts, instability, or resonance phenomena that limit the usable bandwidth. Moreover, interconnects and PCB layout can introduce parasitic effects such as stray capacitance and inductance, degrading signal integrity and imposing upper limits on operating frequency.

External factors like temperature, humidity, and electromagnetic interference can further impact circuit performance, especially at high frequencies. Temperature variations can alter component characteristics, while environmental factors can introduce noise and distortion, reducing the effective operating range. Mitigating frequency limitations often involves careful design considerations and trade-offs. Techniques such as impedance matching, frequency compensation, and signal conditioning can help extend the usable bandwidth of circuits. Additionally, advanced fabrication technologies and materials research continue to push the boundaries of device performance, enabling higher operating frequencies and improved signal integrity.

6.5.1. Small-Signal Equivalent Circuit

A small-signal equivalent circuit is a simplified representation of a nonlinear electronic device or circuit that facilitates analysis under small-signal conditions. In this model, the device or circuit is linearized around a quiescent operating point, assuming that small perturbations from this point can be accurately described by linear relationships.

The small-signal equivalent circuit typically consists of linearized small-signal models of individual components, replacing nonlinear elements such as transistors with linear equivalents. For example, in a transistor amplifier circuit, the transistor may be represented by its small-signal hybrid-π model, consisting of resistors, capacitors, and dependent current and voltage sources.

This simplified model allows engineers to analyze the behavior of the circuit in response to small variations in input signals, such as AC signals superimposed on DC bias voltages. By linearising the circuit, techniques from linear systems theory can be applied to analyze parameters like gain, bandwidth, impedance, and frequency response.

The small-signal equivalent circuit is particularly useful in high-frequency and nonlinear circuit analysis, where nonlinear effects can complicate analysis. By isolating small-signal behavior from large-signal effects, engineers can predict circuit performance accurately within the linear operating range.

However, it's important to note that the small-signal equivalent circuit is only valid for small variations around the quiescent operating point. Large-signal effects, such as saturation, clipping, and distortion, are not accurately captured by the small-signal model and require separate analysis techniques. Overall, the small-signal equivalent circuit is a valuable tool for analyzing linearized behavior in electronic circuits, providing insights into circuit performance under small-signal conditions and facilitating design optimization and troubleshooting.

6.5.2. Frequency Limitation Factors and Cutoff Frequency

Frequency limitation factors in electronic circuits stem from various sources, each imposing constraints on the operational range of the circuit. These factors influence the cutoff frequency, which denotes the frequency at which the circuit's response attenuates by a certain threshold, often -3 dB. Several key factors contribute to frequency limitations:

1. Device Characteristics: Active and passive components within the circuit exhibit frequency-dependent behavior. Active devices like transistors have intrinsic limitations such as transit frequency (f_T), which represents the maximum frequency at which the device can effectively amplify signals. Passive components like capacitors and inductors introduce frequency-dependent impedance, affecting the circuit's frequency response.
2. Circuit Topology: The complexity and configuration of the circuit topology impact its frequency response. Circuits with multiple stages, feedback loops, or complex signal paths introduce phase shifts, resonance, and stability issues that limit the usable bandwidth. High-order filters, amplifiers, and oscillators are particularly susceptible to these limitations.
3. Parasitic Elements: Parasitic

capacitance, inductance, and resistance inherent in components and interconnects affect the circuit's performance at high frequencies. These parasitic elements create unintended impedance paths, signal coupling, and loss mechanisms that degrade signal integrity and limit the circuit's bandwidth.

4. Transmission Line Effects: In high-frequency circuits, transmission line effects become significant, influencing signal propagation and impedance matching. Reflections, standing waves, and signal attenuation in transmission lines impose limitations on the circuit's operating frequency range.
5. Environmental Factors: External factors such as temperature, humidity, and electromagnetic interference can impact circuit performance, especially at high frequencies. Temperature variations alter component characteristics, while environmental noise and interference degrade signal quality, imposing constraints on the circuit's usable bandwidth.

The cutoff frequency (fc) of a circuit is the frequency at which its response attenuates by -3 dB compared to its low-frequency (DC) value. It represents the boundary between the circuit's usable and attenuated frequency range. Engineers often design circuits with cutoff frequencies well above the intended operating frequency to ensure adequate performance within the desired bandwidth.

Understanding and mitigating frequency limitation factors are crucial in designing high-performance electronic circuits, particularly in applications requiring wide bandwidth, low distortion, and high fidelity signal processing. Techniques such as impedance matching, frequency compensation, and component selection help optimize circuit performance and extend the usable frequency range while minimizing unwanted effects.

6.6. THE CMOS TECHNOLOGY

The foundation of contemporary integrated circuits is Complementary Metal-Oxide-Semiconductor (CMOS) technology, which makes it possible to create increasingly high-performing, energy-efficient, and compact electronic devices. The complementary p-type and n-type MOSFETs (Metal-Oxide-Semiconductor Field-Effect Transistors) are the core of CMOS, enabling low-power operation and strong noise immunity. A thin layer of oxide is deposited onto a semiconductor substrate to begin the fabrication process. Next, precise doping and etching steps are used to create the source/drain regions and gate electrodes. A single chip with CMOS technology can have millions or billions of transistors integrated due to its scalability, low static power consumption caused by the tiny current flow that occurs when there are no signal transitions, and high noise margins because of the complementary nature of its logic gates. These characteristics ensure that CMOS is widely used in contemporary electronic systems for a variety of applications, including analog and mixed-signal circuits, microprocessors, memory chips, and more (Sung-Mo Kang & Yusuf Leblebici (2002)).

But for one widely used MOS technology, understanding fundamental properties of these circuits and devices requires taking into account the fundamental fabrication techniques. The complementary MOS, or CMOS, process is the MOS technology that we briefly discuss. The physics of MOSFETs with both n- and p-channel enhancement modes have been examined. A CMOS inverter, the building block of CMOS digital logic circuits, uses both devices. A complementary p-channel and n-channel pair can be used to drastically lower the DC power dissipation in a digital circuit. To accommodate the n- and p-channel transistors in an integrated circuit, electrically isolated p- and n-substrate regions must be formed. For CMOS circuits, the p-well process has been a widely used method. The n-type silicon substrate, on which the p-channel MOSFET will be fabricated, is initially relatively low doped. The process creates a diffused p region known as a p-well, where the n-channel MOSFET will be built. To achieve the intended threshold voltages, the p-type substrate doping level typically needs to be greater than the n-type substrate doping level. To form the p-well, the larger p doping can readily offset the initial n doping. Figure 6.13a depicts a simplified cross-section of the p-well CMOS structure. Field oxide, a comparatively thick oxide that separates the devices, is represented by the notation FOX. The FOX aids in preserving isolation between the two devices by preventing the n or p substrate from inverting. Providing connections to enable the p-well and n-substrate to be electrically connected to the proper voltages is one example of the additional processing steps that must be included in practice. This pn junction will always be reverse biased since the n substrate needs to be at a higher potential than the p-well.

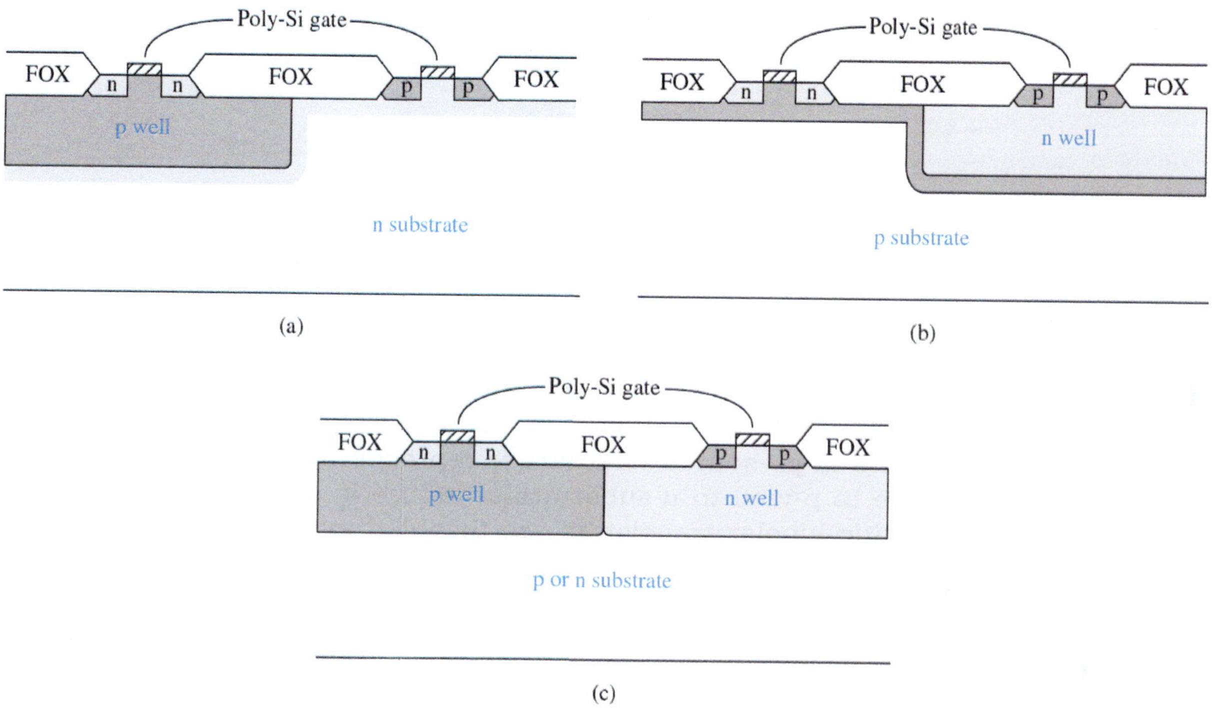

Figure 6.13. CMOS structures: (a) p well, (b) n well, and (c) twin well.

Source: https://www.optima.ufam.edu.br/SemPhys/Downloads/Neamen.pdf.

It is now common practice to use both the n-well and twin-well CMOS processes for threshold voltage control, thanks to the widespread use of ion implantation. An optimized p-type substrate is the first step in the n-well CMOS process, which is depicted in Figure 6.13b, and is used to create n-channel MOSFETs. Given the superior characteristics of n-channel MOSFETs in general, this starting point should result in excellent n-channel devices. After that, the n-well is added, where the p-channel devices are made. Ion implantation is a controllable method for n-well doping.

The dual-well CMOS process, as depicted in Figure 6.13c, enables optimal doping of both the p-well and n-well regions to effectively control the threshold voltage and transconductance of individual transistors. This process also allows for higher packing density due to self-aligned channel stops. A significant issue in CMOS circuits has been latch-up, which involves a high-current, low-voltage scenario that can occur in a four-layer pnpn structure. Figure 6.14a illustrates the CMOS inverter circuit, while Figure 6.14b shows a simplified layout of the integrated circuit. The CMOS layout involves a p^+

source to n substrate to p-well to n^+ source configuration, creating the aforementioned four-layer structure.

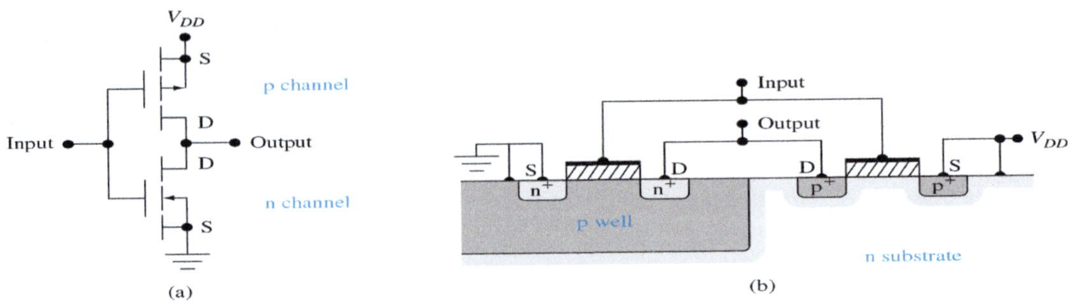

Figure 6.14. (a) CMOS inverter circuit. (b) Simplified integrated circuit cross section of CMOS inverter.

Source: https://www.optima.ufam.edu.br/SemPhys/Downloads/Neamen.pdf.

Figure 6.15 depicts this four-layer structure's equivalent circuit. The parasitic pnp and npn bipolar transistors interact to produce the silicon-controlled rectifier action. The lateral p-well to n-substrate to p+-source structure is represented by the pnp transistor, and the vertical n+-source to p-well to n-substrate structure is represented by the npn transistor. The two parasitic bipolar transistors are disabled in CMOS operation under normal conditions. Avalanche breakdown, on the other hand, could happen in the p-well to substrate junction in specific circumstances, pushing both bipolar transistors to saturation. Positive feedback allows this high-current, low-voltage condition—known as latch-up—to persist. The disorder has the potential to permanently harm and burn out the CMOS circuit in addition to making it impossible for it to function.

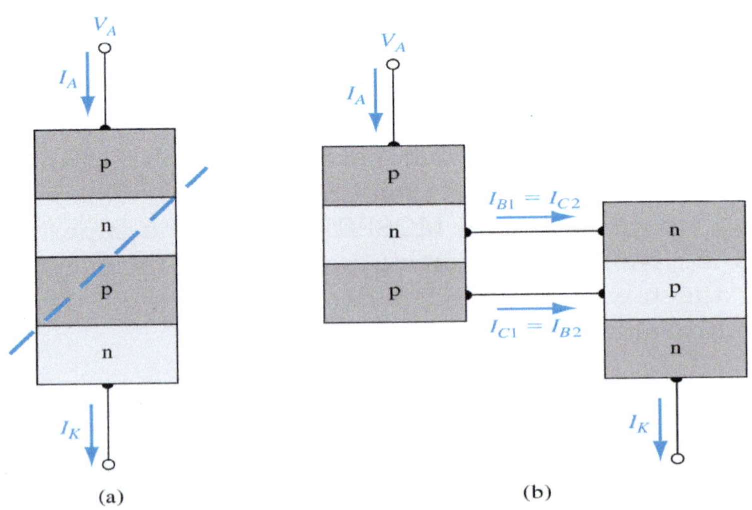

Figure 6.15. (a) The splitting of the basic pnpn structure. (b) The two-transistor equivalent circuit of the four-layered pnpn device.

Source: https://www.optima.ufam.edu.br/SemPhys/Downloads/Neamen.pdf.

Latch-up can be prevented if the product is less than unity at all times, where the common-emitter current gains of the npn and pnp parasitic bipolar transistors are, respectively. One method of preventing latch-up is to "kill" the minority carrier lifetime. Gold doping or neutron irradiation are two methods for achieving minority carrier lifetime degradation, both of which create deep traps inside the semiconductor. The current gain is decreased and the excess minority carrier recombination rate is increased by the deep traps. Latching up can also be avoided by employing appropriate circuit layout techniques. Latch-up can be reduced or avoided if the two bipolar transistors can be successfully decoupled. It is also possible to decouple the two parasitic bipolar transistors by employing an alternative fabrication technique. For example, the n-channel and p-channel MOSFETs can be separated from one another by an insulator thanks to silicon-on-insulator technology. By means of isolation, the parasitic bipolar transistors are separated.

CASE STUDY

METAL–SEMICONDUCTOR CONTACT FOR SCHOTTKY BARRIER TRANSISTORS: FEW-LAYER BLACK PHOSPHORUS

Two-dimensional (2D) materials have shown great potential in the application of nanotransistors, especially for beyond 5 nm node technology. Among thousands of 2D materials, black phosphorus (BP) has triggered intensive research interests owing to its unique material properties. Depending on the number of layers, the band gap of BP varies from 0.35 to 2.0 eV. The hole Hall mobility of BP is as high as 5200 cm^2/V s at room temperature with hexagonal boron nitride (h-BN) passivation. BP has a puckered honeycomb atomic structure, which leads to its highly anisotropic transport characteristics. The moderate direct band gap and high carrier mobility make BP a strong candidate for high-performance transistor applications. However, field-effect transistors (FETs) based on most 2D semiconductors are Schottky barrier transistors. Namely, the transistor characteristics are significantly affected by the source/drain Schottky contacts. This phenomenon is more obvious for short-channel devices, where the contact resistance is even more dominant than the channel resistance. It is thus very important to form a low-resistance metal-semiconductor contact to fully access the intrinsic material properties of the channel. Typically, the contact resistance of 2D semiconductor FETs includes three parts: (i) the Schottky barrier resistance (R_{sb}), which is the result of Fermi-level pinning and the difference between the metal work function and the electron affinity of the semiconductor; (ii) the tunneling resistance (R_t) owing to the existence of a physical gap between the metal and the semiconductor. The physical gap may stem from any interfacial oxide; and (iii) the interlayer resistance (R_{inter}) contacting from top layers to bottom layers. R_{inter} is usually much smaller than R_{sb} and R_t. However, in 2D FETs, R_{inter} cannot be ignored, especially for bottom-gate devices owing to a significantly higher out-of-plane effective mass, which is usually several times larger than the in-plane effective mass.

One disadvantage of BP is that it easily reacts with O_2 and H_2O in an ambient environment, forming an oxide after cleaving from the bulk. The existence of this phosphorus oxide/acid at the surface makes the formation of a clean and low-resistance BP-metal interface complicated. This leads us to believe that much better contacts can be achieved with a clean metal-BP interface by preventing the formation of phosphorus oxide/acid. In this work, we demonstrate how to minimize the total contact resistance ($R_{total} = R_{sb} + R_t + R_{inter}$) using several strategies, including high-work-function contact metal, BN passivation, and top-gate structure. This allows the BP contact resistance to be reduced to a record low value of 0.58 kΩ·μm, resulting in an Ion that exceeds 940 μA/μm. As a consequence, the Ioff is also increased significantly in these low-resistance contact devices. To understand the abnormally high I_{off}, R_t is intentionally increased by adding a thin BN barrier to the contact. It is found that the high I_{off} is due to the reverse electron tunneling current on the drain side, which can be suppressed by the BN tunneling barrier. Finally, by using an asymmetric contact structure with a clean metal-BP contact at the source and BN tunneling barrier at the drain, we have successfully reduced I_{off} by a factor of 120 with only a 20% reduction in Ion.

Device Fabrication

Figure 1 shows the device fabrication flow for fabricating top gate BP FETs. Few-layer BP flakes were exfoliated onto the 90 nm SiO_2/p^{++} Si substrate. The BN thin film was grown on sapphire by metal-organic chemical vapor deposition (MOCVD) with a thickness of 1.6 nm and a root-mean-square roughness of 0.1 nm. The relative dielectric constant of BN is about 3, which is similar to the reported value of CVD BN. MOCVD BN was first transferred onto the BP-SiO_2-Si substrate (details in Materials and Methods). The source/drain region was patterned by electron-beam lithography. After that, BN was etched by a low-power Ar reactive ion etching (RIE) with an etching rate of ~1 nm/10 s. The BN thickness can be controlled by adjusting the Ar etching time. Figure 2 shows the atomic force microscopy (AFM) image of BN etching pattern after Ar RIE for 10 s. The roughness of BN increases from 0.1 nm to about 0.5 nm after RIE. Immediately after the etching process, 5 nm Pt/8 nm Ni/30 nm Al was deposited as the contact metals. A second dielectric layer, 4 nm Al_2O_3, was deposited by atomic layer deposition (ALD) at 200 °C after the metal lift-off process.

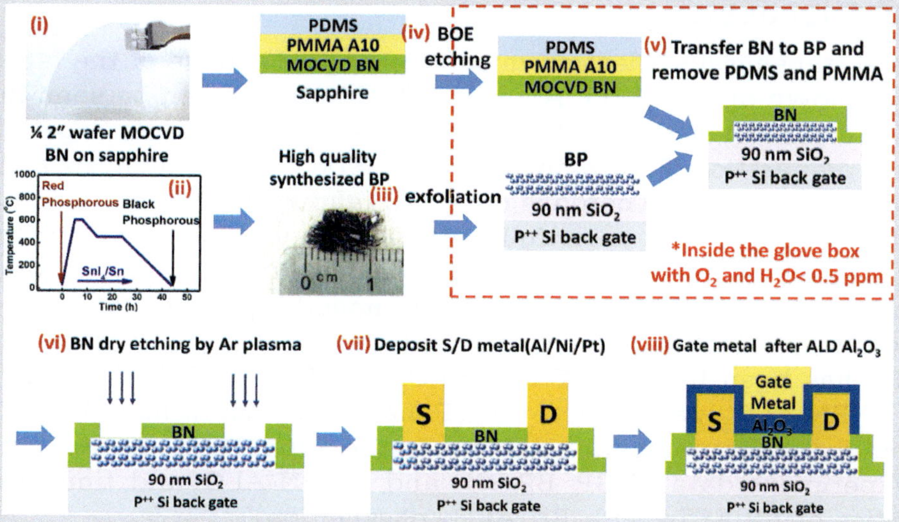

Figure 1. *Fabrication process flow for the top-gate BP FETs with BN–Al_2O_3 gate dielectric/passivation layers. Key steps include (i) MOCVD BN, (ii) bulk BP growth, (iii) BP exfoliation, (iv) BN release, (v) BN transfer, (vi) BN dry etching, (vii) S/D metallization, and (viii) ALD Al_2O_3 and gate metallization. Step (iii) BP exfoliation and step (v) BN transfer were performed in a glovebox with O_2/H_2O concentration less than 0.5 ppm.*

Source: https://engineering.purdue.edu/~yep/Papers/ACS%20Omega_Asymmetric%20Contacts_Lingming%20Yang_2017.pdf

The total equivalent oxide thickness of the BN-Al_2O_3 top-gate dielectric layer is around 4 nm. Finally, 20 nm Ti/50 nm Au was deposited as the top-gate metal. All devices were fabricated with the current flow along the high-mobility armchair direction of BP, which was determined by polarization-dependent Raman spectra using a HORIBA LabRAM HR800 Raman spectrometer with a 532.8 nm wavelength He-Ne laser. All devices have the same channel length of 200 nm and a thickness between 4 and 12 nm. The characterization of FETs was performed in air at room temperature by using a Keithley 4200 semiconductor parameter analyzer.

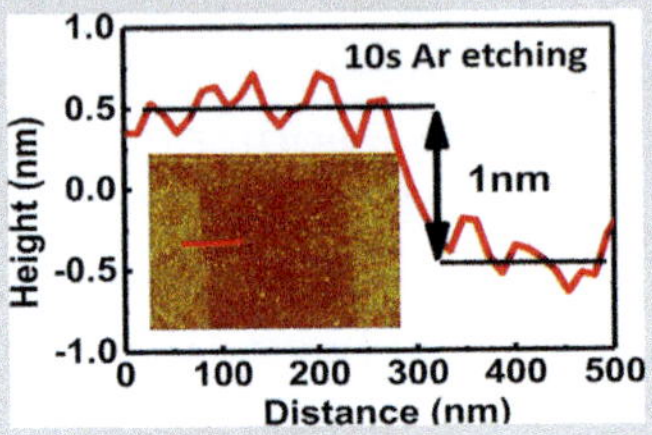

Figure 2. *AFM height profile and inset image of BN etched pattern after Ar RIE for 10 s.*

Source: https://engineering.purdue.edu/~yep/Papers/ACS%20Omega_Asymmetric%20Contacts_Lingming%20Yang_2017.pdf

Results and Discussion

A cross-sectional transmission electron microscopy (TEM) image of the fabricated Al-Ni-Pt-BP contact is shown in Figure 3a. A sharp BP-Pt interface can be seen in the picture, although the 2D crystal structure of BP is vague because of the damage caused during the TEM sample preparation. Figure 3b shows the EDS element analysis along the Al-Ni-Pt-BP-SiO_2 stack. No nitrogen or boron is detected (above the detection limit), confirming that the BN barrier layer has been fully etched away. Along the stack, the carbon signal is negligible showing that the organic residue (polydimethylsiloxane (PDMS) or poly (methyl methacrylate) (PMMA)) has been removed because of the transfer process. However, a significant amount of oxygen is still detected within the phosphorus layer, even though BP has been fully isolated from O_2 and H_2O by BN-Al_2O_3 passivation. The effectiveness of BN-Al_2O_3 passivation on BP has been confirmed by time-dependent Raman and electrical measurements, as shown in the Supporting Information (accelerated degradation experiment data of BP, BP-BN, and BP-BN-Al_2O_3 and time dependence of I-V characteristics of BP FETs with BN-Al_2O_3 passivation).

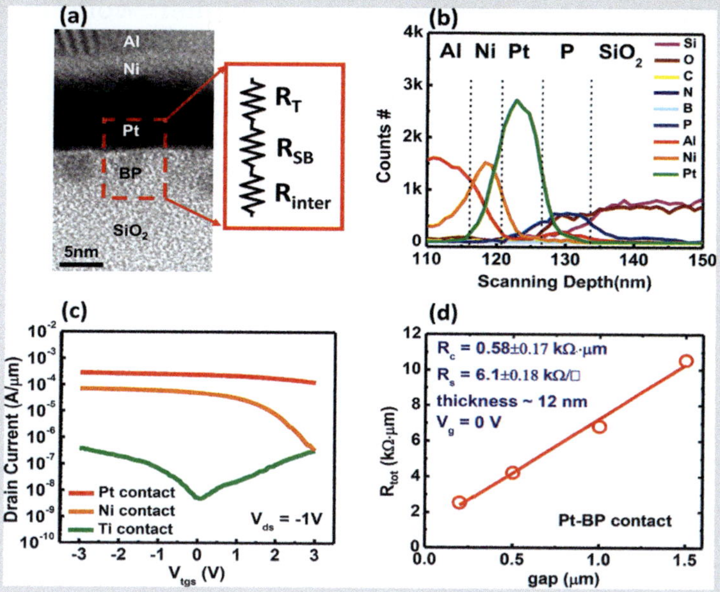

Figure 3. *(a) Cross-sectional TEM image of Pt–BP contact and the diagram of the composition of R_c. (b) Element analysis of Pt–BP contact by EDS. The signal of carbon, boron, or nitrogen is below the detection limit. (c) I–V transfer curves of BP FETs with different contact metals: Pt, Ni, and Ti. (d) TLM resistance of BP with Pt contact at zero back-gate bias. Contact resistance is extracted to be 0.58 kΩ·μm, which is seven times less than the previously reported value.*

Source: https://engineering.purdue.edu/~yep/Papers/ACS%20Omega_Asymmetric%20Contacts_Lingming%20Yang_2017.pdf

Meanwhile, the I-V hysteresis has also been reduced to 0.25 V after BN-Al_2O_3 passivation. As a result, it is very likely that the oxygen is introduced because of exposure to air between the focused ion beam processing and TEM imaging. If oxygen contamination occurred during exfoliating and processing, a higher oxygen concentration would be expected at the Pt-BP interface than at the BP-SiO_2 interface.

The selection of a contact metal with the appropriate work function is the critical first step to reducing R_c. Three contact metals with a range of work functions were investigated: Pt (~5.6 eV), Ni (~5.2 eV), and Ti (~4.33 eV). Figure 3c shows the transfer curves of BP FETs with different source/drain metals. Clearly, the FET with BP-Pt contacts performed the best, whereas BP-Ti performed the worst. Interestingly, the device with Ti contacts shows symmetric electron and hole transport, which indicates that the Fermi level at the BP surface and the work function of Ti are aligned near the middle of the band gap. This result is in good agreement with the theoretical energy band diagram of the BP-Ti contact, in which E_c and E_v of few-layer BP are about 4.1-4.2 and 4.5-4.7 eV, respectively. Notably, there is a big difference in the threshold voltage (Vt) between these devices with different contact metals. This is a result of charge transfer from the floating BP channel to the contact metal, where electrons tend to move from the high-potential region (BP channel) to the low potential region (metal).

Consequently, Vt becomes more positive when a higher work function (i.e., lower potential) metal is used.

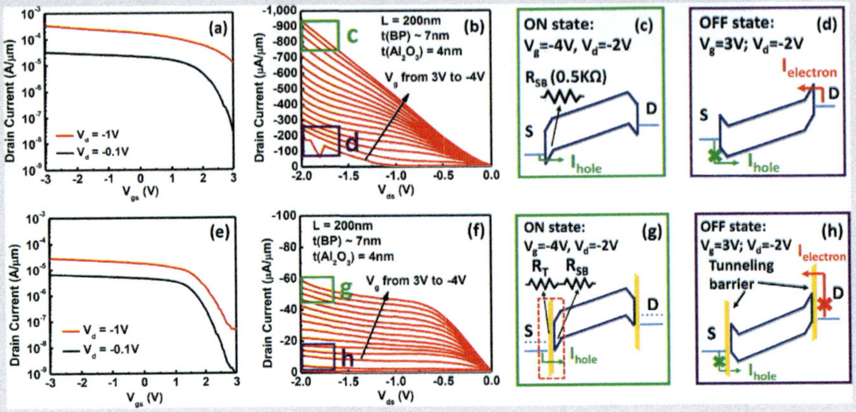

Figure 4. *(a–d) BP FETs without BN tunneling barriers at source/drain contacts and (e–h) BP FETs with BN tunneling barriers at both source/ drain contacts. (a) Transfer curves and (b) output curves of a 7 nm thick BP PMOSFET without BN tunneling barriers. Band diagram of (c) ON state and (d) OFF state for devices without BN barriers. (e) Transfer curves and (f) output curves of a 7 nm thick BP PMOSFET with two-layer BN tunneling barriers. Band diagram of (g) ON state and (h) OFF state for devices with BN barriers.*

Source: https://engineering.purdue.edu/~yep/Papers/ACS%20Omega_Asymmetric%20Contacts_Lingming%20Yang_2017.pdf

The contact resistance of BP-Pt contacts is measured with the transfer length method (TLM) on a 12 nm thick sample. At zero back-gate bias, the extracted Rc is about 0.58 kΩ·μm, and the sheet resistance is about 6.1 kΩ/square as shown in Figure 3d. Encouragingly, the measured R_c of the clean BP-metal contact is 1/7 of the previously reported R_c for BP-Pd contacts. Because Pt and Pd have similar work functions, the Schottky barrier height and R_{sb} of the two contacts should be similar. Thus, the significant reduction of total R_c is due to the contribution of the other resistance factors: R_t and R_{inter}. The transfer and output characteristics of a 7 nm thick BP FET with Pt contact are shown in Figure 4a, b. Owing to the improvement of R_c, the 200 nm channel length device shows Ion as high as 940 μA/μm at a V_{ds} of -2 V. To the best of our knowledge, this is the highest Ion achieved among all 2D semiconductor-based FETs. However, the I_{on}/I_{off} is about ~103 at small drain bias (-0.1 V) and decreases to about ~20 at large drain bias (-1 V). The I_{off} increases to 200 μA/μm at a drain bias of -2 V, and the drain current in Figure 4b increases almost linearly with V_{ds} and does not saturate. The abnormally high I_{off} and nonsaturation phenomena have never been reported for BP FETs. This phenomenon can be explained as follows. First, few-layer BP has a small band gap approaching the bulk value of 0.35 eV, and BP FETs are Schottky barrier transistors, that is, the observed electrical characteristics are the outcomes of both the channel and more importantly the contacts. In principle, the total drain current of a Schottky barrier transistor contains (i) hole current from source to drain and (ii) electron current from drain to source. As can be seen from the band diagram in Figure 4c, in the ON state, the majority of the drain current is the hole current injected from the source terminal. As long as the holes are able to overcome the source Schottky barrier, they can be collected at the drain side. In other words, the total current level is mainly determined by the hole injection at the source. In the OFF state, the channel potential is pulled down by the gate voltage to reduce the hole injection at the source; however, this gate bias also lowers the potential of the channel region near the drain terminal. Meanwhile, the drain potential is lifted up by the negative drain bias, which results in a steep potential gradient. As a result, the electrons at the drain terminal can be injected into the conduction band of a channel by tunneling through the triangular barrier. In other words, by turning off the forward hole current, the reverse electron current is turned on by top-gate bias. Indeed, this phenomenon is universal for all top-gate Schottky barrier transistors. However, the reverse tunneling probability at the drain side is inversely related to the Schottky barrier height for minority carriers. Consequently, a high reverse tunneling current is observed in narrow bandgap semiconductors. For wide bandgap 2D semiconductors, such as MoS_2, the drain-to-channel Schottky barrier for holes is so large that the tunneling current can be ignored in the OFF state. So far, the device characteristics have been well explained by the transport model. However, for narrow bandgap BP devices, this raises important questions: what happens in the BP devices with both low I_{off} and Ion as reported in the literature and where does this inconsistency come from? A possible explanation for the absence of a high reverse tunneling current in BP devices is the presence of a tunneling barrier at the source/drain interface. This additional tunneling barrier may stem from surface phosphorus oxide/acid formed during fabrication. The presence of this tunneling barrier would explain why previous works reported much higher Rc than what we are reporting in this work. On the other hand, this tunneling barrier blocks the reverse electron tunneling current, which leads to a low I_{off}. To verify this hypothesis, we fabricated BP FETs with a thin layer of BN intentionally kept at the source/drain contacts. Figure 4e, f shows the transfer and output characteristics of a BP FET with a thin BN barrier (two layers) added to the source/drain contacts. The geometry of this device is the same as the one without a BN barrier. However, there are several clear differences between the electrical characteristics of these two devices: (i) in the ON state, Ion of the device with BN contacts is 10 times lower than the device without BN. The reduction of Ion is due to the additional tunneling barrier resistance R_t. (ii) In the linear region, the I_d–V_d curve of the device with BN is more linear than the device without BN owing to the Fermi-level depinning. Figure 4g shows the band diagram of the device with BN contacts in the ON state. The Schottky barrier height becomes smaller with Fermi-level depinning, and the Id-Vd curves become more linear. However, Rt increases significantly with insertion of a BN barrier in contacts. As a result, the total R_c (R_t + R_{sb} + R_{inter})of the device with BN is much larger than that of the device without BN. This also tells us that the linear Id-Vd curves do not necessarily indicate that an Ohmic contact has been achieved. (iii) In the saturation region, the current saturates at V_{ds} < -0.5 V. Below this voltage, the drop across the tunneling barrier becomes dominant. Increasing V_d no longer improves the drain current. (iv) In the OFF state, I_{off} is reduced by a factor of 400 at a V_{ds} of -1 V due to the presence of the BN barrier at the drain terminal, which reduces the reverse electron tunneling current. Figure 4h shows the band diagram of the device with BN contacts in the OFF state. In short, the presence of a tunneling barrier at source/drain contacts reduces I_{on}, leads to linear Id-V (Fermi-level depinning), leads to current saturation (V_d drop at contacts), and reduces Ioff (blocks the reverse tunneling current). All of these observations from our BP devices with BN barriers are in good agreement with the characteristics of the previously reported BP devices, which did not use any passivation method to avoid the formation of phosphorus oxidation at contacts. As a result, for Schottky barrier transistors, it is extremely important to distinguish between the contribution from the contacts and the intrinsic channel to the overall electrical characteristics.

Taking advantage of the benefits of having a clean BP-metal contact (high I_{on}) and a BN barrier contact (low I_{off}), we have demonstrated an asymmetric source/drain contact structure to improve the I_{off}, while keeping the I_{on} as high as possible. It is well-known that in FETs I_{on} is mainly controlled by the electrostatics at the source terminal. The asymmetry between source and drain can be used to suppress either the electron or the hole current, depending on the sign of the drain voltage. For BP Schottky barrier transistors, the majority current (i.e., forward hole current) is determined by the potential barrier from metal to the source contact region.

The reverse electron current can be suppressed by adding a tunneling barrier at the drain terminal without losing too much hole current. Figure 5a-c shows the schematic diagram and measurement configuration for a BP FET without BN barriers (device A) and a BP FET with only one BN barrier (~bilayer thickness) at the drain (device B) or the source (device B'). Device A and device B(B') were fabricated from the same BP flake to reduce potential variability between flakes.

The output and transfer curves of device A are shown in Figure 5d, g. The I_{on} and I_{off} are about 638 and 84 μA/μm, respectively, at a V_{ds} of −1.6 V for the device without BN. With a BN tunneling barrier at the drain side, the I_{off} is reduced by a factor of 120 to 700 nA/μm at a V_{ds} of −1.6 V, whereas I_{on} is still about 80% of the device without the BN barrier. Because I_{on} is mainly determined by the source contact, a BN barrier at the drain does not have a significant impact. Figure 5f,i shows the output and transfer curves of device B′ that has a BN barrier at the source terminal. As expected, the device shows a much smaller I_{on} of 153 μA/μm and has current saturation as well. We conclude that the reverse electron tunneling current or high I_{off} current of BP FETs can be significantly reduced by an asymmetric source/drain contact structure and contact engineering.

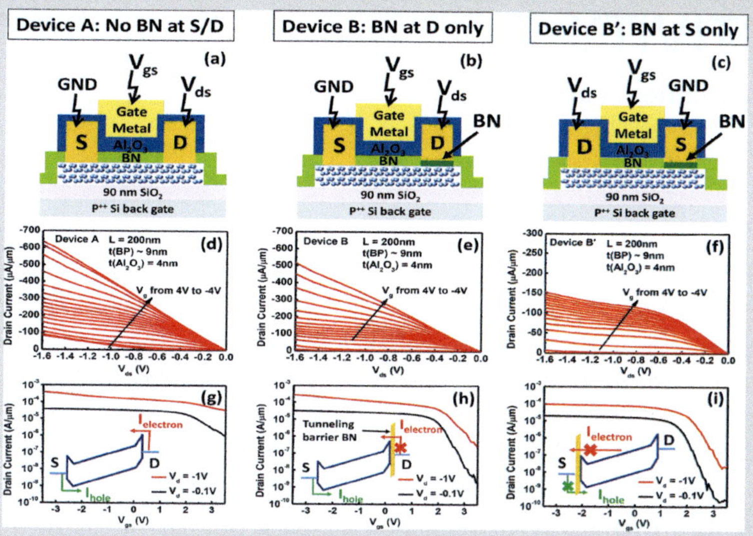

Figure 5. *Device A: no BN tunneling barrier at source/drain. Device B: with bilayer BN tunneling barrier only at drain. Device B′: with BN tunneling barrier only at source. Schematic diagram for (a) device A, (b) device B, and (c) device B′. Output curves for (d) device A, (e) device B, and (f) device B′. Transfer curves and band diagrams for (g) device A, (h) device B, and (i) device B′. Device A and device B(B′) were made from the same BP flake.*

Source: https://engineering.purdue.edu/~yep/Papers/ACS%20Omega_Asymmetric%20Contacts_Lingming%20Yang_2017.pdf

In general, the strategies used in this work can be applied to other Schottky-type FETs. The strategies include BN or other effective passivation for environmentally sensitive materials; choose contact metals with proper work function for either electron or hole transport; and control/suppress the electron or hole current by tuning the contact tunneling barrier. As a case study on BP, we have successfully utilized these methods to improve the electrical performance of BP FETs.

Conclusion

To summarize, the BP–metal contact has been systematically studied and improved, resulting in record low contact resistance. As a result, record high I_{on} of BP FETs has also been obtained. However, an abnormally high I_{off} of BP FETs is also identified owing to the reverse electron tunneling current at the drain terminal. By contact engineering, the I_{off} has also been significantly reduced with minor sacrifice in I_{on}. All of these results have shown the importance of contacts for its electrical characteristics in Schottky barrier transistors.

CLASS ACTIVITY

1. Identify these schematic symbols:

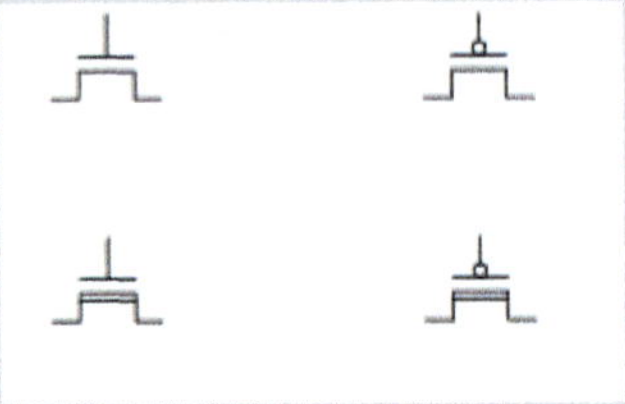

2. Bipolar junction transistors (BJTs) are considered "normally-off" devices, because their natural state with no signal applied to the base is no conduction between emitter and collector, like an open switch. Are insulate-gate field-effect transistors (IGFETs) considering the same? Why or why not?
3. Metal Oxide Field-Effect Transistors (MOSFETs) differ in behavior from Bipolar Junction Transistors (BJTs) in several ways. Address each one of these behavioral aspects in your answer:

SUMMARY

- The heart of the MOSFET is the MOS capacitor. The energy bands in the semiconductor adjacent to the oxide-semiconductor interface bend, depending on the voltage applied to the gate.
- An inversion layer of electrons can be created at the oxide-semiconductor surface in a p-type semiconductor by applying a sufficiently positive gate voltage, and an inversion layer of holes can be created at the oxide-semiconductor surface in an n-type semiconductor by applying a sufficiently negative gate voltage.
- The n-channel MOSFET, both enhancement mode and depletion mode, and the p-channel MOSFET, both enhancement mode and depletion mode.
- The basic transistor action is the modulation of the current at the drain terminal by the gate-to-source voltage.
- The intrinsic Fermi level at the surface is now below the Fermi level. The conduction band at the surface is now close to the Fermi level, whereas the valence band is close to the Fermi level in the bulk semiconductor.
- A depletion region forms instantaneously across a p-n junction. It is most easily described when the junction is in thermal equilibrium or in a steady state: in both of these cases, the properties of the system do not vary in time; they are in dynamic equilibrium.
- N-type semiconductor has an excess of free electrons (in the conduction band) compared to the P-type semiconductor, and the P-type has an excess of holes (in the valence band) compared to the N-type.
- The threshold voltage, commonly abbreviated as V_{th} or $V_{GS(th)}$, of a field-effect transistor (FET) is the minimum gate-to-source voltage (VGS) that is needed to create a conducting path between the source and drain terminals. It is an important scaling factor to maintain power efficiency.
- The current in a MOSFET is due to the flow of charge in the inversion layer or channel region adjacent to the oxide-semiconductor interface.

REVIEW QUESTIONS

1. Describe what is meant by an inversion layer of charge.
2. Describe how an inversion layer of charge can be formed in a MOS capacitor with a p-type substrate.
3. Why does the space charge region in the semiconductor of a MOS capacitor reach a maximum width once the inversion layer is formed?
4. Sketch the energy-band diagram through a MOS structure with a p-type substrate and an n polysilicon gate under zero bias.
5. Define the flat-band voltage. Sketch the energy-band diagram in a MOS capacitor at flat band.

6. What is the effect on the C–V characteristics of a MOS capacitor with a p-type substrate if the amount of positive trapped oxide charge increases?
7. Discuss why the threshold voltage changes when a reverse-biased source-to-substrate voltage is applied to a MOSFET.

REFERENCES

1. Baker, R. J. (2010). *CMOS Circuit Design, Layout, and Simulation* (3rd ed.). New York: Wiley-IEEE.
2. Delaurenti, M. (1999). *Design and Optimization Techniques of High-Speed Vlsi Circuits* (Ph.D. dissertation). Archived 2014-11-10 at the Wayback Machine.
3. Dimitrijev, S. (2006). *Principles of Semiconductor Devices*. New York: Oxford University Press.
4. Hilibrand, J., & Gold, R. D. (1960). Determination of the impurity distribution in junction diodes from capacitance-voltage measurements. *RCA Review, 21*, 245.
5. Hu, C. C. (2010). *Modern Semiconductor Devices for Integrated Circuits*. Upper Saddle River, NJ: Pearson Prentice Hall.
6. Kang, S.-M., & Leblebici, Y. (2002). *CMOS Digital Integrated Circuits: Analysis & Design*. McGraw–Hill Professional.
7. Kano, K. (1998). *Semiconductor Devices*. Upper Saddle River, NJ: Prentice Hall.
8. Muller, R. S., & Kamins, T. I. (1986). *Device Electronics for Integrated Circuits* (2nd ed.). New York: Wiley.
9. Nicollian, E. H., & Brews, J. R. (1982). *MOS Physics and Technology*. New York: Wiley.
10. Nicollian, E. H., & Brews, J. R. (2002). *MOS (Metal Oxide Semiconductor) Physics and Technology*. Wiley.
11. Ong, D. G. (1984). *Modern MOS Technology: Processes, Devices, and Design*. New York: McGraw-Hill.
12. Pierret, R. F. (1996). *Semiconductor Device Fundamentals*. Reading, MA: Addison-Wesley.
13. Roulston, D. J. (1999). *An Introduction to the Physics of Semiconductor Devices*. New York: Oxford University Press.
14. Sansen, W. M. C. (2006). *Analog Design Essentials*. Dordrecht: Springer.
15. Sedra, A. S., & Smith, K. C. (1998). *Microelectronic Circuits* (Fourth ed.). New York: Oxford University Press.
16. Taur, Y., & Ning, T. H. (2009). *Fundamentals of Modern VLSI Devices* (2nd ed.). Cambridge University Press.

CHAPTER

7 Bipolar Junction Transistor

LEARNING OBJECTIVES

After studying this chapter, you will be able to:

- Understand the general configuration and definitions of BJT
- Describe types of bipolar junction transistors
- Discuss the function of bipolar junction transistor
- Define the regions of operation
- Deal with theory and modeling of BJT
- Describe bipolar transistor biasing
- Understand the applications of bipolar junction transistor

KEY TERMS FROM THIS CHAPTER

Baker clamp
Bipolar transistors
Collector-base diode
Doped regions
Electronic amplifiers
Emitter resistors
Bias network
Circuit diagram
Collector–base junction
Ebers–Moll model
Emitter current
Gummel–Poon model

7.1. INTRODUCTION

A bipolar junction transistor (BJT) is a type of transistor in which the charge carriers are electrons and electron holes. On the other hand, a unipolar transistor, like the field-effect transistor (FET), utilizes a single type of charge carrier. A bipolar transistor can control a much larger current flowing between its terminals with a small current injected at one of its terminals, allowing the device to be used for switching or amplification.

BJTs use two p–n junctions between two semiconductor types, n-type and p-type, which are regions in a single crystal of material. Various techniques can be used to create the junctions, including doping the semiconductor material during growth, forming alloy junctions by depositing metal pellets, and diffusing n- and p-type doping substances into the crystal. The original point-contact transistor was quickly replaced by junction transistors due to their superior predictability and performance. Integrated circuits for analog and digital functions consist of diffused transistors and other components. It is very cost-effective to manufacture hundreds of bipolar junction transistors in a single circuit.

A generation of mainframes and minicomputers used integrated circuits with bipolar transistors as their primary active components; however, most computer systems today use complementary metal-oxide-semiconductor (CMOS) integrated circuits, which rely on the field-effect transistor (FET). Bipolar transistors are still used in mixed-signal integrated circuits employing BiCMOS for signal amplification and switching. Specialized types are used in high voltage switches, radio-frequency (RF) amplifiers, and high current switches.

7.2. GENERAL CONFIGURATION AND DEFINITIONS

An example of a solid-state device is the bipolar junction transistor, which has three terminals: the base terminal controls the base current flow, while the emitter and collector terminals handle the device's current flow. It differs from the other type of transistor. The input voltage controls the output current of a field-effect transistor.

There are two types of BJTs, which are three-terminal devices: PNP and NPN BJTs, respectively. Figure 7.1 shows the BJT symbols and their corresponding block diagrams. Three distinct doped regions are used in the fabrication of the BJT. In the PNP device, one n region is located between two p regions, while in the NPN device, one p region is located between two n regions. The BJT has two junctions, which mark the boundaries between the n and p regions. These junctions can be forward-biased or reverse-biased, similar to the diodes' junctions. The three terminals of the BJT are named Base (B), Collector (C), and Emitter (E).

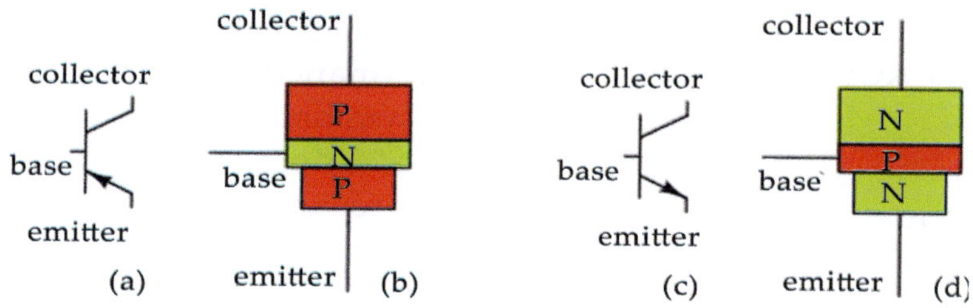

Figure 7.1. Circuit components: BJT transistor: (a) PNP schematic symbol, (b) physical layout (c) NPN symbol, (d) layout.

Source: https://www.researchgate.net/figure/Circuit-components-BJT-transistor-a-PNP-schematic-symbol-b-physical-layout-c_fig18_316098112.

The BJT, with its two junctions, has four possible states of operation because each junction has two possible states of operation (forward or reverse bias).

Both geometrically and in terms of the regions' doping concentration, the n and p regions differ from one another. As an illustration, the doping concentrations in the emitter, base, and collector could be, respectively, 10^{15}, 10^{17}, and 10^{19}. As a result, the device's behavior is not electrically symmetric, and its two ends are incompatible.

The Base-Emitter (B-E) junction is forward biased and the Base-Collector (B-C) junction is reverse biased when the voltage VBE and VC_{CB} are as indicated.

The current through the B-E junction is related to the B-E voltage as:

$$I_E = I_s\left(e^{V_{BE}/V_T} - 1\right) \quad (1)$$

The emitter and base regions have significantly different doping concentrations, causing electrons injected from the emitter region to generate emitter current I_E. Additionally, the number of electrons injected into the collector region depends on the quantity of electrons injected into the base region from the emitter region. (Brar, Sullivan; et al., 2001)

As a result, the emitter current, which depends on the B-E voltage, is connected to the collector current.

This is the basic principle of the BJT.

The collector current and the base current are related by

$$I_C = \beta I_B \quad (2)$$

And by applying KCL we obtain

$$I_E = I_C + I_B \quad (3)$$

And thus from equations (2) and (3) the relationship between the emitter and the base currents is

$$I_E = (1+\beta)I_B \quad (4)$$

And equivalently

$$I_C = \frac{\beta}{1+\beta} I_E \quad (5)$$

The fraction $\frac{\beta}{1+\beta}$ is called α.

For the transistors of interest β = 100 which corresponds to α = 0.99 and $I_C \simeq I_E$.

Unlike other transistors, the bipolar junction transistor is typically not a symmetrical device. This indicates that the transistor enters the reverse mode of operation and exits the forward active mode when the collector and emitter are switched. Since the internal structure of a transistor is typically optimized for forward-mode operation, switching the collector and emitter causes the values of α and β to be significantly smaller in reverse operation than in forward operation; frequently, the α of the reverse mode is less than 0.5. The doping ratios of the emitter and collector are the main cause of the lack of symmetry. Because the collector is lightly doped and the emitter is heavily doped, a high reverse bias voltage can be applied before the collector–base junction fails. Under typical operating conditions, the collector–base junction is reverse biased. To improve emitter injection efficiency—that is, the ratio of carriers injected by the emitter to those injected by the base—the emitter is heavily doped. The emitter must supply the majority of the carriers injected into the emitter-base junction in order to achieve high current gain.

Sometimes CMOS processes use low-performance lateral bipolar transistors that are symmetrically designed, meaning there is no difference in performance between forward and backward operation.

Significant variations in the voltage across the base-emitter terminals result in a change in the current flowing between the emitter and the collector. The input voltage or current can be amplified by using this effect. Although BJTs are more easily described as current-controlled current sources or current amplifiers because of their low base impedance, they can also be thought of as voltage-controlled current sources.

While germanium was used to make early transistors, silicon is now used to make the majority of BJTs. Nowadays, a sizable minority are also made of gallium arsenide, particularly for applications requiring extremely high speeds.

An enhancement of the BJT that can handle signals at very high frequencies up to several hundred GHz is the heterojunction bipolar transistor (HBT). In contemporary ultrafast circuits, primarily RF systems, it is typical.

Although a wide range of semiconductors may be utilized for the HBT structure, silicon–germanium and aluminum gallium arsenide are two HBTs that are frequently used. Typically, MOCVD and MBE epitaxy techniques are used to grow HBT structures.

7.3. TYPES OF BIPOLAR JUNCTION TRANSISTORS

There are two types of junction transistors and they are:

- NPN transistor
- PNP transistor

The majority charge carriers in an n-p-n junction transistor are p-type, while the majority carriers at the other two ends are n-type. However, in a p-n-p transistor, the majority carriers are p-type in the other two hands, and the minority charge carriers are n-type.

7.3.1. NPN Transistor

An NPN transistor is one that has one p-type material sandwiched between two n-type materials. The weak signal that enters the base of the NPN transistor is amplified, producing strong signals at the collector end. An electron in an NPN transistor moves from the emitter to the collector region, which is how the transistor's current is formed. Because electrons have a higher mobility than holes and make up the majority of the transistor's charge carriers, this type of transistor is frequently used in circuits.

7.3.1.1. Construction of NPN Transistor

The NPN transistor consists of two diodes connected in opposite directions. The diode connected to the left side is known as the emitter-base diode, while the diode on the right side is called the collector-base diode. These names are based on the corresponding terminals of the transistor.

The emitter, collector, and base terminals of an NPN transistor are its three terminals. The NPN transistor's middle section, which has a light doping, is the most crucial component for the transistor's operation. The collector is heavily doped, while the emitter is moderately doped.

7.3.1.2. Circuit Diagram of NPN Transistor

Figure 7.2 displays the NPN transistor's circuit diagram. Whereas the emitter and base circuit are connected in forward bias, the collector and base circuit are connected in reverse bias. To control the ON/OFF states of the transistor, the base is always in the negative supply and the collector is always connected to the positive supply.

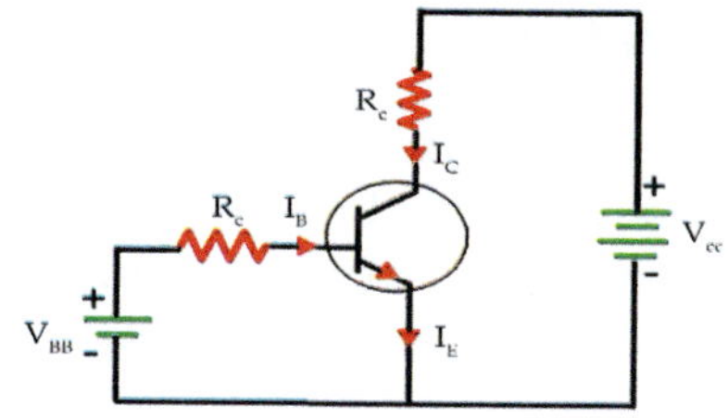

Figure 7.2. Circuit diagram of NPN transistor

Source: https://circuitglobe.com/npn-transistor.html.

7.3.1.3. Working of NPN Transistor

Figure 7.2 displays the NPN transistor's circuit diagram. The emitter-base junction receives the forward biased application, while the collector-base junction receives the reverse biased application. In comparison to the reverse bias voltage VCB, the forward biased voltage VEB is smaller.

The NPN transistor's emitter has a lot of doping. The majority of charge carriers travel toward the base when the forward bias is applied across the emitter. The emitter current IE results from this. The holes and electrons mix as they enter the P-type material.

The NPN transistor has a lightly doped base, resulting in only a small number of electrons combining while the rest form the base current IB. This base current flows into the collector region, where the reverse bias potential attracts the electrons with great force, gathering them at the collector junction.

The emitter current enters the base in its entirety. As a result, the emitter current can be defined as the total of the base and collector currents. (Gummel, H. K.; Poon, H. C. 1970)

7.3.2. PNP Transistor

Basically, the two connected diodes in this kind of PNP transistor construction are the opposite of the ones in the preceding NPN transistor. This results in a configuration of the Positive-Negative-Positive kind, with the arrow in the PNP transistor symbol pointing inward and defining the Emitter terminal.

A PNP transistor additionally has all of its polarities inverted, meaning that, in contrast to an NPN transistor, which "sources" current through its Base, it "sinks" current into its Base. For PNP transistors, holes are the more significant carriers, whereas for NPN transistors, electrons are the more significant carriers. This is the primary distinction between the two types of transistors.

Then, PNP transistors regulate a much higher emitter-collector current with a small base current and a negative base voltage. To put it another way, the Emitter of a PNP transistor is more positive than the Base and Collector combined.

7.3.2.1. A PNP Transistor Configuration

An NPN transistor's construction and terminal voltages are displayed above. With the exception of the fact that, for any one of the three possible configurations—Common Base, Common Emitter, and Common Collector—the polarities (or biasing) of the current and voltage directions are reversed, the PNP transistor's characteristics are very similar to those of their NPN bipolar cousins.

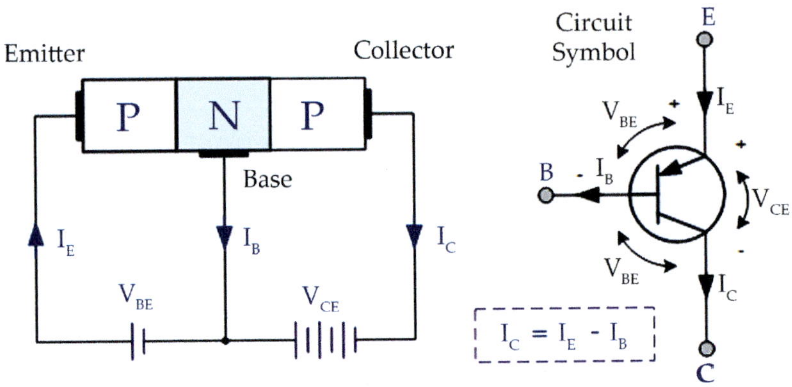

Figure 7.3. A PNP transistor configuration.

Source: https://www.electronics-tutorials.ws/transistor/tran_3.html.

The voltage between the Base and Emitter (V_{BE}) is negative at the Base and positive at the Emitter, as the Base terminal of a PNP transistor is always biased negatively with respect to the Emitter.

Additionally, the voltage between the Emitter and Collector (VCE) is positive. Therefore, in order for a PNP transistor to conduct, the Emitter must always be more positive than the Base and Collector.

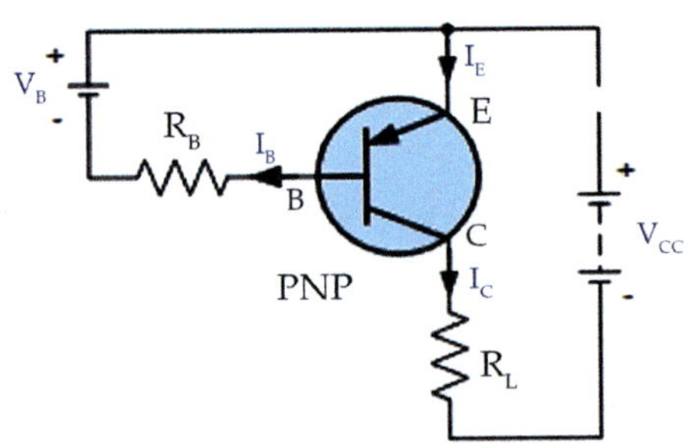

Figure 7.4. PNP transistor connection.

Source: https://www.electronics-tutorials.ws/transistor/tran_3.html

The PNP transistor is connected to voltage sources in the following manner: The Emitter is connected to the supply voltage V_{CC} through the load resistor R_L, which controls the maximum current flowing through the device connected to the Collector terminal. The Base voltage, V_B, is biased negatively relative to the Emitter and is connected to the Base resistor R_B, which again limits the maximum Base current.

To cause the Base current to flow in a PNP transistor, the Base needs to be more negative than the Emitter (current must leave the base) by approximately 0.7 volts for a silicon device or 0.3 volts for a germanium device. The formulas used to calculate the Base resistor, Base current, or Collector current are the same as those used for an equivalent NPN transistor and are given as:

$$I_C = I_E - I_B$$

$$I_C = \beta.I_B \qquad I_B = \frac{I_C}{\beta}$$

Since the current and voltage polarities of an NPN and PNP transistor are always opposite to one another, it is evident that the fundamental distinction between the two types of transistors is the appropriate biasing of their junctions. So for the circuit above: Ic = Ie – Ib as current must leave the Base.

The polarities of the voltages and the directions of the current flow are the only differences between PNP and NPN transistors, which can generally be substituted in most electronic circuits.

7.3.2.2. A PNP Transistor Circuit

PNP transistor output characteristic curves are quite similar to those of an equivalent NPN transistor, with the exception that they have been rotated 180 degrees to account for reverse polarity voltages and currents, which occur when electron current flows from the base and collector of a PNP transistor in the direction of the battery. To determine the operating points of PNP transistors, the same dynamic load line can be drawn onto the I-V curves.

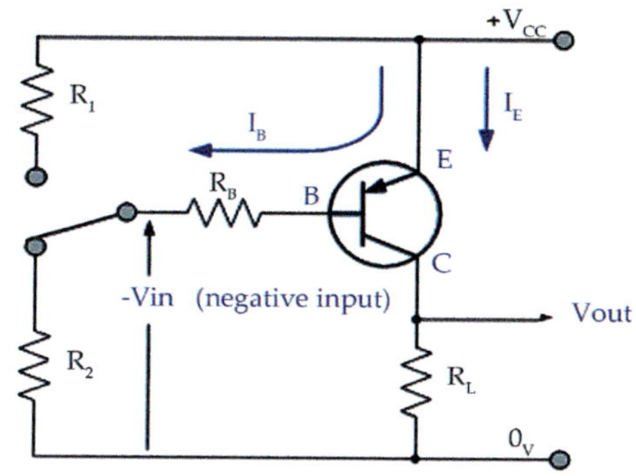

Figure 7.5. A PNP transistor circuit

Source: https://www.electronics-tutorials.ws/transistor/tran_3.html.

7.3.2.3. Transistor Matching

You may wonder what is the point of having a PNP Transistor when there are plenty of NPN Transistors available that can be used as an amplifier or solid-state switch. Having two different types of transistors, PNP and NPN, can be a great advantage when designing power amplifier circuits such as the Class B Amplifier.

When controlling reversible H-Bridge motor control circuits or the output stage of Class-B amplifiers, "Complementary" or "Matched Pair" transistors—a PNP and an NPN connected together—are used. These circuits allow us to regulate the motor's current flow in both directions at different times for forward and reverse motion.

Complementary Transistors are a pair of corresponding NPN and PNP transistors that have nearly identical characteristics to one another. TIP2955 (a PNP transistor) and TIP3055 (an NPN transistor) are two examples of complementary or matched pair silicon power transistors. They are perfect for general motor control or robotic applications because they both have a high collector current of roughly 15A and a DC current gain, Beta (Ic/Ib), matched to within 10%.

Complementary NPN and PNP transistors are also used in the power output stage design of class B amplifiers. Only the positive half of the signal is conducted by the NPN transistor, whereas the negative half is conducted by the PNP transistor.

In both directions, the load loudspeaker can be driven by the amplifier with the necessary power at the nominal impedance and power stated. This results in an output current that is likely to be several amps, evenly divided between the two complementary transistors.

7.3.2.4. Identifying the PNP Transistor

As we saw in the first tutorial in this section on transistors, transistors are essentially just two diodes connected back-to-back.

With this analogy, we can test the resistance between the three separate leads—the emitter, base, and collector—to identify if a transistor is PNP or NPN type.

By testing each pair of transistor leads in both directions with a millimeter will result in six tests in total with the expected resistance values in Ohms given below:

1. Emitter-Base Terminals – The Emitter to Base should act like a normal diode and conduct one way only.
2. Collector-Base Terminals – The Collector-Base junction should act like a normal diode and conduct one way only.
3. Emitter-Collector Terminals – The Emitter-Collector should not conduct in either direction Table 7.1

Table 7.1. Terminal Resistance Values for PNP and NPN Transistors

Between Transistor Terminals **Terminals**		**NPNP**	**NPN**
Collector	Emitter	R_{HIGH}	R_{HIGH}
Collector	Base	R_{LOW}	R_{HIGH}
Emitter	Collector	R_{HIGH}	R_{HIGH}
Emitter	Base	R_{LOW}	R_{HIGH}
Base	Collector	R_{HIGH}	R_{LOW}
Base	Emitter	R_{HIGH}	R_{LOW}

Source: https://www.electronics-tutorials.ws/transistor/tran_3.html

7.4. FUNCTION OF BIPOLAR JUNCTION TRANSISTOR

Depending on the doping types of the three main terminal regions, BJTs can be classified as PNP or NPN types. Two semiconductor junctions that share a thin p-doped region make up an NPN transistor, and two semiconductor junctions that share a thin n-doped region make up a PNP transistor. P-type refers to doping with impurities (like boron) that provide holes that easily accept electrons, whereas N-type refers to doping with impurities (like phosphorus or arsenic) that provide mobile electrons.

Diffusion of charge carriers across a junction between two regions of different charge carrier concentrations is the cause of charge flow in a BJT. The emitter, base, and collector regions make up a BJT. Three leads on a discrete transistor are used to connect to these areas. The collector is typically doped less than the base (collector doping is typically ten times lighter than base doping), and the emitter region is typically heavily doped in comparison to the other two layers. BJTs are categorized as minority-carrier devices because, by design, the majority of the collector current flows from electrons or holes injected from a heavily doped emitter into the base, where they diffuse toward the collector as minority carriers.

During typical operation, the base-emitter junction is forward biased, with the p-doped side at a higher positive potential than the n-doped side, while the base-collector junction is reverse-biased. Applying forward bias to the base-emitter junction disrupts the equilibrium between thermally generated carriers and the repelling electric field of the n-doped emitter depletion region. This allows thermally excited electrons (or holes in a PNP) to flow from the emitter to the base region. These electrons then move through the base from an area of high concentration near the emitter to a region of low concentration near the collector. In the p-type doped base, electrons are considered minority carriers, as holes are the majority carriers. In a PNP device, similar behavior occurs, but with holes being the primary current carriers.

The base region of the transistor must be thin enough for carriers to diffuse across it in a lot less time than the minority-carrier lifetime of the semiconductor in order to reduce the percentage of carriers that recombine before reaching the collector-base junction. Low recombination rates are ensured by having a base that is lightly doped. Specifically, the base's thickness needs to be significantly smaller than the carriers' diffusion length. The reverse bias of the collector-base junction results in minimal carrier injection from the base to the collector. However, carriers injected from the emitter into the base diffuse and reach the collector-base depletion region, where they are swept into the collector by the electric field present in the depletion region. The characteristics that set apart a bipolar transistor from two individual diodes connected in series are the thin shared base and the asymmetric collector-emitter doping.

7.4.1. Voltage, Current, and Charge Control

The current flowing from the collector to the emitter can be influenced either by the current passing through the base to the emitter (current control) or by the voltage across the base to the emitter (voltage control). These two perspectives are connected by the current-voltage relationship of the base-emitter junction, which follows the typical exponential curve of a p-n junction (diode).

The concentration gradient of minority carriers in the base region provides an explanation for collector current. The ambipolar transport rates, where the excess majority and minority carriers flow at the same rate, are essentially set by the excess minority carriers because of low-level injection, which results in significantly fewer excess carriers than normal majority carriers.

In-depth transistor models, like the Gummel-Poon model, precisely consider the distribution of charge to accurately explain transistor behavior. The charge-control perspective effectively explains the operation of phototransistors, where minority carriers in the base are generated by absorbed photons, and also accounts for turn-off dynamics and recovery time based on base region charge recombination. However, since base charge is not directly observable at the terminals, current- and voltage-control perspectives are typically utilized in circuit design and analysis.

Because the current-control view is roughly linear, it is occasionally used in analog circuit design. In other words, the base current is roughly βF times the collector current. Assuming that the collector current is β times the base current and that the base-emitter voltage is roughly constant, some simple circuits can be designed. Nonetheless, the voltage-control (e.g., Ebers–Moll) model is needed in order to precisely and consistently design production BJT circuits. The voltage-control view is frequently preferred because it requires the consideration of an exponential function; however, when this function is linearized, as in the Ebers-Moll model, so that the transistor can be modeled as a transconductance, design for circuits like differential amplifiers again becomes a mostly linear problem. The transistors are typically modeled as voltage-controlled current sources whose transconductance is proportional to their collector current for translinear circuits, where the exponential I-V curve plays a crucial role in the functioning of the circuit. The mathematical model complexity is typically not a major concern for the designer because transistor-level circuit analysis is typically carried out using SPICE or a similar analog-circuit simulator. However, a simplified view of the characteristics enables designs to be created in a logical manner.

7.4.2. Turn-on, Turn-off, and Storage Delay

When bipolar transistors are driven to saturation, especially power transistors, they have long base-storage times; in switching applications, this base storage limits the turn-off time. By keeping the transistor from oversaturating, a Baker clamp can shorten the transistor's switching time by storing less charge in the base.

The term "Baker clamp" refers to a class of electronic circuits that use different types of diodes to apply nonlinear negative feedback that shortens a switching bipolar junction transistor's (BJT) storage time. Saturated BJTs have sluggish turn-off times because of the base's stored charge. Since the storage time of bipolar transistors and IGBTs limits their use in fast switching applications, it must be removed before the transistor turns off. The transistor cannot become saturated and build up a large amount of stored charge because the diode-based Baker clamps keep it from doing so.

The Baker clamp was named after Richard H. Baker, who first introduced it in his 1956 technical report titled "Maximum Efficiency Transistor Switching Circuits." Baker originally referred to the technique as "back clamping," but it is now commonly known as the Baker clamp. Baker's report is often credited with the creation of the two-diode clamp circuit. In addition, Baker included the circuit in a patent application in 1956 and was granted a patent, US 3,010,031, in 1961, which highlighted the use of the clamp in symmetrical flip-flop circuits. Additional clamp circuits exist. A 1959 handbook explains a method known as saturation clamping. A saturation clamp diode is used in that scheme to connect the collector to a saturation clamp supply that is set at roughly 2 volts. The clamp diode activates and provides the additional collector current to prevent the transistor from saturating when it approaches saturation. There must be a significant current supply to the saturation clamp supply. On the other hand, instead of increasing collector current, the Baker clamp lowers the transistor base current.

A single diode clamp is used in another clamp circuit. It employs a resistor divider network but decreases base drive as the transistor gets close to saturation.

Clamp circuits were utilized for accelerating cutoff transitions in electronics. When a transistor is turned off, the output behaves like an RC circuit, gradually decreasing to its ultimate value. As the circuit nears its final value, the available current for charging the capacitor diminishes, causing the rate of change to slow down. It typically takes around 2.3 time constants to reach 90% of the final value. By employing cutoff clamping, the output voltage swing is reduced, leading to quicker transitions. Limiting the collector voltage to 63% of the ultimate value can double the speed of the transition.

7.4.2.1. Basic Idea

By directing base current through the collector, the Baker clamp reduces the voltage differential between the emitter and collector. By lowering the gain close to the saturation point, this nonlinear negative feedback is introduced into a common-emitter stage (BJT switch) in an attempt to prevent saturation. The negative feedback is turned off and the gain is maximum when the transistor is in active mode and sufficiently far from the saturation point; as the transistor gets closer to the saturation point, the negative feedback gradually turns on and the gain rapidly decreases. By connecting a voltage-stable element in parallel to the base-emitter junction, the transistor reduces its gain by functioning as a shunt regulator with respect to its own base-emitter junction.

7.4.2.2. Implementation

By redirecting the excessive input current through the collector to ground, the feedback diode (D_1) between the collector and the input limits the collector voltage to roughly VBE. To increase the effective input voltage, a second silicon diode is connected in series with the base terminal; occasionally, a germanium clamp diode is used in the collector-base feedback to reduce the voltage drop across it. The base diode maintains VCE above VCE(sat) and around a diode drop, enabling the use of a Si diode clamp with a Si transistor. Regretfully, attempting to turn off the transistor results in it turning off and creating a high-impedance return path. It is now harder to extract charge from the base, even though the base charge has decreased.

In order to release the stored base charge in the transistor, a low-impedance return path will be made available by a second base diode that is connected antiparallel to the base diode (D_2 in Baker's schematic). Some sources still refer to this three-diode circuit as a Baker clamp, while others only refer to the two-diode circuit as one.

A single low-voltage diode connected from the collector to the base serves as an easy substitute for the Baker clamp. Low-voltage-drop Schottky and germanium diodes can be used with silicon transistors because they have a much lower forward voltage drop than the VBE bias voltage of the silicon transistor and switch quickly. For a diode to function well, its forward drop must be less than the base-emitter drop. The diode is connected to the intersection of two base-bias resistors using an alternate diode clamp circuit. Combining a Schottky diode and transistor into a single Schottky transistor is the modern solution. This arrangement is also referred to as a Baker clamp in some sources.

Power applications also employ baker clamps, and the selection of diodes is a crucial design consideration. The higher low voltage-output level of the Baker clamp (as in a Darlington transistor) is one of its drawbacks. It raises the dissipated power in power applications and reduces the noise immunity in logic circuits.

7.4.3. Transistor Characteristics: Alpha (α) and Beta (β)

The BJT efficiency is determined by the percentage of carriers that are able to pass through the base and reach the collector. A greater number of electrons are injected from the emitter into the base than holes are injected from the base into the emitter due to the heavy doping of the emitter region and the light doping of the base region. The majority of the minority carriers injected into the base will diffuse to the collector and not recombine because of the thin and lightly doped base region.

7.4.3.1. Common-Emitter Current Gain

The common-emitter current gain, βF (hFE), is the ratio of the DC collector current to the DC base current, not the other way around. In small-signal transistors, it is usually greater than 50; however, in transistors intended for high-power applications, it may be less. The BJT gain is decreased by both base recombination and injection efficiency.

7.4.3.2. Common-Base Current Gain

A further valuable feature is the common-base current gain, or αF. In the forward-active region, the current gain from emitter to collector is roughly equal to the common-base current gain. Typically, this ratio falls between 0.980 and 0.998, or very near to unity. Because of charge carrier recombination as they traverse the base region, it is less than unity.

Alpha and beta are related by the following identities:

$$\alpha_F = \frac{I_C}{I_E}, \beta_F = \frac{I_C}{I_B},$$

$$\alpha_F = \frac{\beta_F}{1+\beta_F} \Leftrightarrow \beta_F = \frac{\alpha_F}{1-\alpha_F}.$$

Although it is a useful metric for characterizing a bipolar transistor's performance, beta is not a fundamental feature of the device. Bipolar transistors are classified as voltage-controlled devices because the base current, which is determined by recombination in the base and the characteristics of the base-emitter junction, may be viewed as a defect. Essentially, the collector current is controlled by the base-emitter voltage. In a lot of designs, beta is taken to be high enough that base current barely affects the circuit. Certain circuits (usually switching circuits) have enough base current supplied to ensure that the necessary collector current flows even with the lowest possible beta value for that specific device.

7.5. REGIONS OF OPERATION

BJT junction biases distinguish four different operating regions for bipolar transistors.

7.5.1. Forward-Active (Or Simply Active)

There is forward bias at the base-emitter junction and reverse bias at the base-collector junction. In forward-active mode, the majority of bipolar transistors are engineered to provide the maximum common-emitter current gain, βF. If so, the collector-emitter current is roughly proportional to the base current, but for small variations in the base current, it is much larger.

7.5.2. Reverse-Active (Or Inverse-Active Or Inverted)

A bipolar transistor enters reverse-active mode by flipping the biasing conditions of the forward-active region. The roles of the emitter and collector regions alternate in this mode. In inverted mode, the βF is several times smaller than in forward-active mode (2-3 times for an ordinary germanium transistor), because most BJTs are designed to maximize current gain. This transistor mode is rarely utilized; it is typically reserved for failsafe scenarios and specific kinds of bipolar logic. In this area, the reverse bias breakdown voltage to the base might be ten times lower.

7.5.3. Saturation

A BJT in saturation mode allows high current conduction from the emitter to the collector (or, in the case of an NPN, the opposite direction, with negatively charged carriers flowing from emitter to collector) when both junctions are forward biased. This mode is equivalent to a closed switch or logical "on."

7.5.4. Cut-off

Biasing conditions opposite of saturation—that is, both junctions reverse biased—are present in the cut-off. Very little current means that the switch is logically "off," or open.

These regions overlap slightly for small biases (less than a few hundred millivolts), despite being well defined for sufficiently large applied voltages. This end of the forward active region can be considered the cutoff region because, in the typical grounded-emitter configuration of an NPN BJT used as a pull-down switch in digital logic, the base voltage never drops below ground, so there is never a reverse-biased junction in the off state. However, the forward bias is close enough to zero that practically no current flows.

7.5.5. Active-mode Transistors in Circuits

A schematic representation of an NPN transistor coupled to two voltage sources is shown in the diagram. The same explanation holds true for a PNP transistor with applied voltage and current flow directed in the opposite directions. The lower P-N junction becomes forward biased as a result of this applied voltage, enabling an electron flow from the emitter into the base. Most of these electrons will cross the upper P-N junction into the collector to form the collector current (I_C) in the active mode due to the electric field generated by the V_{CE} between the base and collector. The base current, or I_B, is created when the remaining electrons recombine with the holes, which are the majority carrier in the base. This current flows through the base connection. As shown in the diagram, the emitter current, IE, is the total transistor current, which is the sum of the other terminal currents (i.e., $I_E = I_B + I_C$).

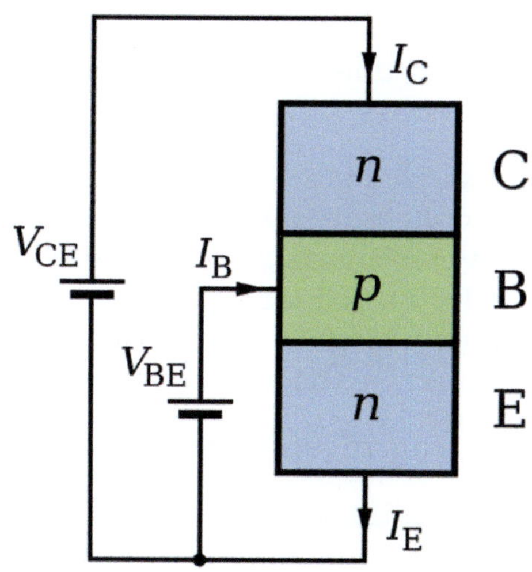

Figure 7.6. Structure and use of NPN transistor. Arrow according to schematic.

Source: https://en.wikipedia.org/wiki/Bipolar_junction_transistor#/media/File:NPN_BJT_-_Structure_&_circuit.svg.

Because electrons carry a negative electric charge, the flow of electrons in the diagram is in the opposite direction from the arrows representing current, which point in the direction of conventional current. The DC current gain in an active mode is defined as the collector current divided by the base current. Typically, this gain is 100 or higher, but reliable circuit designs don't rely on the precise amount. The terms "h_{FE}" and "h_{fe}" refer to the values of this gain for DC signals and small signals, respectively.

7.6. THEORY AND MODELING

BJTs are conceptualized as two diodes (P–N junctions) that share a common area through which minority carriers are able to pass. When it comes to operation, a PNP BJT is similar to two diodes sharing an N-type cathode region and an NPN is similar to two diodes sharing a P-type anode region. A BJT cannot be created by connecting two diodes with wires because the wire will prevent minority carriers from passing through one P-N junction to the other.

A small current input to the base of either type of BJT controls an amplified output from the collector. As a result, the BJT produces an effective switch that is managed by its base input. Because it can increase a weak input signal to roughly 100 times its original strength, the BJT also functions well as an amplifier. BJT networks are used to create strong amplifiers with a wide range of uses.

In the discussion below, focus is on the NPN BJT. In what is called active mode, the base–emitter voltage V_{BE} and collector–base voltage V_{CB} are positive, forward biasing the emitter–base junction and reverse-biasing the collector–base junction. In this mode, electrons are injected from the forward biased n-type emitter region into the p-type base where they diffuse as minority carriers to the reverse-biased n-type collector and are swept away by the electric field in the reverse-biased collector–base junction.

For an illustration of forward and reverse bias, see semiconductor diodes.

7.6.1. Large-signal Models

In 1954, Jewell James Ebers and John L. Moll introduced their mathematical model of transistor currents:

7.6.1.1. Ebers–Moll model

The DC emitter and collector currents in active mode are well modeled by an approximation to the Ebers–Moll model

$$I_E = I_{ES}\left(e^{\frac{V_{BE}}{V_T}} - 1\right)$$

$$I_C = \alpha_F I_E$$

$$I_B = (1-\alpha_F)I_E$$

The base internal current is mainly by diffusion and

$$J_{n(base)} = \frac{1}{W} qD_n n_{bo} e^{\frac{V_{EB}}{V_T}}$$

Where: VT is the thermal voltage (approximately 26 mV at 300 K ≈ room temperature). IE is the emitter current. IC is the collector current. αF is the common base forward short-circuit current gain (0.98 to 0.998).

IES is the reverse saturation current of the base–emitter diode (on the order of 10–15 to 10–12 amperes)

VBE is the base–emitter voltage

Dn is the diffusion constant for electrons in the p-type base

W is the base width

The α and forward β parameters are as described previously. A reverse β is sometimes included in the model.

The unapproximated Ebers–Moll equations used to describe the three currents in any operating region are given below. These equations are based on the transport model for a bipolar junction transistor.

$$i_C = I_S\left[\left(e^{\frac{V_{BE}}{V_T}} - e^{\frac{V_{BC}}{V_T}}\right) - \frac{1}{\beta_R}\left(e^{\frac{V_{BC}}{V_T}} - 1\right)\right]$$

$$i_B = I_S\left[\frac{1}{\beta_F}\left(e^{\frac{V_{BE}}{V_T}} - 1\right) + \frac{1}{\beta_R}\left(e^{\frac{V_{BC}}{V_T}} - 1\right)\right]$$

$$i_E = I_S\left[\left(e^{\frac{V_{BE}}{V_T}} - e^{\frac{V_{BC}}{V_T}}\right) + \frac{1}{\beta_F}\left(e^{\frac{V_{BE}}{V_T}} - 1\right)\right]$$

where: iC is the collector current. iB is the base current. iE is the emitter current. βF is the forward common emitter current gain (20 to 500). βR is the reverse common emitter current gain (0 to 20). IS is the reverse saturation current (on the order of 10–15 to 10–12 amperes). VT is the thermal voltage (approximately 26 mV at 300 K ≈ room temperature).. VBE is the base–emitter voltage. VBC is the base–collector voltage.

7.6.1.2. Base-Width Modulation

As the collector–base voltage $V_{CB} = V_{CE} - V_{BE}$ varies, the collector–base depletion region varies in size. For instance, a higher voltage across the collector-base junction results in a larger reverse bias, which widens the collector-base depletion region and reduces the base's width. Because of its discoverer, James M. Early, this variation in base width is frequently referred to as the "Early effect."

Narrowing of the base width has two consequences:

- There is a lesser chance for recombination within the "smaller" base region.
- The charge gradient is increased across the base, and consequently, the current of minority carriers injected across the emitter junction increases.

When the collector-base voltage rises, the transistor's collector or "output" current increases as a result of both factors.

In the forward-active region, the Early effect modifies the collector current (i_C) and the forward common emitter current gain (β_F) as given by:

$$i_C = I_S e^{\frac{V_{BE}}{V_T}} \left(1 + \frac{V_{CE}}{V_A}\right)$$

$$\beta_F = \beta_{F0} \left(1 + \frac{V_{CB}}{V_A}\right)$$

$$r_o = \frac{V_A}{I_C}$$

where: V_{CE} is the collector–emitter voltage. V_A is the Early voltage (15 V to 150 V). β_{F0} is forward common-emitter current gain when V_{CB} = 0 V. r_o is the output impedance. IC is the collector current

7.6.1.3. Punchthrough

The base-collector depletion region boundary meets the base-emitter depletion region boundary when the base-collector voltage reaches a specific (device-specific) value. The transistor essentially has no base when it is in this state. Thus, in this state, the device loses all gains.

7.6.1.4. Gummel–Poon charge-control model

Others have adapted and expanded the Gummel–Poon model, a comprehensive charge-controlled model of BJT dynamics, to provide a more thorough explanation of transistor dynamics than terminal-based models usually can. In this model, the β-values of transistors are also dependent on their direct current levels, whereas in the Ebers-Moll model, the transistors are assumed to be current-independent.

7.6.2. Small-signal Models

7.6.2.1. Hybrid-pi model

A well-liked circuit model for studying the small signal and AC behavior of field-effect and bipolar junction transistors is the hybrid-pi model. It is also referred to as the Giacoletto model, named after L.J. Giacoletto, who introduced it in 1969. If the proper inter-electrode capacitances and other parasitic components are added, the model can be readily modified for higher-frequency circuits and is quite accurate for low-frequency circuits.

7.6.2.2. h-parameter model

The h-parameter model, also called the hybrid equivalent model, is another model that is frequently used to analyze BJT circuits. It is similar to the hybrid-pi model and the y-parameter two-port model, but it uses input current and output voltage as independent variables instead of input and output voltages. This two-port network is especially well-suited for BJTs since it makes it simple to analyze circuit behavior and can be used to create more precise models. As can be seen, depending on the topology chosen, the term "x" in the model denotes a different BJT lead.

For common-emitter mode the various symbols take on the specific values as:

- Terminal 1, base
- Terminal 2, collector
- Terminal 3 (common), emitter; giving x to be e
- i_i, base current (i_b)
- i_o, collector current (i_c)
- V_{in}, base-to-emitter voltage (V_{BE})
- V_o, collector-to-emitter voltage (V_{CE})

and the h-parameters are given by:

- $h_{ix} = h_{ie}$ for the common-emitter configuration, the input impedance of the transistor (corresponding to the base resistance r_{pi}).
- $h_{rx} = h_{re}$, a reverse transfer relationship, it represents the dependence of the transistor's (input) I_B–V_{BE} curve on the value of (output) V_{CE}. It is usually very small and is often neglected (assumed to be zero) at DC.
- $h_{fx} = h_{fe}$, the "forward" current-gain of the transistor, sometimes written h_{21}. This parameter, with lower case "fe" to imply small signal (AC) gain, or more often with capital letters for "FE" (specified as h_{FE}) to mean the "large signal" or DC current-gain (β_{DC} or often simply β), is one of the main parameters in datasheets, and may be given for a typical collector current and voltage or plotted as a function of collector current. See below.
- $h_{ox} = 1/h_{oe}$, the output impedance of transistor. The parameter h_{oe} usually

corresponds to the output admittance of the bipolar transistor and has to be inverted to convert it to an impedance.

Because the h-parameters' subscripts are in lowercase, they denote AC conditions oranalysis. The conditions are stated in uppercase for DC conditions. An approximate h-parameter model is frequently used for the CE topology, which further streamlines the circuit analysis. The h_{oe} and h_{re} parameters are disregarded for this, meaning they are set to zero and infinity, respectively. The presented h-parameter model is appropriate for small-signal, low-frequency analysis. The inter-electrode capacitances that are significant at high frequencies need to be added for high-frequency analysis.

7.6.2.3. Etymology of hFE

The h stands for "h-parameter," which is a class of parameters called after the hybrid equivalent circuit model in which they originated. The choice of capital or lower case letters for the letters that follow the h is important, just like it is for all other h parameters. Lower case letters indicate small signal parameters, such as the slope of a given relationship, while upper case letters indicate large signal or DC values, such as the ratio of voltages or currents.

In the case of the very often used h_{FE}:

- F is from Forward current amplification also called the current gain.
- E refers to the transistor operating in a common Emitter (CE) configuration.

Therefore, h_{FE} is dimensionless and equals the (total; DC) collector current divided by the base current. It is a parameter that is typically specified at a typical collector current and voltage, or it is graphed as a function of collector current. It varies somewhat with collector current, but it is frequently approximated as a constant. If the subscript had not been written in capital letters, i.e., if the parameter were written as "small signal (AC) current gain," that is, the graph's slope at a given point of collector current versus base current, which, unless the test frequency is high, is frequently near to the h_{FE} value.

7.7. BIPOLAR TRANSISTOR BIASING

For bipolar transistors to function properly, they need to be biased correctly. Resistors are frequently used in biasing networks for discrete circuits, or circuits built with individual devices. Integrated circuits, such as bandgap voltage references and current mirrors, use far more complex biasing configurations. The voltage divider configuration uses resistors in specific patterns to produce the desired voltages. It is possible to achieve stable current levels with minimal temperature variations and transistor characteristics like β by choosing the appropriate resistor values.

The point on the output characteristics that displays the DC collector-emitter voltage (V_{ce}) and the collector current (I_c) in the absence of an applied input signal is called the operating point of a device, also referred to as the bias point, quiescent point, or Q-point.

7.7.1. Bias Circuit Requirements

A bias network is selected to stabilize the operating point of the transistor by reducing the following effects of device variability, temperature, and voltage changes:

- Transistor gain can differ greatly between batches, leading to widely disparate operating points for successive units in serial production or following transistor replacement.
- Due to the Early effect, the current gain is affected by the collector–emitter voltage.
- Both gain and base–emitter voltage depend on the temperature.
- The leakage current also increases with temperature.

Based on the range of expected operating conditions, a bias circuit may consist solely of resistors or may also include components like diodes, temperature-dependent resistors, or additional voltage sources.

7.7.1.1. Signal Requirements

The Q-point is set for analog operation of a class-A amplifier such that the transistor operates continuously in the active mode throughout the input signal's range, without shifting to operate in the saturation or cut-off regions. In order to permit comparable signal swings in both positive and negative directions, the Q-point of a transistor characteristic is frequently set close to the center of the active region.

Instead, the Q-point is selected for digital operation, causing the transistor to flip between the on (saturation) and "off" (cutoff) states.

7.7.1.2. Thermal Considerations

For every degree Celsius that the temperature rises (with 25 °C as the reference), the voltage across the emitter-base junction V_{BE} of a bipolar transistor falls by 2 mV for silicon and 1.8 mV for germanium when the current is constant. According to the Ebers-Moll model, an increase in temperature will result in an increase in the current flowing through the base-emitter junction I_B, which will raise the collector current I_C, if the base-emitter voltage V_{BE} remains constant. The power dissipated in the transistor may also increase depending on the bias point, which will raise its temperature and make the issue worse. The outcome of this harmful positive feedback is thermal runaway. To lessen the risk of bipolar transistor thermal runaway, there are various methods.

For example,

- The biasing circuit can be designed with negative feedback to ensure that a rise in collector current results in a fall in base current. Therefore, its source is throttled by the rising collector current.
- It is possible to use heat sinks to remove surplus heat and stop the temperature of the base-emitter from rising.
- Collector-emitter power dissipation is maximized when the transistor is biased such that its collector is typically less than half of the power supply voltage. Because of the half-voltage principle, which states that an increase in collector current results in a decrease in dissipated power, runaway is thus impossible. The circuits below serve as main examples of how negative feedback is used to stop thermal runaway.

7.7.2. Types of Bias Circuit For Class-A Amplifiers

The following discussion treats five common biasing circuits used with class-A bipolar transistor amplifiers:

- Fixed bias
- Collector-to-base bias
- Fixed bias with emitter resistor
- Voltage divider bias or potential divider
- Emitter bias

7.7.2.1. Fixed Bias (Base Bias)

This form of biasing is also called base bias or fixed resistance biasing. In the given fixed bias circuit,

$$I_b = \frac{V_{cc} - V_{be}}{R_b}.$$

For a given transistor, V_{be} doesn't vary significantly during use. And since R_b and the DC voltage source V_{cc} are constant, the base current I_b also doesn't vary significantly. Thus this type of biasing is called fixed bias.

The common-emitter current gain of a transistor (specified as a range on its data sheet as h_{FE} or β), allows us to obtain c as well:

$$I_c = \beta I_b.$$

Now V_{ce} can be determined:

$$V_{ce} = V_{cc} - I_c R_c.$$

Thus, an operating point (V_{ce}, I_c) for a transistor can be set using R_b and R_c.

Advantages

The operating point is set by two resistors and the calculation is very simple.

Disadvantages

- Since the base current establishes the bias, the collector current is directly related to β. As a result, the operating point is unstable to temperature changes and will fluctuate greatly when transistors are switched.
- In the case of small-signal transistors (i.e., not power transistors), β values that are relatively high (i.e., between 100 and 200), thermal runaway will be likely in this setup. Specifically, the stability factor—a measurement of how the collector current changes in response to variations in the reverse saturation current—approximately β+1. Small-signal transistors have high stability factors because it is desirable for the amplifier to have a stability factor of less than 25.

Usage

Fixed bias is rarely utilized in linear circuits (i.e., circuits that use transistors as current sources) because of the aforementioned intrinsic disadvantages. Rather, it is frequently employed in circuits that employ transistors as switches. But one use for fixed bias is to feed a DC signal from the AC output of a subsequent stage to the base resistor of the transistor in order to achieve rudimentary automatic gain control.

7.7.2.2. Collector feedback bias

Negative feedback is used in this configuration to stop thermal runaway and maintain the operating point. Instead of connecting the base resistor R_b to V_{cc}, this biasing method connects it to the collector. Therefore, any thermal runaway will cause a voltage drop across the R_c resistor, which will reduce the base current of the transistor.

From Kirchhoff's voltage law, the voltage V_{R_b} across the base resistor R_b is

$$V_{R_b} = V_{cc} - \overbrace{(I_c + I_b)R_c}^{\text{Voltage drop across } R_c} - \overbrace{V_{be}}^{\text{Voltage at base}} .$$

By the Ebers–Moll model, I_c=b I_b, and so

$$V_{R_b} = V_{cc} - (\overbrace{\beta I_b}^{I_c} + I_b)R_c - V_{be} = V_{cc} - I_b(\beta + 1)R_c - V_{be}.$$

From Ohm's law, the base current $I_b = V_{R_b} / R_b$, and so

$$\overbrace{I_b R_b}^{V_{R_b}} = V_{cc} - I_b(\beta + 1)R_c - V_{be}.$$

Hence, the base current I_b is

$$I_b = \frac{V_{cc} - V_{be}}{R_b + (\beta + 1)R_c}$$

If V_{be} is held constant and temperature increases, then the collector current I_c increases. However, a larger I_c causes the voltage drop across resistor R_c to increase, which in turn reduces the voltage V_{R_b} across the base resistor R_b. A lower base-resistor voltage drop reduces the base current I_b, which results in less collector current I_c. Because an increase in collector current with temperature is opposed, the operating point is kept stable.

Advantages

Circuit stabilizes the operating point against variations in temperature and β (i.e., replacement of transistor).

Circuit stabilizes the operating point (as a fraction of V_{cc}) against variations in V_{cc}.

Disadvantages

Modest variations in β are acceptable, but significant variations will significantly alter the operating point. Once β is known fairly precisely (perhaps within ~25%), Rb must be selected; however, the variability of β between identical parts is frequently greater than this.

Usage

This setup, referred to as voltage-shunt feedback, senses the output voltage and applies the feedback signal—a current—in a shunt fashion, meaning it runs parallel to the input. This indicates a real reduction in the input impedance looking into the base. Miller's Theorem can be applied to easily verify this. This scenario is comparable to an inverting op-amp circuit, in which the external series resistor controls the overall input impedance and the amplifier's input impedance is almost zero at the virtual earth. This biasing form is used only when the trade-off for stability is justified because of the feedback's reduction in gain. The input impedance of this circuit will rise with the addition of an emitter resistor.

7.7.2.3. Fixed bias with emitter resistor

The emitter of the fixed bias circuit is connected to an external resistor to alter it.

This resistor introduces negative feedback that stabilizes the Q-point. From Kirchhoff's voltage law, the voltage across the base resistor is

$$V_{R_b} = V_{cc} - I_e R_e - V_{be}$$

From Ohm's law, the base current is

$$I_b = \frac{V_{R_b}}{R_b}$$

Feedback regulates the bias point in the following ways. Emitter current rises when temperature rises while maintaining a constant V_{be}. But a higher $_{Ie}$ results in an increase in the emitter voltage $V_e = I_e R_e$, which lowers the voltage V_{Rb} across the base resistor. Because Ic = β I_b, a lower base-resistor voltage drop lowers the base current, which in turn lowers the collector current. Since collector current and emitter current are related by $I_c = \alpha\, I_e$, where $\alpha \approx 1$, the operating point is maintained stable and the increase in emitter current with temperature is opposite.

Similarly, there might be a change in I_c (corresponding to a change in β-value, for example) if a different transistor is used in its place. The change is reversed and the operating point is maintained by following the same procedure as before.

For the given circuit,

$$I_b = \frac{V_{cc} - V_{be}}{R_b + (\beta + 1)R_e}$$

Advantages

The circuit has the tendency to stabilize operating point against changes in temperature and β-value.

Disadvantages

- In this circuit, to keep Ic independent of β the following condition must be met:

$$I_c = \beta I_b = \frac{\beta(V_{cc} - V_{be})}{R_b + (\beta + 1)R_e} \approx \frac{(V_{cc} - V_{be})}{R_e}$$

which is approximately the case if

$$(\beta+1)R_e \gg R_b.$$

- As the β-value is fixed for a given transistor, this relation can be satisfied either by keeping R_e very large or making R_b very low.
- If R_e has a large value, a high V_{cc} is necessary. This increases the cost as well as the precautions necessary while handling.
- If R_b is low, a separate low-voltage supply should be used in the base circuit. Using two supplies of different voltages is impractical.

Usage

When viewed from the base, the feedback also raises the amplifier's input impedance, which has certain benefits. Due to the aforementioned drawbacks, this kind of biasing circuit should only be employed after carefully weighing the associated trade-offs.

7.7.3. Class-B and AB Amplifiers

7.7.3.1. Signal requirements

Two active devices are used by class B and AB amplifiers to cover the full 360 degrees of input signal flow. As a result, each transistor is biased to operate across roughly 180 degrees of the input signal. When there is no signal, the collector current I_c just conducts (roughly 1% of the maximum value that can be reached), which is known as class B bias. In the case of class-AB bias, the collector current I_c is approximately 1/4 of its maximum value. A moderate-power audio amplifier could be built using the class-AB push-pull output amplifier circuit below.

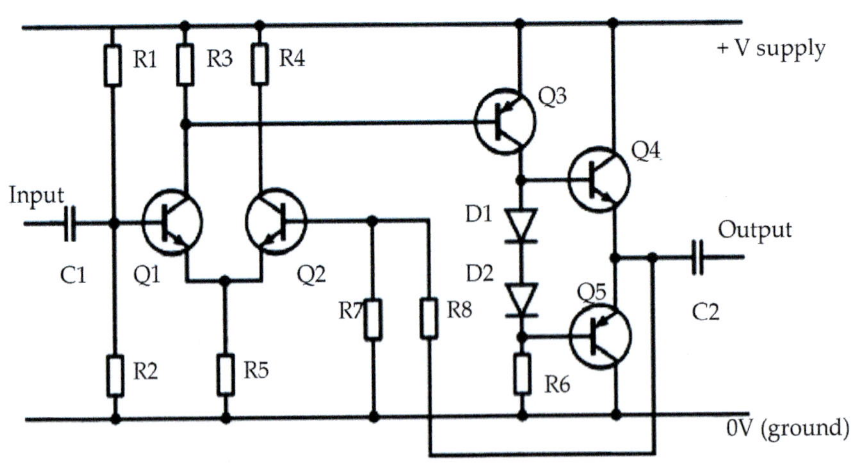

Figure 7.7. A practical amplifier circuit.

Source: https://upload.wikimedia.org/wikipedia/commons/thumb/f/fd/Amplifier_Circuit_Small.svg/500px-Amplifier_Circuit_Small.svg.png.

Using D1 and D2, the common emitter stage Q3 amplifies the signal and supplies DC bias current to create a bias voltage for the output devices. The output pairs, also known as complementary pairs, are arranged in a class-AB push-pull configuration. To reduce crossover distortion, the output pair is slightly biased into the conducting state by the diodes D1 and D2, which supply a steady voltage bias. In other words, if heat dissipation lowers the base-emitter drop of the output transistors, the diodes force the output stage into class-AB mode.

Since the overall feedback in this design operates internally from DC up through the audio range and beyond, it automatically stabilizes its operating point. The diodes must be thermally and electrically matched to the output transistors in order to use fixed diode bias. Since the power supply's total current is unrestricted at this point, excessive conductivity by the output transistors could quickly lead to overheating and their own destruction.

A common fix is to add some emitter resistors—usually a few ohms or so—to help stabilize the output device operating point. Based on the parts used and the amplifier's intended use, the resistor and capacitor values in the circuit are determined.

7.8. APPLICATIONS

Because of its high transconductance and output resistance in comparison to MOSFETs, as well as the large variety of BJT types available, the BJT continues to be an excellent device for certain applications, such as discrete circuit design.

Demanding analog circuits, particularly those involving very high frequencies, like radio-frequency circuits for wireless systems, are best served by the BJT.

7.8.1. High-Speed Digital Logic

By utilizing a BiCMOS wafer fabrication process, bipolar transistors and MOSFETs can be combined in an integrated circuit to build circuits that capitalize on the respective application strengths of both types of transistors.

7.8.2. Amplifiers

The BJT's current gain is characterized by the transistor parameters α and β. Because of their gain, BJTs can be utilized as the basic components of electronic amplifiers. The three main BJT amplifier topologies are:

- Common emitter
- Common base
- Common collector

7.8.3. Temperature Sensors

By subtracting two voltages at two distinct bias currents in a known ratio, the BJT can be used to measure temperature because of the forward-biased base-emitter junction voltage's known temperature and current dependence.

7.8.4. Logarithmic Converters

A BJT can also be used to calculate logarithms and anti-logarithms since the base-emitter voltage varies as the logarithm of the base-emitter and collector-emitter currents. These nonlinear tasks can also be carried out by a diode, but the transistor offers greater circuit flexibility.

7.8.5. Avalanche Pulse Generators

A lower collector-to-emitter breakdown voltage than the collector-to-base breakdown voltage can be purposefully incorporated into transistor construction. It is possible to maintain the collector-emitter voltage at a voltage slightly below breakdown if the emitter-base junction is reverse biased. The transistor turns fully on as soon as the base voltage is allowed to rise and current flows, causing an avalanche and impact ionization in the collector-base depletion region. This quickly floods the base with carriers. This effect can be used to create very sharp falling edges, provided that the pulses are short enough and rare enough to avoid damaging the device.

CASE STUDY

AC Characteristics of van der Waals Bipolar Junction Transistors Using an $MoS_2/WSe_2/MoS_2$ Heterostructure

Bipolar junction transistors (BJTs), as some of the important semiconductor devices, have attracted great attention in the past decades. They contain three separately doped regions, which are defined as the collector, base, and emitter. Two-dimensional (2D) materials offer significant opportunities for the construction of high-performance BJTs owing to their unique interlayered van der Waals (vdW) bonding characteristics. By employing mechanical exfoliation or chemical vapor deposition (CVD), monolayer or multilayer 2D materials, including hexagonal boron nitride (h-BN), transition metal dichalcogenides (TMDs), and graphene, can be easily obtained. These ultrathin 2D materials provide possibilities for fabricating vdW BJTs with base regions of atomic thicknesses.

In the past few years, several 2D-material-based BJTs, such as $BP/MoS_2/BP$, $Cu_9S_5/PtS_2/WSe_2$, and $MoS_2/WSe_2/MoS_2$, have been fabricated successively. The static characteristics of these devices have been extensively studied, and they show promising application potential in the fields of photodetection, gas sensing, and biosensing. However, the AC characteristics of vdW BJTs fabricated from 2D materials have not been reported yet, which is crucial for determining a device's ability to process alternating signals.

Materials and Methods

Device Fabrication

A controlled multistep dry transfer process was employed to fabricate the vdW BJT. Firstly, the WSe_2 and MoS_2 flakes were exfoliated from the bulk crystals supplied by HQ Graphene Company (Groningen, The Netherlands). Then, using the dry transfer technique, MoS_2, WSe_2 and MoS_2 sheets were stacked onto a clean 300 nm SiO_2/Si substrate in sequence. In this case, the top and bottom MoS_2 sheets were separated by the middle WSe_2 flake. Thirdly, maskless lithography was utilized to define the locations of the metal electrodes, and thermal evaporation was employed to deposit Cr/Ag metals with thicknesses of 10 nm and 100 nm. Finally, the device underwent a two-hour annealing process in an argon atmosphere at 300 °C to eliminate the photoresist residues and potentially facilitate Ag diffusion into the underlying MoS_2 flakes, thereby reducing the contact resistance.

Characterization

AFM (NTEGRA Spectra, NT-MDT, Moscow, Russia) and Raman spectroscopy (InVia Reflex, Renishaw, Wotton-under-Edge, Gloucestershire, UK) instruments were employed to characterize the height profile and composition of the vdW BJT. A semiconductor parameter analyzer (B1500A, Agilent Technologies, Santa Clara, CA, USA) was used to investigate the static characteristics of the device. The AC performance of the vdW BJT was measured using an oscilloscope (DPO 7354C, Tektronix, Portland, OR, USA) and an arbitrary waveform generator (DG4062, RIGOL, Beijing, China).

Results and Discussion

Figure 1a,b show the schematic diagram and the optical image of the vertically stacked $MoS_2/WSe_2/MoS_2$ BJT. Here, the top MoS_2 sheet acts as the collector (C) while the bottom MoS_2 sheet serves as the emitter (E). The multilayer WSe_2 sheet was designed for the base (B) region. Figure 1c shows the height profile of the device. Apparently, the thicknesses of bottom MoS_2, middle WSe_2, and top MoS_2 are 13 nm, 3.5 nm, and 63.9 nm, respectively. To analyze the composition of the device, Raman spectra were obtained for the individual 2D materials as well as their overlap regions, as shown in Figure 1d. From the bottom MoS_2, Raman peaks at 383.7 and 408.8 cm^{-1} can be observed. The two Raman peaks have a relatively large separation of 25.1 cm^{-1}, confirming the multilayer nature of the MoS_2 material. The peaks at 249.8 and 258.3 cm^{-1} for the WSe_2 flake are ascribed to the E^1_{2g} mode and the A_{1g} mode. In addition, the Raman spectra of the three flake overlap region are the sum of the Raman peaks of the MoS_2 and WSe_2 flakes, thereby confirming the successful fabrication of the vertically stacked heterostructure.

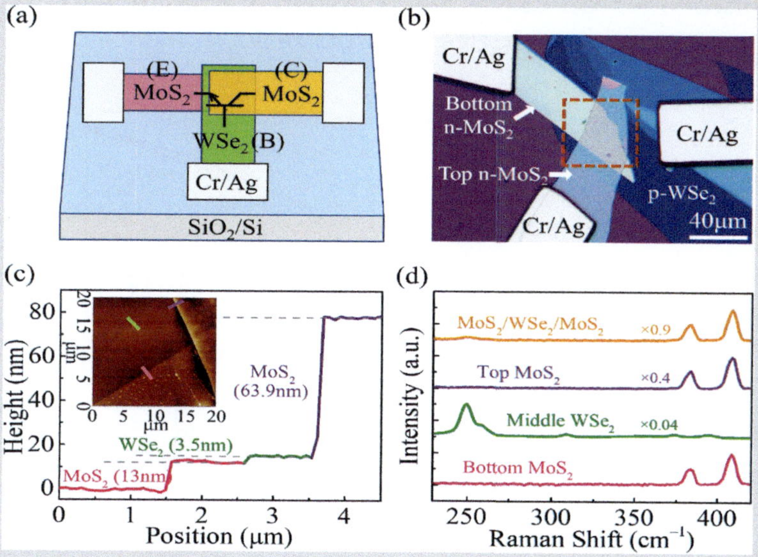

Figure 1. *(a) Schematic illustration of the vdW BJT. (b) Optical microscope image of the vdW BJT. The inside of the dashed brown square represents the three-flake overlapped region. (c) Height profile of the device. The dashed lines indicate the horizontal position of the steps. The inset illustrates the corresponding AFM image. (d) Raman spectra of the bottom MoS_2, middle WSe_2, top MoS_2, and $MoS_2/WSe_2/MoS_2$ three flake overlap regions.*

The static performance of the vdW BJT in common base mode was initially investigated. In this case, the base was grounded, whereas the base–collector and base–emitter junctions were separately reverse-biased and forward-biased. The band diagram of the vdW BJT operating in the forward-active operating mode is depicted in Figure 2a. Figure 2b shows the relationship between the base–emitter voltage (V_{BE}) and the emitter current (I_E) at various fixed collector–base voltages (V_{CB}). With the increase in V_{BE}, the depletion region of the base–emitter junction narrows, facilitating an enhanced diffusion of electrons from emitter to base. Hence, I_E increased with a larger V_{BE}. Figure 2c illustrates the output characteristic of the vdW BJT. The V_{BE} can effectively affect the collector current (I_C), since it can influence the electrons diffusing from the emitter. The electrons were transferred into the collector, constituting the main component of the I_C. The common base current gain (α) was determined to be around 1.01 at V_{BE} = 5 V by calculating the ratio of I_C and I_E. Figure 2d shows the output performance of the vdW BJT operating in common emitter mode. At low V_{CE} values, the collector current shows an approximately linear increase with the V_{CE}, indicating the saturation region of the device. Beyond the saturation region, changes in V_{CE} have minimal impact on the I_C. Instead, the I_C is primarily influenced by variations in V_{BE}. This region is defined as the active region of the device. A maximum current gain ($\beta = I_C/I_B$) of approximately 9 can be obtained at V_{BE} = 0.4 V under the common emitter configuration, as shown in Figure 2d. It is noteworthy that the $MoS_2/WSe_2/MoS_2$ BJT has a relatively low on/off ratio, which may be attributed to the ultrathin base region. The negatively biased base–collector junction introduces an extra electric field perpendicular to the base–emitter junction, thereby diminishing the device's on/off ratio.

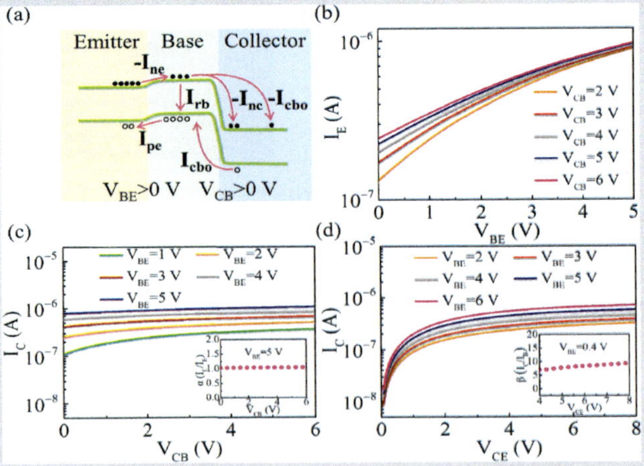

Figure 2. *(a) Band diagram of the vdW BJT operating in forward-active operating mode. The orange, green, and blue areas indicate the emitter, base, and collector region of the device, respectively. (b) The relationship between I_E and V_{BE} at various fixed V_{CB} values. (c) The relationship between I_C and V_{CB} at various values of V_{BE}. Inset shows α as a function of V_{CB} at a fixed V_{BE} = 5 V. (d) The relationship between I_C and V_{CE} at various fixed V_{BE} values. Inset shows β as a function of V_{CE} at a fixed V_{BE} = 0.4 V.*

Since the device showed excellent static performance, the AC characteristics of the vdW BJT operating in common emitter mode were investigated. The schematic diagram of the electrical connection is shown in Figure 3a. Here, a DC voltage (V_{BE} = 5.8 V) and a small AC voltage (v_i) were applied to the base–emitter junction. The base–collector junction was reverse-biased by connecting the collector to another power supply (V_{CE} = 34 V) through a load resistor (R_L = 22 MΩ). An oscilloscope was used to monitor the input and output waveforms of the device in real time. Figure 3b illustrates the time domain characteristics of the device operating at 1 Hz. Here, an AC voltage v_i with an amplitude of 0.2 V is superimposed on the V_{BE}, which causes the voltage applied on the base–emitter junction to fluctuate sinusoidally above and below its DC bias level. The resulting variation in I_B causes the output current change. Therefore, an output AC signal with an amplitude approximately 3.5 times higher than the input AC signal can be observed in the collector region.

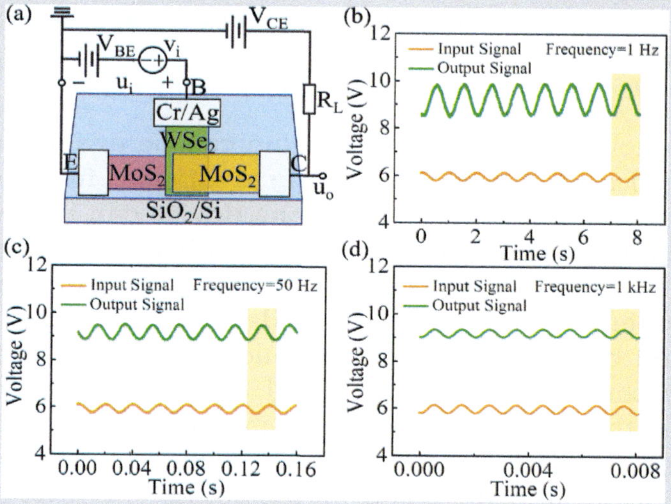

Figure 3. *(a) Schematic diagram of the electric connection of the vdW BJT in common emitter mode. (b–d) Time domain characteristics of the vdW BJT operating at 1 Hz, 50 Hz, and 1 kHz, respectively. The yellow area demonstrates the relative phase between the input and output signals during one cycle of the sinusoidal signal.*

The time domain characteristics of the vdW BJT operating at other frequencies were also investigated. The representative results, such as for the device operating at 50 Hz and 1 kHz, are shown in Figure 3c,d. It is worth mentioning that the voltage in the collector region was higher than that in the base region throughout the experiment, which indicates that the BJT remained in forward-active operating mode at all times. Figure 4a summarizes the common emitter voltage gain versus the operating frequency. With the increase in operating frequency, the amplitude of the output voltage signal gradually decreases until the voltage gain falls to unity at 200 Hz, indicating that the device has voltage amplification capability in the 0–200 Hz region. Figure 4b illustrates the phase response of the device. The output signal at the collector region is 180° out of phase with the input signal in the low-frequency range. As the operating frequency increases, the relative phase between the input and output signals begins to shift until the output signal is in phase with the input signal at 2.3 kHz.

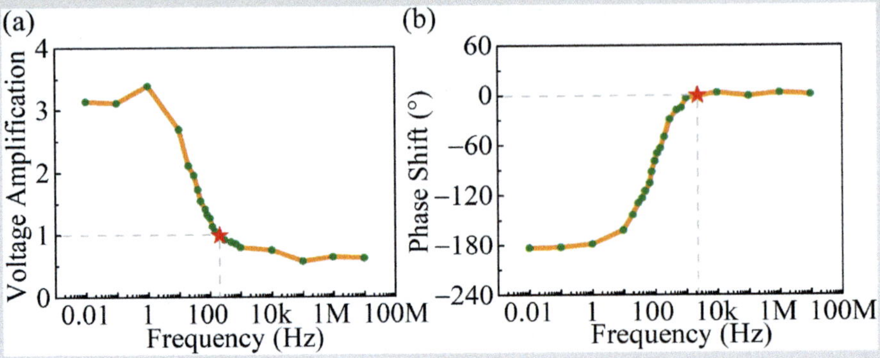

Figure 4. *(a) The frequency response of the device. The red star represents the frequency corresponding to a voltage amplification of 1. (b) The phase response of the device. The red star represents the frequency when the input and output signals are in phase. The green dots represent experimental test results and are connected by orange lines.*

Further, the frequency response of the β value can be evaluated according to the time domain characteristics of the device operating at different frequencies, as illustrated in Figure 3b–d. Here, the output current (i_C) of the device can be determined from the output signal (u_o) by applying the formula $i_C = (V_{CE} - u_o)/R_L$. As the operating frequency continuously increases, the amplitude of i_C is unchanged at first and then gradually decreases when the operating frequency exceeds 1 Hz. However, the input current (i_B) is almost unaffected by the operating frequency. Therefore, the β value changing with the frequency is consistent with the trend of i_C changing with the frequency; that is, it decreases as the operating frequency increases. In addition, to investigate the repeatability of the AC performance of the device, several other vdW BJTs with similar configurations were fabricated and investigated. All devices exhibited similar AC characteristics.

Conclusion

In summary, a vdW BJT was fabricated by vertically stacking MoS_2, WSe_2, and MoS_2 flakes in sequence. The static characteristics of the device were investigated in common emitter and common base modes, demonstrating excellent current modulation and saturation characteristics. The AC performance of the device in common emitter mode was also investigated. A phase inversion from output to input with a maximum voltage gain of around 3.5 was obtained in the low-frequency range. As the operating frequency increases, the voltage gain gradually decreases to unity at 200 Hz and the relative phase between the input and output signals gradually changes to 0° at 2.3 kHz. This work demonstrates the AC characteristics of the vdW BJT and experimentally proves the device's ability to process alternating signals. If the issues of device array fabrication and device-to-device variation can be further addressed, this will significantly promote the application of vdW BJTs as neuromorphic devices and wearable healthcare devices.

ACTIVITY

Consider the figure, the 2N3904 transistor is a general-purpose transistor with a typical β_{DC} value of 200.

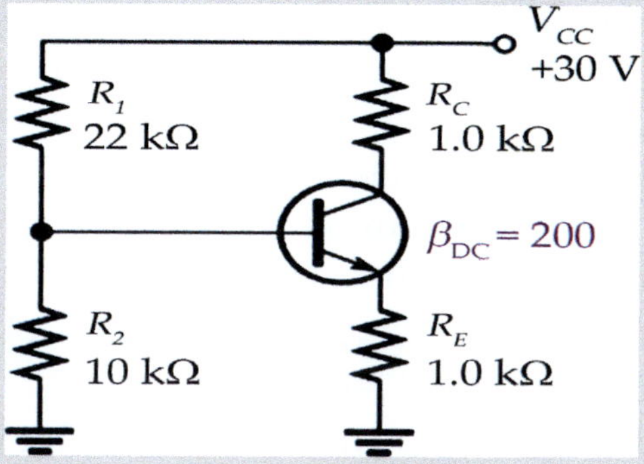

Calculate:

- The base voltage V_B
- The emitter voltage V_E
- The emitter current I_E
- The collector current I_C
- The base current I_B
- The collector voltage V_C
- The collector-to-emitter voltage $_V$CE

SUMMARY

- A bipolar junction transistor (BJT) is a type of transistor that uses both electrons and electron holes as charge carriers.
- BJTs use two p–n junctions between two semiconductor types, n-type and p-type, which are regions in a single crystal of material.
- The bipolar junction transistor is a solid-state device. In these transistors, the current flows in two terminals: the emitter and collector, and the flow of current is controlled by the third terminal, which is the base terminal.
- The bipolar junction transistor, unlike other transistors, is usually not a symmetrical device. This means that interchanging the collector and the emitter makes the transistor leave the forward active mode and start to operate in reverse mode.
- The transistor in which one p-type material is placed between two n-type materials is known as an NPN transistor.
- A pair of corresponding NPN and PNP transistors with near identical characteristics to each other are called Complementary Transistors. For example, a TIP3055 (NPN transistor) and the TIP2955 (PNP transistor) are good examples of complementary or matched pair silicon power transistors.
- Bipolar transistors, and particularly power transistors, have long base-storage times when they are driven into saturation; the base storage limits turn-off time in switching applications.
- Baker clamp is a generic name for a class of electronic circuits that reduce the storage time of a switching bipolar junction transistor (BJT) by applying nonlinear negative feedback through various kinds of diodes.
- The common-emitter current gain is represented by β_F or the h-parameter h_{FE}; it is approximately the ratio of the DC collector current to the DC base current in forward-active region

REVIEW QUESTIONS

1. Explain the basic structure of a Bipolar Junction Transistor (BJT) and the function of each of its regions.
2. Describe the difference between NPN and PNP BJTs in terms of their structure and current flow.
3. How does the current gain (β) of a BJT relate to the collector current (I_C) and the base current (I_B)?
4. How can a BJT be used as a switch? Describe the conditions necessary for switching between the 'on' and 'off' states.
5. What is bipolar junction transistor biasing? Discuss.

REFERENCES

1. Ahn, S. H., Sun, G. M., & Baek, H. (2022). Turn-off time improvement by fast neutron irradiation on pnp Si Bipolar Junction Transistor. *Nuclear Engineering and Technology, 54*(2), 501–506. https://doi.org/10.1016/j.net.2021.11.008.
2. Bowers, D. F. (2014). A fast precision operational amplifier featuring two separate control loops. In *2014 IEEE Bipolar/BiCMOS Circuits and Technology Meeting (BCTM)* (pp. 72–75). https://doi.org/10.1109/BCTM.2014.6987370.
3. Brar, B., Sullivan, G. J., & Asbeck, P. M. (2001). Herb's bipolar transistors. *IEEE Transactions on Electron Devices, 48*(11), 2473–2476. https://doi.org/10.1109/16.959334.
4. Bullis, W. M., & Runyan, W. R. (1967). Influence of mobility and lifetime variations on drift-field effects in silicon-junction devices. *IEEE Transactions on Electron Devices, 14*(2), 75–81. https://doi.org/10.1109/T-ED.1967.15912.
5. De Langen, K.-J., & Huijsing, J. H. (1998). Compact low-voltage power-efficient operational amplifier cells for VLSI. *IEEE Journal of Solid-State Circuits, 33*(10), 1482–1496. https://doi.org/10.1109/4.720304.
6. Gilasgar, M., Barlabé, A., & Pradell, L. (2020). High-Efficiency Reconfigurable Dual-Band Class-F Power Amplifier with Harmonic Control Network Using MEMS. *IEEE Microwave and Wireless Components Letters, 30*(7), 677–680. https://doi.org/10.1109/LMWC.2020.3003992.
7. Gummel, H. K., & Poon, H. C. (1970). An Integral Charge Control Model of Bipolar Transistors. *Bell System Technical Journal, 49*(5), 827–852. https://doi.org/10.1002/j.1538-7305.1970.tb04209.x.
8. Hosseini, S. E., & Dehrizi, H. G. (2012). A new BJT-transistor with ability of controlling current gain. In *International Multi-Conference on Systems, Signals & Devices* (pp. 1–4). https://doi.org/10.1109/SSD.2012.6207993.
9. Kong, L., Liu, H., Zhu, X., Boon, C. C., Li, C., Liu, Z., & Yeo, K. S. (2020). Design of a wideband variable-gain amplifier with self-compensated transistor for accurate db-linear characteristic in 65 nm CMOS technology. *IEEE Transactions on Circuits and Systems I: Regular Papers, 67*(12), 4187–4198. https://doi.org/10.1109/TCSI.2020.3006992.
10. Liu, Y., Hiblot, G., Furuhashi, T., Lin, H., Velenis, D., & De Wolf, I. (2019). Study of out-of-plane mechanical stress impact on Si BJT and diffusion resistor using in-situ nanoindentation probing. *Microelectronics Reliability, 100–101*, 113367. https://doi.org/10.1016/j.microrel.2019.113367.
11. Nam, H., Nguyen, D.-A., Kim, Y., & Seo, C. (2023). Design of 6 GHz variable-gain low-noise amplifier using adaptive bias circuit for radar receiver front end. *Electronics, 12*(9), 2036. https://doi.org/10.3390/electronics12092036.
12. Ozeren, E., Cahskan, C., Davulcu, M., Kayahan, H., & Gurbuz, Y. (2014). 4-Bit SiGe phase shifter using distributed active switches and variable gain amplifier for X-band phased array applications. In *2014 9th European Microwave Integrated Circuit Conference* (pp. 257–260). https://doi.org/10.1109/EUMIC.2014.6997851.

13. Perez-Verdu, B., Huertas, J. L., & Rodriguez-Vazquez, A. (1988). A new nonlinear time-domain op-amp macromodel using threshold functions and digitally controlled network elements. *IEEE Journal of Solid-State Circuits, 23*(4), 959–971. https://doi.org/10.1109/4.323.

14. Sivonen, P., Kangasmaa, S., & Parssinen, A. (2003). Analysis of packaging effects and optimization in inductively degenerated common-emitter low-noise amplifiers. *IEEE Transactions on Microwave Theory and Techniques, 51*(4), 1220–1226. https://doi.org/10.1109/TMTT.2003.809623.

15. Um, J.-Y. (2022). A compact variable gain amplifier with continuous time-gain compensation using systematic predistorted gain control. *IEEE Transactions on Circuits and Systems II: Express Briefs, 69*(2), 274–278. https://doi.org/10.1109/TCSII.2021.3143446.

CHAPTER

8 The Junction Field-Effect Transistor

LEARNING OBJECTIVES

After studying this chapter, you will be able to:

- Understand JFET concepts
- Know the device characteristics
- Explain the JFET biasing
- Focus on high electron mobility transistor

KEY TERMS FROM THIS CHAPTER

Aluminium gallium arsenide
Bipolar transistor
Current Source Bias
Drain terminal
Gallium arsenide
Gate terminal
Microwave amplifier
Bipolar junction transistors
Current
Depletion layers
Electrons
Gate Bias
High Electron Mobility Transistor
n-channel electrons

8.1. INTRODUCTION

One of the most basic varieties of field-effect transistor is the junction field-effect transistor (JFET). JFETs are semiconductors with three terminals that can be used to create amplifiers, electronically controlled switches, or resistors.

Because they don't require a biasing current, JFETs are only voltage-controlled, in contrast to bipolar junction transistors. Between the source and drain terminals, there is a semiconducting channel that conducts electricity. A reverse bias voltage is applied to a gate terminal to pinch the channel, which inhibits or stops the flow of electricity. When there is no voltage between the gate and source terminals of a JFET, the device is typically conducting. Less current will flow in the channel between the source and drain terminals of the JFET if a potential difference of the correct polarity is applied between its gate and source terminals. This is because the JFET will be more resistant to current flow.

Because JFETs operate on the idea of a depletion region—a region without majority charge carriers—they are sometimes referred to as depletion-mode devices. For current to flow, the depletion region needs to close.

An n-type or p-type channel can be found in JFETs. The current will be lowered in the n-type if the voltage applied to the gate is negative relative to the source (and in the p-type if the voltage applied to the gate is positive relative to the source). Little current is pulled from circuits used as input to the gate because a JFET in a common source or common drain configuration has a large input impedance (sometimes on the order of 10^{10} ohms).

8.2. JFET CONCEPTS

The bipolar (PNP/NPN) transistor, despite revolutionizing electronic equipment design, still has a major flaw with its low input impedance at the base-emitter junction. This causes impedance matching problems between interstage amplifiers. Scientists aimed to find a solution that would offer the high input impedance of a vacuum tube along with the benefits of a transistor. The result of this research is the field-effect transistor (FET). Unlike the bipolar transistor that uses bias current to control conductivity, the FET uses voltage to control an electrostatic field within the transistor.

In the figure below, the components of one kind of FET, the junction field-effect transistor (JFET), are contrasted with those of a bipolar transistor. The JFET is a three-element device similar to the other one, as the figure illustrates. In terms of operation, the JFET's "gate" element and the bipolar transistor's base are extremely similar. The emitter and collector of a bipolar transistor are represented by the "source" and "drain" elements of the JFET.

8.2.1. History

Julius Lilienfeld patented a number of devices in the 1920s and 1930s that were similar to FETs. However, decades of progress in materials science and fabrication technology would be necessary before FETs could be produced.

Heinrich Welker initially received a patent for JFET in 1945. Researchers Walter Houser Brattain, William Shockley, and John Bardeen made several unsuccessful attempts to construct an FET in the 1940s. While attempting to identify the causes of their failures, they came across the point-contact transistor. Following Shockley's theoretical treatment on JFET in 1952, a working practical JFET was made in 1953 by George C. Dacey and Ian M. Ross. Japanese engineers Jun-ichi Nishizawa and Y. Watanabe applied for a patent for a similar device in 1950 termed static induction transistor (SIT). The SIT is a type of JFET with a short channel.

After Silicon carbide (SiC) wide-bandgap devices were commercially introduced in 2008, high-speed, high-voltage switching using JFETs became technically possible. SiC JFETs were initially a niche product with high costs because of early manufacturing challenges, specifically inconsistencies and low yield. These manufacturing problems had largely been fixed by 2018. By that time, SiC JFETs were frequently utilized in addition to traditional low-voltage Silicon MOSFETs. The advantages of wide band-gap devices and the simple gate drive of MOSFETs are combined in SiC JFET + Si MOSFET devices.

8.2.2. JFET Structure

The figure below illustrates how a JFET is put together. The device's main body is a solid bar composed of either P-type or N-type material. Two deposits of material that differ from the bar material, which together form the "gate," are diffused into each side of the bar. A "channel" connecting the source and the drain is formed by the section of the bar with a smaller cross section than the rest of the bar that is located between the deposits of gate material. A P-type material gate and an N-type material bar are depicted in the figure 8.1 below. The device is referred to as an N-channel JFET because the channel's material is N-type. P-type material makes up the channel and N-type material makes up the gate of a P-channel JFET.

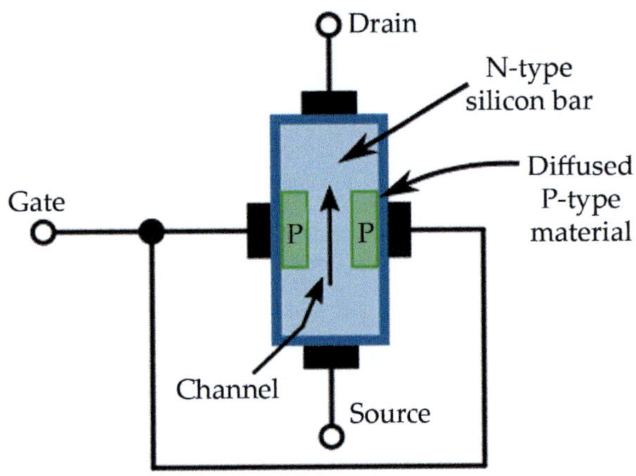

Figure 8.1. JFET structure.

Source: https://ecstudiosystems.com/discover/textbooks/basic-electronics/field-effect-transistors/junction-field-effect-transistors/.

8.2.3. JFET Symbols

The NPN and PNP bipolar transistor schematic symbols are compared with those of the two types of JFET in the figure 8.2 below. Similar to bipolar transistor types, the only differences between the two JFET types are in the required bias voltage configuration and the symbol's arrow direction. The arrow in a JFET symbol always points in the direction of the N-type material, just like it does in bipolar transistor symbology. As a result, the N-channel JFET symbol displays an arrow pointing in the direction of the drain/source channel, while the P-channel symbol displays an arrow pointing in the opposite direction, toward the gate.

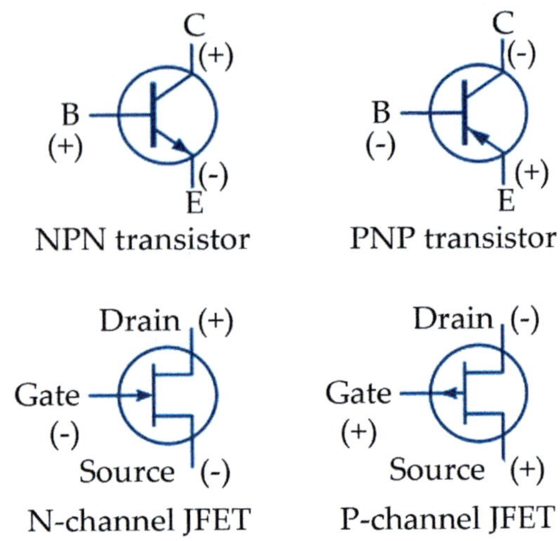

Figure 8.2. Symbols and bias voltages for bipolar transistors and JFET.

Source: Source: https://ecstudiosystems.com/discover/textbooks/basic-electronics/field-effect-transistors/junction-field-effect-transistors/.

8.2.4. Functions

The way a garden hose operates is comparable to how a JFET operates. A hose's water flow can be adjusted by squeezing it to decrease its cross section, and a JFET's electric charge flow can be adjusted by narrowing its current-carrying channel. Similar to the pressure differential on either end of the hose, the electric field between the source and drain determines the current as well. The characteristics displayed in the diagram above an applied voltage do not support this current dependency. The JFET is typically operated in this saturation region, which is a constant-current zone where the drain-source voltage essentially has no effect on the device current. The JFET, junction transistors, and thermionic tube (valve) tetrodes and pentodes all have this constant-current feature.

By applying a voltage between the gate and the source to reverse bias the gate-source pn-junction, the conducting channel's cross-sectional area is constrained and its depletion layer widens, encroaching upon it. This technique is known as the field effect. The depletion layer gets its name from the fact that it is practically electrically non-conductive due to the exhaustion of mobile carriers.

Pinch-off occurs and drain-to-source conduction ceases when the depletion layer crosses the conduction channel's width. At a specific reverse bias (VGS) of the gate-source junction, pinch-off happens. Even among devices of the same type, there are significant differences in the pinch-off voltage (Vp), also referred to as the threshold voltage or cut-off voltage.

For example, $V_{GS(off)}$ for the Temic J202 device varies from −0.8 V to −4 V. Typical values vary from −0.3 V to −10 V. (Confusingly, the term pinch-off voltage is also used to refer to the V_{DS} value that separates the linear and saturation regions.)

A negative gate-source voltage (VGS) is needed to turn off an n-channel device. Conversely, positive VGS is needed to turn off a p-channel device. During regular operation, the gate's electric field partially prevents source-drain conduction. A few JFET devices have symmetric source and drain configurations.

8.2.5. Junction Field Effect Transistor Construction

JFETs are made up of a semiconducting material channel that allows current to flow. There are two different kinds of JFETs: n-channel and p-channel. Since electrons make up the majority of carriers in n-channel JFETs and electrons are known to be more mobile than holes, n-channel JFETs are generally preferred over p-channel JFETs.

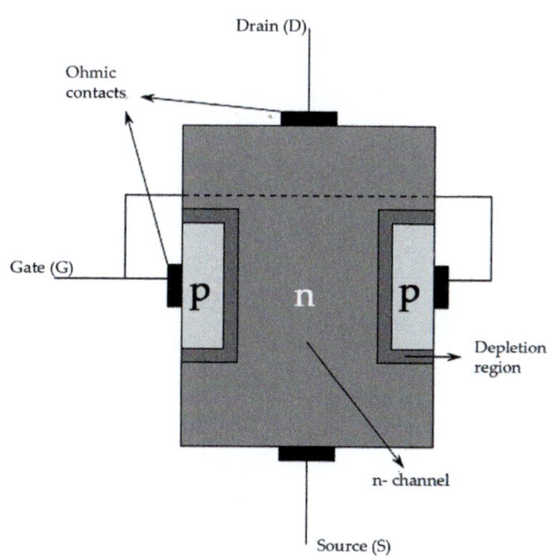

Figure 8.3. N-channel JFET.

Source: By Hermitage17, 6 February 2007, public domain, https://commons.wikimedia.org/wiki/File:N-channel_JFET.JPG.

An N-channel JFET has an n-type silicon bar that is considered the channel and two smaller pieces of p-type silicon material that diffuse on opposite sides of its middle section to form a p-n junction. Similarly, in a p-channel JFET, a p-type silicon bar is considered the channel and smaller pieces of n-type silicon material diffuse on the opposite sides of its middle section to form a p-n junction. Internally, the p-n junction creates a terminal or gate known as a gate terminal.

It establishes ohmic contacts at both ends of the channel—the source terminal and the drain terminal—using a battery. The gate terminal is used to control the current flow from the source to the drain. A p-n junction diode is formed by the gate-source, and another by the drain. These diodes are referred to as the gate-source and gate-drain diodes.

8.2.6. JFET Working Operation

The operation of p-channel and N-channel JFETs is identical, despite the charge carriers being inverted, i.e., in the n-channel, electrons make up the majority of carriers, whereas in the p-channel, holes do. Since n-channel JFET operation is more desirable, we will explain it here.

The amount of bias applied to the gate terminal and source-drain terminal determines how wide the channel is. We describe how the channel width changes when a positive or negative voltage is applied across it in the figure 8.3 above. Let's examine it in more detail.

Depending on the voltage applied across the terminals, a JFET can operate in three different conditions.

No voltage

When neither any voltage is applied across the source to drain terminal i.e., V_{ds} = 0 nor any bias is applied to the gate terminal i.e., V_{gs}= 0 the depletion region around the p-n junction is of equal thickness and symmetrical in nature.

8.2.6.1. Negative Voltage

The p-n junction becomes reverse biased and forms a depletion region when the drain is applied with a positive bias with respect to the source, and the gate is negatively biased with respect to the source. There is a voltage drop along the channel's length as the drain current passes through it. As a result, the width of the depletion layer is greater at the drain because the reverse bias there is greater than it is at the source. The channel narrows as a result of rising resistance and falling current I_d.

The drain current cuts off entirely, and the depletion layers meet at the center if the negative voltage across the gate is increased further. In a similar vein, when the negative voltage across the gate is decreased, the depletion layers begin to shrink, which lowers resistance and raises the drain current I_d.

8.2.6.2. Positive Voltage

In contrast to conventional drain current I_d, which flows through the channel from the drain to the source, electrons begin to move from the source terminal to the drain terminal when a positive voltage is applied to the drain terminal with respect to the source terminal without connecting the gate terminal to the supply. The diode is reverse biased as a result of the uniform voltage drop that results from this current flow across the channel resistance. The channel's points closest to the drain than the source have a more negative gate.

As a result, at locations closer to the drain than the source, depletion layers enter the channel more deeply. Consequently, when V_{ds} is applied across the terminals, the wedge-shaped depletion regions are created.

8.2.7. Basic MESFET Operation

A MESFET, also known as a metal-semiconductor field-effect transistor, is a semiconductor device that is comparable to a JFET, but with a Schottky junction for the gate instead of a p-n junction.

8.2.7.1. Construction

Compound semiconductor MESFETs, like gallium arsenide, indium phosphide, and silicon carbide, lack high-quality surface passivation. Despite being faster, they are more costly than silicon-based JFETs or MOSFETs. With a production limit of around 45 GHz, MESFETs are commonly used for microwave frequency communications and radar [Lepkowski, W.; Wilk, S.J.; Thornton, T.J. (2009)]. The first MESFETs were created in 1966, and within a year, their exceptional RF microwave performance was showcased.

8.2.7.2. Functional Architecture

The MESFET, like the JFET, is distinct from the typical insulated-gate FET or MOSFET because it lacks an insulator beneath the gate in the active switching area. This means that the gate of the MESFET must be biased in such a way that a reversed-biased depletion zone regulates the channel below, rather than a forward-conducting metal-semiconductor diode.

While this restriction limits certain circuit possibilities due to the need for the gate to remain reverse-biased and not exceed a certain level of forward bias, MESFETs' analog and digital devices function well within design limits. The key aspect of design is the extent of the gate metal over the switching region. A narrower gate modulated carrier channel generally improves frequency handling capabilities. The spacing of the source and drain in relation to the gate, as well as the lateral extent of the gate, are important but slightly less critical design parameters. Increasing the lateral length of the gate can improve MESFET current handling ability, but there may be limitations due to phase shift along the gate caused by the transmission line effect. Therefore, most production MESFETs use a built-up top layer of low-resistance metal on the gate, often resulting in a mushroom-like profile in cross-section [Lepkowski, W.; Wilk, S.J.; Thornton, T.J. (2009)].

8.2.7.3. Advantages and Disadvantages

MESFET offers an advantage over MOSFET by having carriers with higher mobility in the channel, although it is still less than half of the mobility of bulk counterpart material. The carriers are located in the inversion layer, extending into the oxide layer. As the depletion region separates the charge carriers from the surface, their mobility increases and approaches that of bulk material, resulting in higher current, transconductance, and transit frequency for the MESFET. The presence of a Schottky metal gate limits the forward bias voltage on the gate to the turn-on voltage of the Schottky diode, which is a disadvantage of the MESFET structure. GaAs Schottky diodes have a turn-on voltage of approximately 0.7 V, with the threshold voltage needing to be lower. It is more challenging to fabricate circuits with a large number of MESFETs in the enhancement mode. The higher transit frequency of the MESFET allows for superior microwave amplification. MESFET in depletion mode is preferred for its larger current and transconductance, with GaAs being the most common material choice due to its higher electron mobility and velocity compared to silicon, as well as its ability to resolve the issue of absorbing microwave power.

8.2.7.4. Applications

A multitude of MESFET fabrication options have been investigated for a broad range of semiconductor systems. Military communications, commercial optoelectronics, satellite communication, military radar devices, as a front-end low noise amplifier of microwave receivers, as a power amplifier for the output stage of microwave links, and as a power oscillator are some of the primary application areas.

8.2.8. Advantages of Junction Field Effect Transistor (JFET)

Transistors are essential to modern technology because of their many benefits, some of which are as follows:

- Stability: They provide good stability under a variety of operating conditions.
- Low power consumption: They are energy-efficient because they use little power.
- High impedance: JFETs are known for having high input impedances, which make them ideal for use in amplifier circuits.
- Simplicity: JFETs don't require the intricate biasing configurations that are frequently present in other transistors, making them comparatively easy to use.
- No Gate Current: Because JFETs don't have gate current flow, circuit design is made easier in situations where current flow must be prevented.

8.2.9. Disadvantages of Junction Field Effect Transistor (JFET)

Despite their widespread use, transistors still have certain drawbacks. Some of these include being unipolar devices, such as JFETs, where the current flow can only be controlled by one type of charge carrier. JFETs can also exhibit gate-source leakage currents, which are necessary for certain applications. Additionally, JFETs have limited availability, making it challenging to find specific ones with desired characteristics. They also have low gain compared to other types of transistors, making them unsuitable for high-gain applications. Lastly, their cost is relatively high, which can impact the overall cost of electronic devices.

8.3. THE DEVICE CHARACTERISTICS

As will be covered below, the JFET characteristics can be examined for both N-channel and P-channel.

8.3.1. N-Channel JFET Characteristics

The figure 8.4 below displays the transconductance curve, or N-channel JFET characteristics, as a function of gate-source voltage and drain current. The transconductance curve has several regions, including the ohmic, saturation, cutoff, and breakdown regions.

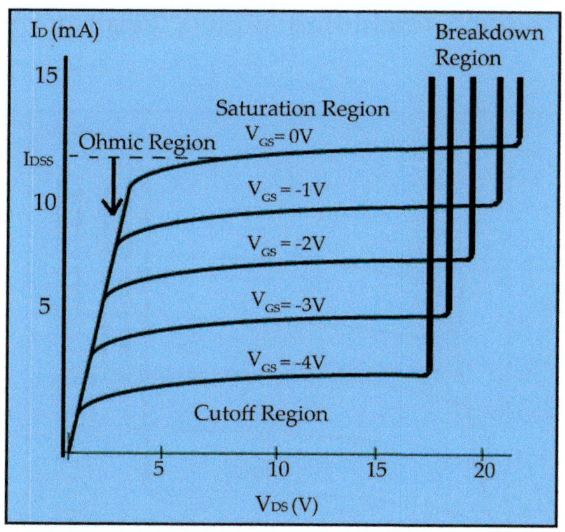

Figure 8.4. N-Channel JFET Characteristics.

Source: https://www.elprocus.com/junction-field-effect-transistor-jfet/.

8.3.1.1. Ohmic Region

The term "Ohmic region" refers to the area where the drain current is opposed by the JFET transistor resistance, and the transconductance curve exhibits a linear response.

8.3.1.2. Saturation Region

Due to the applied gate-source voltage, the N-channel junction field-effect transistor is in the saturation region and operating at maximum current.

8.3.1.3. Cutoff Region

The N-channel JFET is in the off position because there won't be any drain current flowing in this cutoff region.

8.3.1.4. Breakdown Region

If the VDD voltage applied to the drain terminal is higher than the required maximum voltage, the transistor is unable to control the current and allows it to flow from the drain terminal to the source terminal, causing the transistor to enter the breakdown region.

8.3.2. P-Channel JFET Characteristics

The figure 8.5 below illustrates the characteristics of a P-channel JFET, showing the transconductance curve graphed between drain current and gate-source voltage. The curve consists of several regions, including ohmic, saturation, cutoff, and breakdown regions.

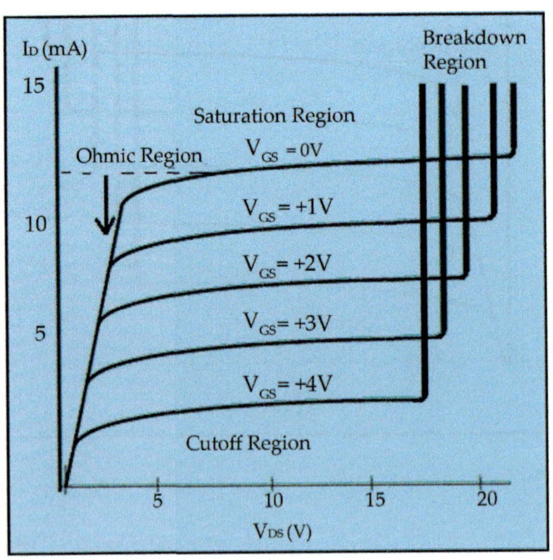

Figure 8.5. P-Channel JFET Characteristics.

Source: https://www.elprocus.com/junction-field-effect-transistor-jfet/.

8.3.2.1. Ohmic Region

The term "Ohmic region" refers to the only area where the drain current is opposed by the JFET transistor resistance and the transconductance curve exhibits a linear response.

8.3.2.2. Saturation Region

Because of the applied gate-source voltage, the N-channel junction field effect transistor is in the saturation region and operating at maximum current.

8.3.2.3. Cutoff Region

The N-channel JFET is in the off position because there won't be any drain current flowing in this cutoff region.

8.3.2.4. Breakdown Region

If the voltage applied to the drain terminal (VDD) exceeds the required maximum voltage, the transistor will not be able to handle the current and will allow it to flow from the drain terminal to the source terminal. This will cause the transistor to enter the breakdown region.

8.4. JFET BIASING

We are aware that a JFET's input resistance is significantly higher than that of a bipolar transistor. Therefore, it is important to note that the JFET's input resistance should never decrease during biasing. As a result, the JFET's gate-source diode is always reverse-biased to maintain this characteristic. For biasing an N-channel JFET, a negative V_{gs} is required, while for a P-channel JFET, a positive V_{gs} is needed. Conversely, the high input resistance of a JFET will decrease abruptly if the gate-source is forward-biased instead of reverse-biased, causing it to lose the advantage that makes it superior to a BJT.

Biasing can occur in the active or ohmic regions of a JFET. The resistance is equivalent to the resistance when biased in the ohmic region, but it behaves like a current source when biased in the active region. Below are some common techniques for JFET biasing:

1. Gate Bias
2. Self-Bias
3. Voltage Divider Bias
4. Source Bias
5. Current Source Bias

8.4.1. Gate Bias

In this biasing scheme, the JFET's source is earth, and the biasing resistor (R_g) supplies the gate with the negative gate voltage ($-V_{gg}$) (figure 8.6). This biasing technique is known as gate biasing since the bias voltage is applied to the JFET gate. When a negative bias voltage is applied to the gate—a voltage that is less than Id_{ss}—drain current builds up. The drain voltages that are produced in parallel to this drain current as it travels from Rd are as follows.

$$V_d = V_{dd} - I_d R_d$$

The reason for the unpopularity of this JFET biasing technique in an active region is that gate biasing makes point Q extremely unstable and vulnerable. Nonetheless, the technique is favored for biasing in the ohmic region as opposed to the active region since there is no problem with Q point stability there.

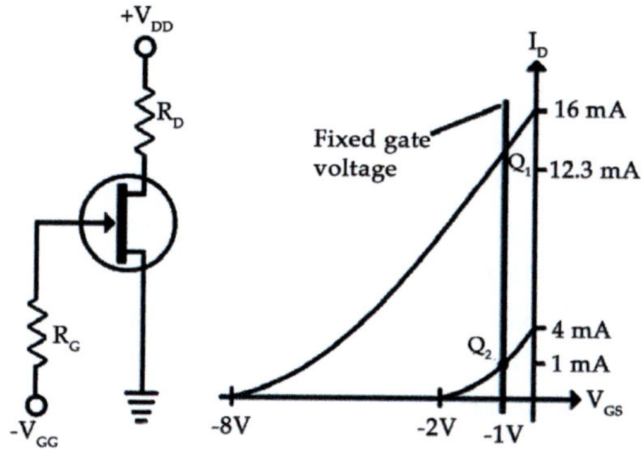

(a) (b)

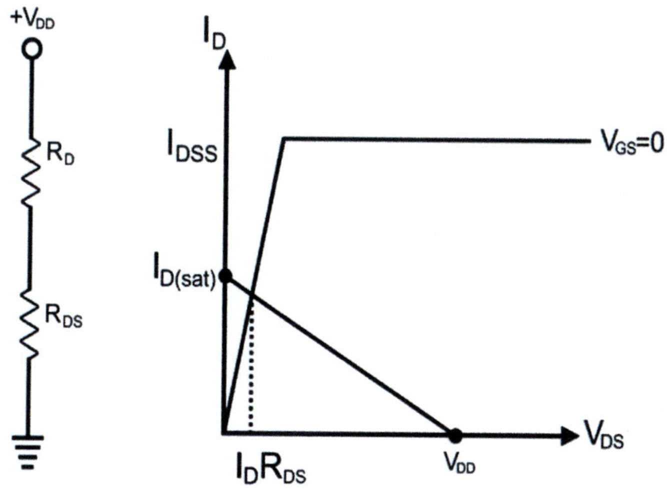

Figure 8.6. (a) Gate bias. (b) Q point unstable in active region. (c)biased in ohmic region. (d) JFET is equivalent to resistance.

Source: https://www.electroniclinic.com/biasing-of-jfet-gate-bias-self-bias-voltage-divider-bias-source-bias-current-source-bias/.

The minimum and maximum trans-semiconductor curves of a JFET 2N5459 operating in its inactive region are displayed in figure (b). In this case, the values of I_{dss} and $V_{gs\,(off)}$ range from 4 to 16 mA and -2 to -8, respectively. As shown in the figure, we obtain the minimum and maximum Q points if the gate bias value of this JFET drops to -1V. As a result, in this case, Q1's drain current is 12.3 mA, while Q2's drain current is only 1 mA. A JFET's bias in the ohmic region is demonstrated in figure (c). The DC load line's upper end has a drain saturation value that can be ascertained as follows.

$$I_{D\,(sat)}=V_{DD}/R_D$$

Remember, when a JFET is biased in the ohmic region, the value of V_{gs} is retained at zero (i.e., V_{gs}=0) and $I_{D\,(sat)}$ value is far lower compared to I_{DSS} i.e.

$$I_{D\,(sat)} << I_{DSS}$$

The equation reflects that the value of drain saturation current has to be significantly lower compared to maximum drain currents' value i.e., if a JFET's I_{DSS} value is 10mA, hard saturation results in case V_{gs} value is considered as zero and $I_{D\,(SAT)}$ value 1mA.When a JFET is biased in the ohmic region, t is reflected vide a resistance R_{DS} (i.e., JFET is converted to R_{DS}) as shown in figure (d). The figure indicates that JFET is fixed in R_D series, which treats JFET as if it were resistant to R_{DS}. Drain voltages can be computed using such equivalent circuits. The value of the drain voltages approaches zero when the R_{DS} value is lowest relative to the R_D.

It is evident from the discussion above that this kind of bias is thought to be very unstable. Another significant flaw in it is that additional power supplies will become necessary if the biasing value of the V_{GS} is not zero. As a result, using this biasing method results in very little.

8.4.2. Self-Bias

In this biasing technique, a power supply is connected only to the drain of the JFET, with no power supply at the gate (see figure 8.7). The gate resistor R_G has no impact on biasing as there is no voltage drop across it. Since the gate is connected to ground through resistor R_G, it remains at zero voltage. It is important to use R_{GS} to isolate AC signals from ground when using JFET as an amplifier. The current source resistance passes through R_s, leading to a voltage drop across R_s. Therefore, the voltage drops across R_s in this biasing method serve as the gate-source voltage.

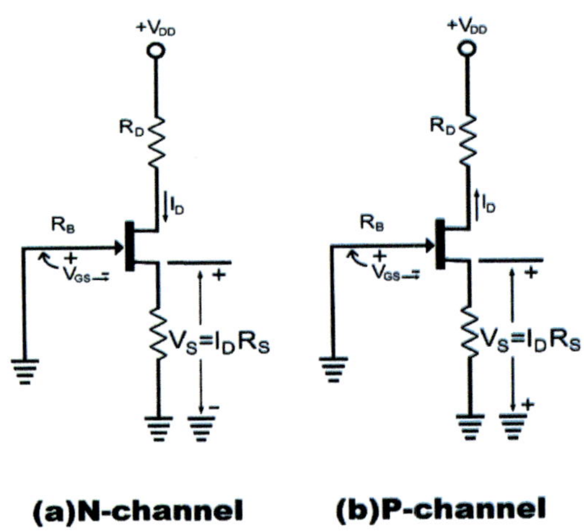

Figure 8.7. Self Bias Circuit.

Source: https://www.electroniclinic.com/biasing-of-jfet-gate-bias-self-bias-voltage-divider-bias-source-bias-current-source-bias/.

Voltage drops parallel to Rs when source current flows through the N-channel JFET's source resistance, as seen in figure (a), making the source positive with respect to ground. As drain current passes through source resistor $R_{s,}$ and values of source current and drain current are equal (i.e., $I_S=I_D$), further as the value of gate voltage (V_g) is zero thus, in this case, source voltage V_S equals the product of drain current and source resistance (R_S)i.e., $V_S=I_D\ R_S$. But the gate-to-source voltage is as follows:

$$V_{GS} = V_G - V_S = 0 - I_D R_S$$

$$V_{GS} = -I_D\ R_S \text{ ...... (N-Channel)}$$

It is clear from the equation that the gate-source voltage is equal to the parallel source resistor's negative voltage. Basically, this circuit creates bias for itself by applying reverse bias to the gate in response to negative voltages generated parallel to R_S. In the case of a P-channel JFET, this is demonstrated in figure b, where the current flowing through R_S results in a negative voltage on the source. As a result, V_{GS} has the following value.

$$V_{GS} = +I_D\ R_S \text{ .... (P-Channel)}$$

Drain voltage value with respect to ground is as under

$$V_D = V_{DD} - I_D\ R_D$$

As, $V_S = I_D\ R_S$, thus, drain to source voltage equals the following

$$V_{DS} = V_D - V_S$$

Entering the values of V_D and V_S in the above equation

$$V_{DS} = V_{DD} - I_D\ R_D - I_D R_S$$

$$V_{DS} = V_{DD} - I_D\ (R_D + R_S) \text{ ...(N-Channel)}$$

$$V_{DS} = -V_{DD} + I_D\ (R_D + R_S) \text{ ...(P-Channel)}$$

Recall that self-interest serves as feedback as well. As draining current rises, source current rises as well (because $I_S = I_D$), which causes the drop in R_S to rise as well. Reverse bias on the gate-source diode increases with an increase in drop parallel to R_S. Since this bias supplies reverse bias voltage to the diode, which lowers drain current. Only small-signal amplifiers can be self-biased because the Q point is not as stable as it is with other biasing techniques; for high signals, a JFET is biased with alternative techniques. When a tiny signal is present, the front portion of communication receivers is where self-bias JFET circuits are primarily utilized.

8.4.3. Voltage Divider Bias

Thevenin voltages are applied to the gate using this biasing technique by fixing two resistors on the gate. A voltage divider bias circuit is depicted in Figure 8.8. As the name implies, bias voltage, also known as gate voltage, is divided into two parts with the assistance of two resistors, and its value is reduced to such an extent that it is only a small portion of the supply voltage (i.e., only a small portion of the supply voltage is equalized in the gate voltage).

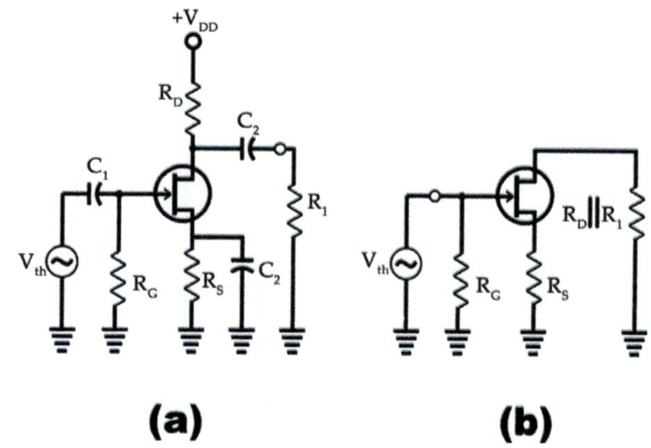

Figure 8.8. JFET amplifier and its equivalent.

Source: https://www.electroniclinic.com/biasing-of-jfet-gate-bias-self-bias-voltage-divider-bias-source-bias-current-source-bias/.

As per the diagram, R_1 and R_2 are used to set the gate voltage (V_G). The voltage divider formula gives the following values for R_1 and R_2.

$$V_G = V_{1h} [R_2/R_1+R_2] V_{DD}$$

Thus, we get the following voltage parallel to the source resistor or between ground and source.

$$V_S = V_G - V_{GS}$$

As VGS is negative, therefore, source voltages are relatively higher as compared to the gate voltage. When source voltages are divided by source resistance, we get drain current (which is equal to source current) i.e.

$$I_D = V_G - V_{GS} / R_S = V_{1h} - V_{GS} / R_S$$

Almost all JFETs have normalized drain currents if the value of V_{GS} is significantly higher than V_G or V_{ih} (see diagram 8.9 b). Nevertheless, since V_G varies greatly between transistors, I_D may take on a variety of values. Recall that the presence of a large gate voltage relative to V_{GS} (the Q point, which is the middle point of intersection between a straight line and a trans-conductance curve) can cause voltage divider bias to establish a strong Q point. The voltage divider bias technique is not as balanced as the bipolar transistor, despite being more stable than self-bias.

For clarification, separate voltage divider bias circuits of N-channel and P-channel JFET have been shown in Figure 8.9.

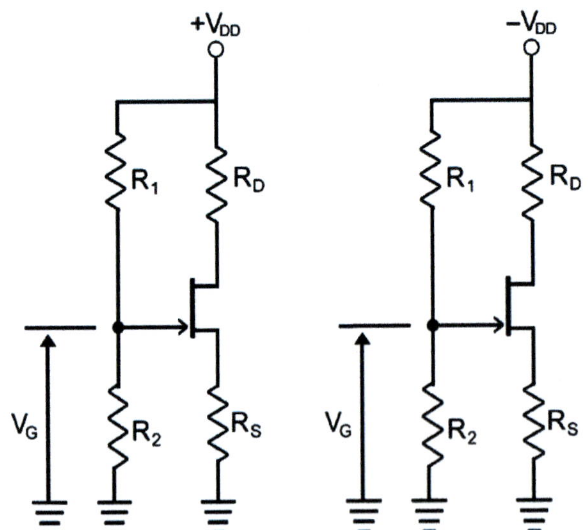

Figure 8.9. Biasing the gate using a voltage divider rather than a separate VGG.

Source: https://www.electroniclinic.com/biasing-of-jfet-gate-bias-self-bias-voltage-divider-bias-source-bias-current-source-bias/

8.4.4. Source Bias

In order to eliminate variations in V_{GS} as much as possible, the source bias method—also known as the two supply source bias—is used. This method results in an excessive increase in the V_{SS} value relative to the V_{GS}. A source bias circuit is depicted in Figure 8.10. In actuality, the drain current value will be as follows

$$I_D = V_{SS} - V_{GS}/ R_S$$

In a typical situation, drain current is achieved by dividing supply voltage (V_{SS}) by source resistance (R_S). i.e.

$$I_D = V_{SS}/ R_S$$

In such a scenario, temperature and JFET changes, but the drain current value stays nearly constant. Since the VSS value is significantly higher than the V_{GS} value, which typically ranges from -1V to -5V, source bias can produce somewhat better results.

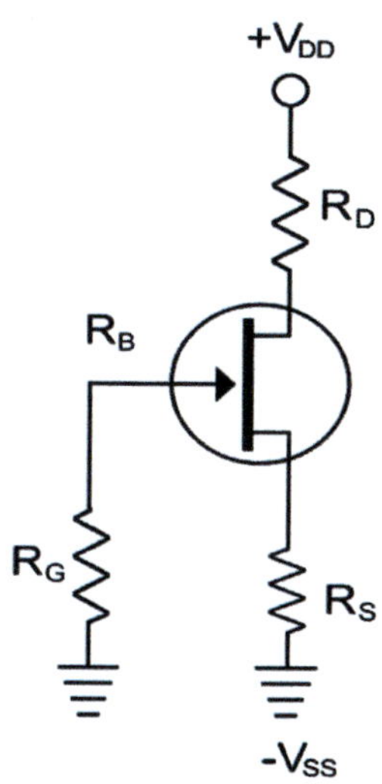

Figure 8.10. Two Supply Source Bias.

Source: https://www.electroniclinic.com/biasing-of-jfet-gate-bias-self-bias-voltage-divider-bias-source-bias-current-source-bias/.

8.4.5. Current Source Bias

A bipolar transistor generates a fixed value of drain current in current source bias by acting as a constant current source. Additionally, a suitable gate voltage value for preventing variations in VGS does not exist when the drain supply voltage is not too high. Under such circumstances, the circuit designer's preference is to apply current source bias, as shown in Figure 8.11a.

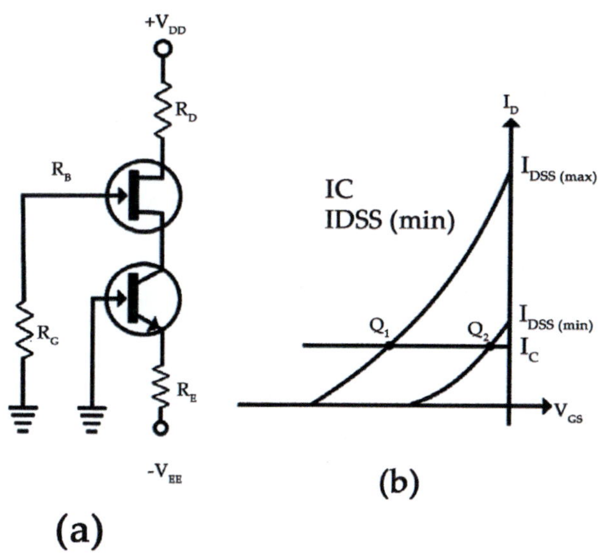

Figure 8.11. Current source bias.

Source: https://www.electroniclinic.com/biasing-of-jfet-gate-bias-self-bias-voltage-divider-bias-source-bias-current-source-bias/.

The circuit makes use of a bipolar transistor, on whose emitter bias is applied. JFET is therefore subjected to a fixed current. As a result, in this case, the drain current value will be as follows.

$$I_D = V_{EE} - V_{BE}/ R_E$$

The goal of the junction transistor is to equalize the collector current and drain current of the JFET because it functions as a current source (i.e., $I_D = I_C$). The drain current value stays constant when the IC value is constant. As a result, the effects of variations on V_{GS} caused by current source bias are negated, and JFET functions in a balanced manner.

Figure (b) illustrates the effectiveness of current source bias. The diagram indicates that both Q points have the same current capacity even though their V_{GS} values differ. It suggests that V_{GS} has no appreciable effect on the drain current value. Therefore, in contrast to other biasing techniques, current source biasing creates the most stable Q points since drain currents maintain a constant value despite changes in JFET characteristics.

8.5. HIGH ELECTRON MOBILITY TRANSISTOR

High Electron Mobility Transistor is what the name HEMT stands for. The device is a type of field-effect transistor, or FET, that can function at extremely high frequencies because it makes use of an uncommon property of a very narrow channel. The HEMT offers an extremely appealing low noise performance in addition to its very high frequency performance.

The apparatus functions essentially as a field-effect transistor by integrating a junction between two materials that have distinct band gaps (i.e., a heterojunction) as the channel as opposed to the conventional MOSFET's use of a doped region. The HEMT may also be referred to as a heterojunction FET, HFET, or modulation-doped FET, MODFET, depending on the context of its structure.

8.5.1. HEMT Development

Despite being essentially, a type of field-effect transistor, the HEMT uses an uncommon electron mobility mechanism. The first experiments with this mode of carrier transport were conducted in 1969, but the first experimental devices were not available for investigation and initial use until 1980. They were first used in the 1980s, but because of their initial exorbitant cost, their use was fairly restricted.

Since they are now somewhat less expensive, they are being used more frequently. They can even be found in a wide range of microwave radio communications links, mobile telecommunications, and many other RF design applications.

8.5.2. HEMT Structure and Fabrication

The unique PN junction that a HEMT employs is its essential component. It is referred to as a heterojunction and is made up of two different types of materials on either side of the junction. Gallium arsenide (GaAs) and aluminum gallium arsenide (AlGaAs) are the most often utilized materials. Because gallium arsenide offers a high degree of fundamental electron mobility, which is essential to the device's functioning, it is typically used. Because silicon has much less electron mobility, it is not used.

A HEMT can contain a wide range of structures, all of which essentially employ the same production techniques. An intrinsic layer of gallium arsenide is first placed on top of the semi-insulating gallium arsenide layer in the process of manufacturing a HEMT. This has a thickness of just one micron. On top of this is then applied a very thin layer of intrinsic aluminum gallium arsenide, ranging from 30 to 60 Angstroms. Its goal is to guarantee that the doped aluminum gallium arsenide region and the heterojunction interface remain apart. If high electron mobility is to be attained, this is crucial.

Above this is the doped layer of aluminum gallium arsenide, which is roughly 500 Angstroms thick. It is necessary to precisely control this layer's thickness, and this requires the use of specialized techniques. Two primary structures are employed. These are the recess gate structure and the self-aligned ion-implanted structure. The gate, drain, and source of the self-aligned ion-implanted structure are typically metallic contacts, though germanium can occasionally be used for the source and drain contacts. Usually constructed of titanium, the gate creates a tiny reverse-biased junction resembling the GaAsFET.

To facilitate the creation of the drain and source connections, an additional layer of n-type gallium arsenide is applied to the recess gate structure. Because it determines the FET's threshold voltage, the thickness beneath the gate is also extremely important. The channel is extremely small due to the gate's tiny size. Because the gate is typically only 0.25 microns or smaller, the device performs exceptionally well at high frequencies.

8.5.3. HEMT Operation

Compared to other FET types, the HEMT operates somewhat differently.

Many of the electrons from the n-type region stay near the hetero-junction as they travel through the crystal lattice. A two-dimensional electron gas is created by these electrons forming a layer that is only one electron thick. Because the mobility of the electrons in the gas is very high and there are no other donor electrons or objects with which the electrons will collide, the electrons can move freely within this region.

The device's conductivity is controlled by modulating the number of electrons in the channel formed by the 2D electron gas through the application of a bias to the gate, which functions as a Schottky barrier diode. This is comparable to the more conventional FET types, in which the gate bias modifies the channel's width [Kasamatsu, A; Kasai, K; Hikosaka, K; Matsui, T; Mimura, T (2004)].

Using HEMT devices has a number of benefits.

- High gain: At microwave frequencies, HEMTs have a high gain because the majority carrier is nearly always the charge carrier and the minority carrier is not heavily involved.
- Low noise: Compared to other field effect devices, HEMTs have very little current variation, which results in very quiet operation.

8.5.4. Applications

The original purpose of the HEMT's development was high-speed applications. It wasn't until the initial devices were manufactured that their extremely low noise figure was discovered. This is connected to the two-dimensional electron gas's characteristics and the decrease in electron collisions.

Owing to their exceptional noise performance, these devices find extensive application in low-noise small signal amplifiers, power amplifiers, oscillators, and mixers that operate at frequencies exceeding 60 GHz. It is projected that these devices will eventually be widely accessible for frequencies up to approximately 100 GHz. As a matter of fact, HEMT devices find extensive usage in radio frequency (RF) design applications such as radio astronomy, radar, DBS, cellular telecommunications, and any other application requiring high-frequency performance with low noise. Worldwide, a large number of semiconductor device manufacturers produce HEMTs. Although they can still be found as discrete transistors, integrated circuits are now more often where you'll find them. These MMICs, or monolithic microwave integrated circuit chips, are widely used for RF design applications. HEMT-based MMICs are particularly popular because they offer the necessary level of performance in a variety of applications [Ajayan, J.; Nirmal, D.; Mathew, Ribu; Kurian, Dheena; Mohankumar, P.; Arivazhagan, L.; Ajitha, D. (2021)].

8.5.5. Other HEMT-Based Devices

The fundamental HEMT device is available in several variations. In certain areas, the performance of these additional devices is enhanced.

- pHEMT: The Pseudomorphic High Electron Mobility Transistor, or PHEMT, is the source of its name. These gadgets are widely utilized in LNA and wireless communication applications. PHEMT transistors have good low noise figures and performance along with high power added efficiency. Consequently, PHEMTs find widespread application in satellite communication systems across various formats, such as DBS-TV (direct broadcast satellite television), where they are utilized in low noise boxes and LNBs that are connected to satellite antennas. Radar and microwave radio communications systems, as well as general satellite communication systems, also use them. When exceptionally high speed is needed, PHEMT technology is also employed in high-speed digital and analog integrated circuits.
- mHEMT: The metamorphic HEMT, also known as the mHEMT, is an advancement over the pHEMT. In order to match the lattice constant of the GaAs substrate and the GaInAs channel, the indium concentration in the AlInAs buffer layer is graded. This has the benefit of allowing for the utilization of almost any indium concentration in the channel, allowing the devices to be tailored for various uses. It is discovered that while a high indium concentration yields more gain, a low indium concentration offers superior low noise performance.

Although these HEMT variations are not as well-known, they can still offer certain features that are required in certain specialized applications.

CLASS ACTIVITY

JFET GATE TRANSFER CHARACTERISTIC: CURVE TRACER

Obtain the detailed transfer characteristic of your JFET with the computerized JFET Transfer Tracer option on the curve tracer. Plug your JFET into the Terminal Block provided with the curve tracer. You will find that the lead ordering on the JFET is reversed relative to the lead ordering on the Terminal Block. This would seem to require that the JFET leads be twisted. This is probably not a good thing to do; you will be using this JFET repeatedly in this lab, and you don't want to mangle the JFET leads. The tracer can be used without mangling the leads by using one or both of these solutions:

1. The 2N4392 JFET is a symmetric JFET; the Source and Drain are technically interchangeable (though we do not generally advise you to do this). If you plug the JFET Source into the Terminal Block Drain, and vice versa, you will get the same characteristic curves as if you plugged the JFET in properly, even though the leads have been effectively reversed from the Tracer's perspective.
2. If you swap the leads as in solution 1 immediately above, you can then rotate the Terminal Block by 180 degrees so that the Terminal Block Source is plugged into the Tracer Drain, and vice versa. From the Tracer's perspective, this will swap the leads yet again; this is one of those rare cases where two wrongs (two swaps) do make a right (no net swap).

Use the Tracer's analysis option to fit the transfer characteristic to a parabola, and to find the transconductance gm and source resistance, rs as a function of VGS. How close is the characteristic to a parabola? Is it at least a parabola over some limited range?

Plot all your data, and add the points that you took by hand to the transfer characteristic curve. For a gate voltage of -1V, find the transconductance directly by differentiating the transfer characteristic curve, and check that your value agrees with the value automatically calculated by the Curve Tracer. The 2N4392 JFET is designed to be operated as a switch, and its transfer characteristic is far from ideal.

Finally, find the complete output characteristic for your JFET with the JFET Output Tracer option. Take scans in both the linear and saturated regimes.

SUMMARY

- Although it has brought about a revolution in the design of electronic equipment, the bipolar (PNP/NPN) transistor still has one very undesirable characteristic. The low input impedance associated with its base-emitter junction causes problems in matching impedances between interstage amplifiers.
- A succession of FET-like devices was patented by Julius Lilienfeld in the 1920s and 1930s. However, materials science and fabrication technology would require decades of advances before FETs could actually be manufactured.
- JFET operation can be compared to that of a garden hose. The flow of water through a hose can be controlled by squeezing it to reduce the cross section and the flow of electric charge through a JFET is controlled by constricting the current-carrying channel. The current also depends on the electric field between source and drain (analogous to the difference in pressure on either end of the hose).
- JFET consists of the channel of semiconducting material through which current flows. JFETs are of two types; one is n-channel JFET and another is p-channel JFET. Generally, n-channel JFETs are more preferred than p-channel JFETs because in n-channel electrons are the majority carriers and as we know electrons are more mobile than holes.
- Both N-channel JFET and P-channel JFET are operated in the same way, although the charge carriers are inverted, i.e., electrons are majority carriers in the N-channel and holes are majority carriers in the P-channel. Here we explain N-channel JFET operation because it is more preferable.
- A MESFET (metal–semiconductor field-effect transistor) is a field-effect transistor semiconductor device similar to a JFET with a Schottky (metal–semiconductor) junction instead of a P–N junction for a gate.
- A JFET can be biased in the ohmic or active regions. When it is biased in the ohmic region, it is equal to the resistance. However, when it is biased in an active region, it becomes equivalent to a current source.
- The name HEMT stands for High Electron Mobility Transistor. The device is a form of field-effect transistor, FET, that utilizes an unusual property of a very narrow channel enabling it to operate at exceedingly high frequencies.
- The key element within a HEMT is the specialized PN junction that it uses. It is known as a heterojunction and consists of a junction that uses different materials on either side of the junction. The most common materials used are aluminum gallium arsenide (AlGaAs) and gallium arsenide (GaAs).

REVIEW QUESTIONS

1. What is meant by field effect transistor?
2. What is the function of the JFET?
3. What is the principle of JFET?

4. What is the JFET symbol?
5. What are the characteristics of FET transistor?
6. How do JFETs act as voltage-controlled resistors?
7. What is the drain current in JFET?
8. What is the main difference between N-channel JFET and P-channel JFET?
9. What are the limitations of field effect transistor?

REFERENCES

1. Ajayan, J., Nirmal, D., Mathew, R., Kurian, D., Mohankumar, P., Arivazhagan, L., & Ajitha, D. (2021). A critical review of design and fabrication challenges in InP HEMTs for future terahertz frequency applications. *Materials Science in Semiconductor Processing, 128*, 105753. https://doi.org/10.1016/j.mssp.2021.105753.
2. Asif Khan, M., Bhattarai, A., Kuznia, J. N., & Olson, D. T. (1993). High electron mobility transistor based on a GaN-AlxGa1−xN heterojunction. *Applied Physics Letters, 63*(9), 1214–1215. https://doi.org/10.1063/1.109775.
3. Bargieł, K., Bisewski, D., & Zarębski, J. (2020). Modelling of dynamic properties of silicon carbide junction field-effect transistors (JFETs). *Energies, 13*(1), 187. https://doi.org/10.3390/en13010187.
4. Blalock, B. J., Cristoloveanu, S., Dufrene, B. M., Allibert, F., & Mojarradi, M. M. (2002). The multiple-gate MOS-JFET transistor. *International Journal of High-Speed Electronics and Systems, 12*(02), 511–520. https://doi.org/10.1142/S0129156402001423.
5. Casady, J. B., Sheridan, D. C., Ritenour, A., Bondarenko, V., & Kelley, R. (2010). High temperature performance of normally-off SiC JFET's compared to competing approaches. *Additional Conferences (Device Packaging, HiTEC, HiTEN, and CICMT) 2010, HITEC*, 000152–000159. https://doi.org/10.4071/hitec-jcasady-tp23.
6. Kasamatsu, A., Kasai, K., Hikosaka, K., Matsui, T., & Mimura, T. (2004). 60nm gate-length Si/SiGe HEMT. *Applied Surface Science, 224*(1), 382–385. https://doi.org/10.1016/j.apsusc.2003.08.064.
7. Lepkowski, W., Wilk, S. J., & Thornton, T. J. (2009). 45 GHz silicon MESFETs on a 0.15 μm SOI CMOS process. In *2009 IEEE International SOI Conference* (pp. 1–2). https://doi.org/10.1109/SOI.2009.5318754.
8. Marcoux, J., Orchard-Webb, J., & Currie, J. F. (1987). Complementary metal oxide semiconductor-compatible junction field-effect transistor characterization. *Canadian Journal of Physics, 65*(8), 982–986. https://doi.org/10.1139/p87-156.
9. Pattnaik, G., & Mohapatra, M. (2021). Design of AlGaAs/InGaAs/GaAs-Based PHEMT for High Frequency Application. In S. K. Sabut, A. K. Ray, B. Pati, & U. R. Acharya (Eds.), *Springer Singapore* (pp. 329–337). ISBN 978-981-33-4866-0.
10. Perez, S., Francis, A. M., Holmes, J., & Vrotsos, T. (2021). Silicon carbide junction field effect transistor compact model for extreme environment integrated circuit design. *Additional Conferences (Device Packaging, HiTEC, HiTEN, and CICMT) 2021, HiTEC*, 000118–000122. https://doi.org/10.4071/2380-4491.2021.hitec.000118.

11. Ye, P. D., Yang, B., Ng, K. K., Bude, J., Wilk, G. D., Halder, S., & Hwang, J. C. M. (2004). GaN MOS-HEMT using atomic layer deposition Al_2O_3 as gate dielectric and surface passivation. *International Journal of High-Speed Electronics and Systems, 14*(3), 791–796. https://doi.org/10.1142/S0129156404002843.

CHAPTER

9 Optical Devices

LEARNING OBJECTIVES

After studying this chapter, you will be able to:

- Learn about the optical absorption
- Examine the basic function and features of solar cells
- Discuss on photodetectors

KEY TERMS FROM THIS CHAPTER

Absorption Coefficient
Chemical analysis
Holography
Laser communication systems
Optical fibers
Photonic integrated circuits
Semiconductor
Atmospheric turbulence
Electron-hole pair (EHP)
Interference patterns
Optical devices
Photon Energy
Remote sensing
Urbach model

9.1. INTRODUCTION

Optical devices are a wide variety of instruments and tools, from simple lenses to intricate optical systems that manipulate light for different purposes. Fundamentally, these apparatuses depend on the concepts of optics, which is the study of light and its properties. Refracting light means bending its path to focus or diverge it. Lenses are one of the basic optical components. They have different optical properties in each of their forms, which include convex, concave, and compound lenses. In devices like cameras, telescopes, microscopes, and eyeglasses, these lenses are essential for creating images and adjusting vision. In addition to lenses, optical devices can also be mirrors, which reflect light and change its direction without modifying its characteristics. Mirrors can be spherical, curved, or flat, and they can be used for a variety of purposes. For example, they can form images in laser systems and makeup mirrors, or they can reflect light in periscopes and telescopes. Another important class of optical devices are optical fibers, which are made of flexible, thin glass or plastic strands that transmit light through total internal reflection. In order to facilitate high-speed data transfer over extended distances and to enable minimally invasive procedures in medical endoscopes, optical fibers are essential to telecommunications. With the use of optical instruments called spectroscopes, one can learn more about an object's emission or interaction with light by examining its characteristics, such as wavelength and intensity. They are extensively utilized in astronomy, science, and chemical analysis, allowing for the identification of chemical elements, the description of celestial objects, and the investigation of the composition of materials.

Moreover, cameras and other imaging systems that record and capture optical images for use in photography, videography, and surveillance are included in the category of optical devices. Advances in imaging technologies have resulted in the development of advanced imaging devices, such as thermal imagers and digital cameras, which have revolutionized various fields, including security, medical, and remote sensing. Light Amplification by Stimulated Emission of Radiation, or lasers, are among the most effective and versatile optical instruments. They generate extremely focused coherent light beams that possess coherence, directionality, and monochromaticity. Surgical procedures, cutting and welding, barcode scanning, and optical storage devices like CD and DVD players are just a few of the many industries that use lasers. They are essential instruments in industrial manufacturing, scientific research, and medical procedures due to their controllability and precision. Technological advancements have continued to shape optical devices, giving rise to breakthroughs like adaptive optics, photonic integrated circuits, and holography. Photonic integrated circuits manipulate light signals for information processing and communication, while holography records patterns of light wave interference to create three-dimensional images. Adaptive optics improve the resolution of astronomical telescopes and laser communication systems by compensating for distortions brought on by atmospheric turbulence through the use of deformable mirrors and other methods.

9.2. OPTICAL ABSORPTION

In physics, the process by which matter (usually electrons confined in atoms) absorbs photon energy and converts it into internal energy of the absorber (such as thermal energy) is known as absorption of electromagnetic radiation. Attenuation, which is the progressive loss of light wave intensity as it travels through the medium, is a prominent result of electromagnetic radiation absorption. Saturable absorption, also known as nonlinear absorption, happens when the medium's transparency varies in relation to wave intensity under specific conditions (optics), despite the fact that the absorption of waves is typically independent of their intensity (linear absorption).

When optical energy is transformed into electrical energy, optical absorption—a basic process—is utilized. The best examples of converting electrical energy into optical energy are solar cells and photodetectors. If the energy of each photon in the light hitting a semiconductor sample is higher than the band gap in the semiconductor, the photons can be absorbed and converted to an electron. By this process, the electron moves up into the conduction band from the valence band. Because the absorption process leaves a hole in the valence band and an electron in the conduction band, it consequently produces an electron-hole pair (EHP). The electron is free to move within the semiconductor's crystal lattice when the covalent bond is broken (it is now in the conduction band). Two formulas are typically used to analyze the optical absorption edges of amorphous semiconductors, like silicon, which are distinguished by an optical gap. The absorption follows the Tauc formula at photon energies above the optical gap.

$$(\omega\alpha)^{1/2} \propto \hbar\omega - E_T$$

where E_T is known as the Tauc gap.

This relationship assumes that both the conduction and valence bands are continuous with a parabolic DOS function. Below the optical gap, the absorption behaves according to the Urbach formula, that is,

$$\alpha \propto \exp\left(\frac{\hbar\omega - \hbar\,\omega_0}{E_0}\right)$$

where $\hbar_\omega$ is called the Urbach parameter and E_o represents the width of the absorption edge.

The electrons must travel in a Gaussian-correlated random potential profile in order for this relationship to hold. When it comes to DLC films, the optical gap calculated using the Tauc formula ranges from nearly 0 to 5 eV, depending on the deposition parameters that cause significant changes in the sp^3–sp^2 ratio, for instance. The areas of Tauc and Urbach behavior, however, are not clearly distinguished in a-C; furthermore, the parameters obtained have not consistently demonstrated consistency, nor has it been evident what their physical significance is. In the case of a-C, whose structure has been described as nonhomogeneous and composed of clusters of sp^2-bonded material in a sp^3-bonded skeleton or with even more disorder than this, the fundamental presumptions underlying these models are invalid. According to this model, the material's band gap fluctuates significantly, resulting in localized states brought on by electron confinement inside the clusters. Rather than between continuous bands, near-edge transitions occur in the sp^2 clusters between these localized states. Additionally, fluctuations at the location of the sp^2 clusters invalidate the Urbach model's assumptions due to the large band gap. To account for these issues, a quantum well model with a stronger physical foundation was created. Nonetheless, the Tauc band gap is still frequently used as a comparison tool for different films and as a gauge for the optical gap due to its ease of measurement. The optical gap is also measured as the photon energy at which the optical absorption coefficient has a value of 10^4 cm^{-1} E_{04}.

The effects of low levels of nitrogen incorporation into films with an (undoped) sp^3–sp^2 ratio of roughly 0.2–0.35, which may approximate a "doping" situation rather than a structural rearrangement, have revealed that the optical gap, estimated from Tauc plots or from the 10^4 level, initially increases with nitrogen concentration and peaks at about 4% or 7%, although there is a significant amount of scatter in the data, as seen in particular. Figure 9.1 displays these outcomes. Two distinct processes were used to produce the films: one was RF-PECVD, and the other was plasma beam deposition. In both cases, the films had a notable proportion of hydrogen (19% and 14%, respectively), which typically raises the sp^3–sp^2 ratio. Measurements on ion beam-deposited films showing that the nitrogen content remained constant but the hydrogen content changed serve to illustrate this. These findings demonstrate the significant impact of hydrogen on the optical gap. Figure 9.2 displays a plot of this data. The Urbach energy has been measured as a function of nitrogen flow during plasma beam deposition, despite doubts about the validity of the Urbach model. The Urbach energy stays close to its level in carbon films for low flows (and correspondingly low nitrogen content in the films), which is consistent with a high level of defects of about 10^{20} cm^{-3}. The decrease in the optical gap depicted in Figure 9.1 is caused by a larger energy at larger flows, which is consistent with the band tail states widening.

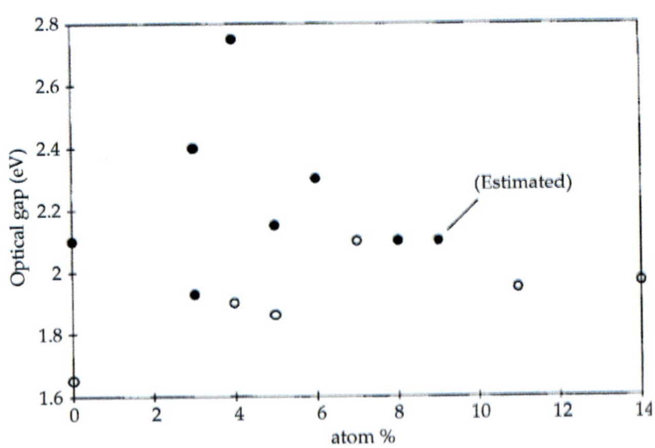

Figure 9.1. Optical gap as a function of nitrogen content.

Source: https://ars.els-cdn.com/content/image/3-s2.0-B9780125139045500090-f07-30-9780125139045.jpg.

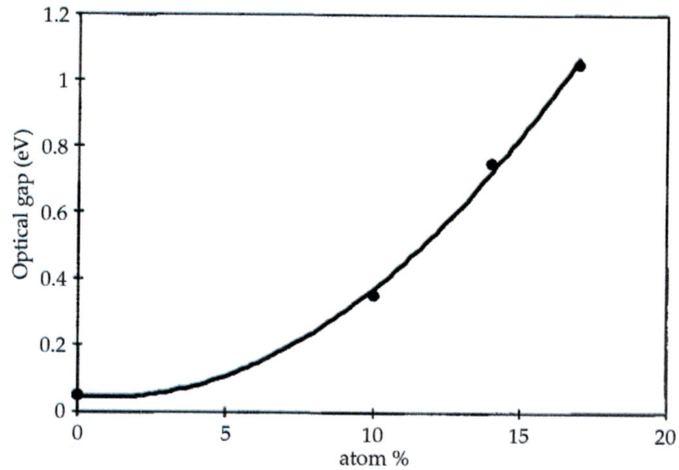

Figure 9.2. Variation of optical gap with hydrogen concentration.

Source: https://ars.els-cdn.com/content/image/3-s2.0-B9780125139045500090-f07–31–9780125139045.jpg.

The optical gap is significantly lower for films deposited primarily under hydrogen-free conditions, such as those deposited by DC magnetron sputtering, laser ablation (LA), or ion beam deposition (IBD), due to the larger sp^2 content; in films containing hydrogen, the H atoms can terminate some carbon bonds and promote the sp^3 configuration. But the behavior can be highly contradictory with increased N incorporation. According to results published by Monclus and colleagues, initial nitrogen incorporation dramatically reduces the Tauc gap from approximately 0.5 eV for carbon films to approximately 0 eV for 25% content, and further nitrogen incorporation increases the optical gap. Alternatively, a reduction in Tauc gap from approximately 0.6 eV for nitrogen-free films to approximately 0 eV has been reported, and this could be attributed to broadening of the π–π^* bands by increased sp^2 cluster size. Lu and colleagues also discovered that the amount of nitrogen increased the optical gap. The band gap can be extremely large if the sp^3–sp^2 ratio is very high, as it is in films deposited by FCA deposition. Chen and colleagues observed that the optical gap decreased initially with nitrogen content and then increased, reaching a maximum of 3.8 eV at approximately 24% content. In contrast, the behavior of hydrogen-containing films showed that the optical gap decreased from 2.5 to 1.6 eV under the same conditions. This disparity in the properties highlights the fact that it is very challenging to draw any broad conclusions about the impact of nitrogen doping without having a thorough understanding of the film structure.

9.2.1. Physical Process

The valence electrons of an atom can move to higher electronic energy levels when a material is illuminated by photons. In the process, the photon is destroyed and the radiant energy that was absorbed is converted to electric potential energy. Subsequently, the absorbed energy can undergo various outcomes. It could be re-emitted as radiant energy by the electron (resulting in a scattering of light overall in this scenario), dissipated to the remaining material (i.e., converted into heat), or the electron may even leave the atom (as in the case of the Compton and photoelectric effects). Since the energy of the incident photon must be comparable to an authorized electronic transition, the amount of absorption varies with the wavelength of light for the majority of substances. This causes pigments that absorb certain wavelengths but not others to become colored. When an object is exposed to white light, it will appear red because it has absorbed blue, green, and yellow light.

9.2.2. Quantitative Measurements

An object's absorbance indicates how much of the incident light it absorbs; some photons are reflected or refracted in addition to being absorbed. The Beer-Lambert law may link this to additional characteristics of the object. Using absorption spectroscopy, which involves lighting a sample from one side and measuring the amount of light that exits the sample in all directions, precise measurements of the absorbance at various wavelengths enable the identification of a substance.

9.2.3. Earth surface

The particular phenomena of electromagnetic radiation absorption at the surface of the Earth have multiple significant features. Earth's crust, surface waters, and lower atmosphere all have temperature regulation mechanisms. The amount and wavelength selectivity of electromagnetic absorption at the Earth's surface will inevitably change due to changes in the Earth's crust, such as glaciation, deforestation, and melting of polar ice. As a result, variations in the albedo, or inverse of electromagnetic absorption, may precede variations in climate, such as global warming. The effects of both local albedo and total solar insolation have been identified through an analysis of the regulation of surface water temperature with regard to electromagnetic radiation absorption.

9.2.4. Photon Absorption Coefficient

Photon Absorption Coefficient: Depending on the photon energy (E) and the band gap energy (E_G), photons may either be absorbed or propagate through a semiconductor when light strikes it. When the photon energy is lower than E_G, light passes through the material and the semiconductor appears transparent because the photons are not easily absorbed.

$$\lambda = \frac{c}{f} = \frac{hc}{E} = \frac{1.24}{E} \mu m \qquad (1)$$

When $E = h_f > E_G$, there is a high probability of interaction between the photons and a valence electron. The result of this interaction is an electron in the valence band and a hole in the conduction band, which are referred to as an electron-hole pair (EHP). The surplus energy gives the hole or electron more kinetic energy, which causes the semiconductor to heat.

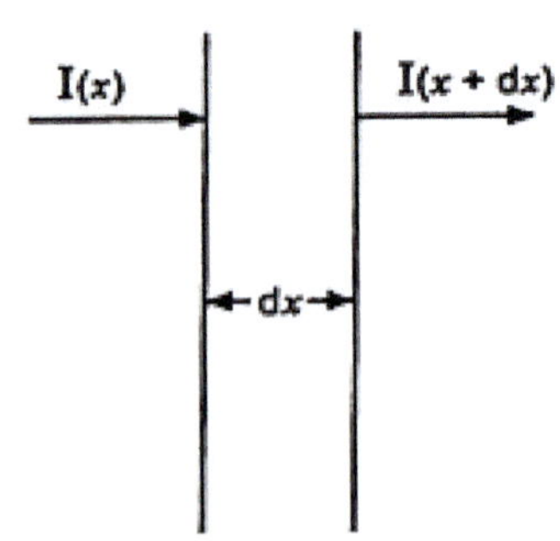

Figure 9.3. Optical absorption in a differential length.

Source: https://www.eeeguide.com/wp-content/uploads/2022/11/Photon-Absorption-Coefficient-06.jpg.

The intensity of photon flux I(x) is expressed in terms of energy/cm^2-s. Figure. 9.3 depicts an incident photon intensity at a position x and the photon flux emerging at a distance x + dx. The energy absorbed per unit time in the distance dx is given by

$$\alpha I(x)dx \tag{2}$$

where α is the photon absorption coefficient (the relative number of photons absorbed per unit distance, given in units of cm^{-1})

From Fig. 9.3, we may write

$$I(x+dx)-I(x)=\frac{dI(x)dx}{dx}=-\alpha I(x)dx \tag{3}$$

$$\text{or } \frac{dI(x)}{dx}=-\alpha I(x) \tag{4}$$

From initial condition

$$I(0)=I_0$$

$$I(x)=I_0e^{-\alpha x} \tag{5}$$

If It be the intensity of transmitted beam after traversing the sample thickness t, then

$$I_t=I_0e^{-\alpha t}$$

$$\text{or } \frac{I_t}{I_0}=e^{-\alpha t} \tag{6}$$

The intensity of photon flux falls exponentially with distance through the semiconductor material as illustrated in Figure. 9.4. If the photon absorption coefficient is large, the photons are absorbed over a relatively short distance.

The absorption coefficient in the semiconductor is a very strong function of photon energy and band gap energy. This coefficient varies with the photon wavelength and with material.

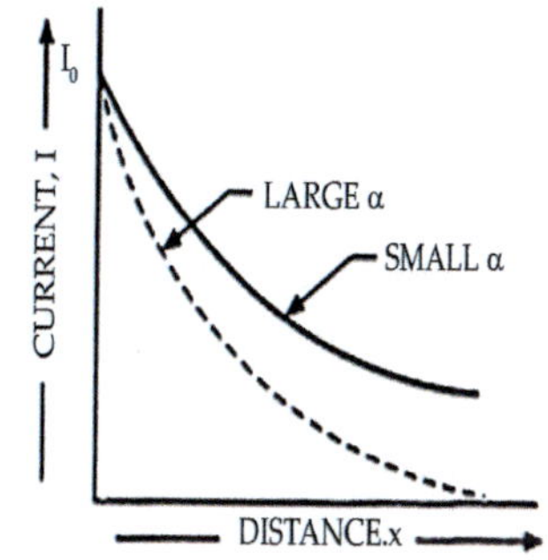

Figure 9.4. Photon intensity vs distance for two absorption coefficients.

Source: https://www.eeguide.com/wp-content/uploads/2022/11/Photon-Intensity-Vs-Distance-for-Two-Absorption-Coefficients-07.jpg.

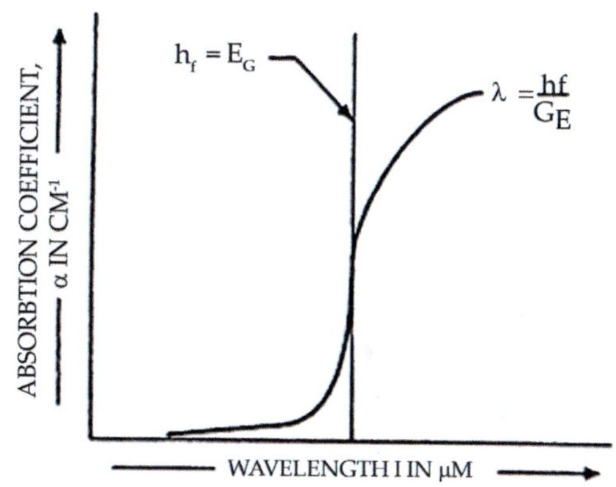

Figure 9.5. Plot of absorption coefficient α versus wavelength

Source: https://www.eeguide.com/wp-content/uploads/2022/11/Photon-Absorption-Coefficient-08.jpg.

In a typical plot of absorption coefficient α versus wavelength, depicted in Figure. 9.5, there is negligible absorption at long wavelengths (hf small) and considerable absorption of photons with energies larger than E_G. According to Eq. (1) the relation between photon energy and wavelength is

$$\lambda=\frac{1.24}{E}\mu m \text{ or } E=\frac{1.24}{\lambda} \tag{7}$$

9.2.5. Electron–Hole Pair Generation Rate

In the fields of optoelectronics and semiconductor physics, the electron-hole pair generation rate—which indicates the speed at which photons are absorbed and produce electron-hole pairs inside a material—is a crucial metric. The functioning of numerous semiconductor devices, such as light-emitting diodes (LEDs), solar cells, and photodetectors, is dependent on this process. A hole can be left in the valence band of a semiconductor material when an electron in the valence band is excited to the conduction band by absorbing a photon with energy greater than or equal to the bandgap energy of the material.

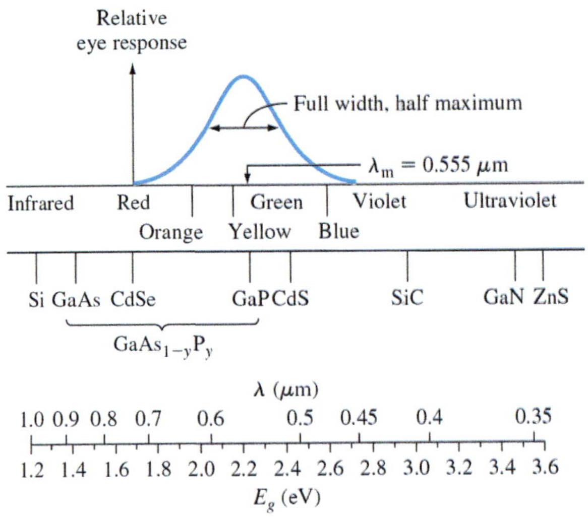

Figure 9.6. Light spectrum versus wavelength and energy. Figure includes relative response of the human eye.

Source: https://www.optima.ufam.edu.br/SemPhys/Downloads/Neamen.pdf.

We have shown that photons with energy greater than Eg can be absorbed in a semiconductor, thereby creating electron–hole pairs. The intensity I(x) is in units of energy/cm^2 -s and $\alpha I_v(x)$ is the rate at which energy is absorbed per unit volume. If we assume that one absorbed photon at an energy h creates one electron–hole pair, then the generation rate of electron–hole pairs is

$$g' = \frac{\alpha I_\nu(x)}{h\nu} \quad (8)$$

which is in units of #/cm^3 -s. We may note that the ratio $I_v(x)/hv$ is the photon flux. If, on the average, one absorbed photon produces less than one electron–hole pair, then Equation (8) must be multiplied by an efficiency factor.

9.3. SOLAR CELLS: BASIC FUNCTION AND FEATURES

A solar cell is a crucial component in photovoltaic energy conversion, which transforms light energy into electrical energy. Semiconductors are typically used as the material for solar cells. The process of converting light energy into electron-hole pairs in a semiconductor and charge carrier separation is known as energy conversion. Most of the time, charge carrier separation is accomplished using a p-n junction. To comprehend both the conventional and novel types of solar cells, it is crucial to grasp the fundamentals of semiconductors and the workings of the conventional p-n junction solar cell. Understanding p-n junction solar cells is essential for making improvements to solar cells in terms of cost, energy consumption during fabrication, and efficiency. The fundamental principles of p-n junction solar cells are then described, which are dependent on an understanding of basic semiconductor physics. The concepts for solar cells made of nanocrystalline materials are presented at the end. The fundamental phenomena are reviewed because, in comparison to the conventional p-n junction solar cell, solar cells based on nanocrystalline materials are more complicated.

9.3.1. Solar Photovoltaic Cell Basics

A photovoltaic (PV) cell, sometimes referred to as a solar cell, can absorb, reflect, or let light flow straight through it. Since semiconductor material makes up the PV cell, it can conduct electricity more effectively than an insulator but not as effectively as a metal, which is a better conductor of electricity. PV cells are made of a variety of semiconductor materials. The semiconductor absorbs light energy and transfers it to electrons, which are negatively charged particles found in the material. The electrons can move through the material as an electrical current thanks to this additional energy. The grid-like lines on solar cells, or conductive metal contacts, are where this current is extracted and used to power the rest of the electric grid and your house. A photovoltaic cell's efficiency can be defined as the difference between the electrical power it produces and the energy it receives from the light it receives. This ratio shows how well the cell converts energy from one form to another. The qualities of the light that is available, including its intensity and wavelength, as well as the cell's various performance characteristics, determine how much electricity is generated by PV cells. The bandgap of photovoltaic semiconductors is a crucial characteristic that describes the range of light wavelengths that the material can absorb and transform into electrical energy. The PV cell can effectively use all of the available energy if the semiconductor's bandgap matches the wavelengths of light shining on it. Find out more about the semiconductor materials that are most frequently used in PV cells below.

Silicon. With silicon accounting for nearly 95% of solar modules on the market today, silicon is by far the most widely used semiconductor material in solar cells. In addition, it is the most prevalent semiconductor used in computer chips and the second most abundant material on Earth (after oxygen). The building blocks of crystalline silicon cells are silicon atoms joined together to create a crystal lattice. This lattice offers a structured framework that improves the efficiency of turning light into electricity. Currently, silicon-based solar cells offer a long lifespan, high efficiency, and low cost. It is anticipated that modules will continue to produce over 80% of their initial power for at least 25 years after that point.

Thin-Film Photovoltaics. A thin-film solar cell is created by covering a supporting material, like glass, plastic, or metal, with one or more thin layers of photovoltaic material. Cadmium telluride (CdTe) and copper indium gallium diselenide (CIGS) are the two primary kinds of thin-film photovoltaic semiconductors available on the market today. One can directly deposit the materials onto the front or rear surface of the module. The second most popular PV material after silicon is cadmium telluride (CdTe), and low-cost manufacturing techniques can be used to create CdTe cells. They are still not as efficient as silicon, but they are still a more affordable option. The transition from the lab to manufacturing is more difficult with CIGS cells due to the intricacy of combining four elements, even though they have excellent efficiencies and ideal properties for a PV material. For long-lasting operation outside, CdTe and CIGS both need more protection than silicon.

Perovskite Photovoltaics. Thin-film solar cells, or perovskite solar cells, get their name from the peculiar crystal structure of these cells. The substrate is the underlying support layer on which layers of materials are printed, coated, or vacuum-deposited to create perovskite cells. They can achieve efficiencies comparable to crystalline silicon and are generally simple to assemble. Perovskite solar cells' efficiency increased in the lab more quickly than that of any other PV material, going from 3% in 2009 to over 25% in 2020. Perovskite PV cells need to stabilize enough to withstand 20 years outdoors in order to be commercially viable, so researchers are trying to increase their durability and create low-cost, large-scale manufacturing processes.

Organic Photovoltaic. Composed of carbon-rich (organic) compounds, organic photovoltaic (OPV) cells can be engineered to improve a particular PV cell function, like bandgap, transparency, or color. Although OPV cells have shorter operating lifetimes and are currently only about half as efficient as crystalline silicon cells, they may be less expensive to produce in large quantities. Moreover, they can be applied to a range of supporting materials, including flexible plastic, which expands the range of applications for OPV.

Quantum Dots. Small semiconductor particles, known as quantum dots, that are only a few nanometers wide are used in quantum solar cells to conduct electricity. While they offer a novel approach to the processing of semiconductor materials, quantum dots are currently not very effective because of the difficulty in establishing an electrical connection between them. But creating solar cells out of them is simple.

A spin-coat technique, a spray, or roll-to-roll printers similar to those used to print newspapers can all be used to deposit them onto a substrate. The tunable bandgap and range of sizes of quantum dots allow them to be combined with other semiconductors, such as perovskites, to maximize the efficiency of a multijunction solar cell and to collect light that is otherwise hard to capture.

Multijunction Photovoltaic. Multijunction solar cells, which are created by stacking multiple semiconductors, are another method for increasing PV cell efficiency. Unlike single-junction cells, which contain a single semiconductor, these cells are essentially stacks of various semiconductor materials. Since each layer's bandgap varies, it can absorb a distinct portion of the solar spectrum, allowing it to absorb more light than single-junction cells. Because a layer below the first semiconductor layer absorbs light that is not absorbed by it, multijunction solar cells are able to achieve record levels of efficiency. A solar cell with precisely two bandgaps is referred to as a tandem solar cell, whereas all solar cells with more than one bandgap are multijunction solar cells. Although multijunction solar cells have shown efficiencies of over 45%, space exploration is the only application for them due to their high cost and manufacturing complexity. Researchers are looking into other applications for III-V solar cells where high efficiency is essential, in addition to the military's use of them in drones.

Concentration Photovoltaic. Concentration photovoltaics, or CPV, uses a mirror or lens to focus sunlight onto a solar cell. Less PV material is needed when sunlight is concentrated on a small area. The highest overall efficiencies are achieved with CPV cells and modules because PV materials become more efficient as light concentration increases. However, proving the required cost advantage over today's high-volume silicon modules has become difficult due to the requirement for more expensive materials, manufacturing processes, and the ability to track the movement of the sun.

9.3.2. Solar Cell Structure and Operation

All solar cells share the same fundamental structure, regardless of whether they are found in a calculator, satellite, or central power plant. The optical coating, or antireflection layer, reduces light loss from reflection, allowing more light to pass through to the energy-conversion layers below and trapping light within the solar cell. Usually formed on the cell surface by spin-coating or vacuum deposition, the antireflection layer is an oxide of silicon, tantalum, or titanium. Beneath the antireflection layer are three energy-conversion layers: the back junction layer, the absorber layer (which forms the device's core), and the top junction layer. To complete an electric circuit, two extra electrical contact layers are required to transfer electricity from an external load back into the cell. The electrical contact layer, which is made of a good conductor like metal, is typically arranged in a grid pattern on the cell face where light enters. Since metal absorbs light, the grid lines are as thin and widely separated as they can be without compromising the cell's ability to collect current. These diametrically opposed limitations do not apply to the back electrical contact layer. It covers the entire back surface of the cell structure and only needs to serve as an electrical contact. The back layer is always made of metal since it needs to be an excellent electrical conductor as well.

Since visible electromagnetic radiation makes up the majority of the energy in both artificial and solar light, a solar cell absorber should be effective at absorbing radiation at those wavelengths. Materials that are classified as semiconductors are those that absorb visible light very well. All incident visible light can be absorbed by semiconductors with thicknesses of one hundredth of a centimeter or less; the thickness of a solar cell is essentially the absorber because the junction-forming and contact layers are much thinner. Semiconductor materials used in solar cells include silicon, copper indium selenide, gallium arsenide, and indium phosphide. Electrons in the absorber layer of a solar cell are excited by light to move from their lower-energy "ground state," where they are confined to particular atoms in the solid, to a higher "excited state," where they can travel throughout the solid. There cannot be an direct current because these "free" electrons are moving randomly in the absence of the junction-forming layers. On the other hand, the photovoltaic effect is generated by the introduction of junction-forming layers, which creates an inherent electric field. The electrons that pass through the electrical contact layers and enter an external circuit where they can perform beneficial tasks are, in essence, given collective motion by the electric field.

For the two junction-forming layers to generate the internal electric field and transport the electric current, their materials must differ from that of the absorber. Therefore, these could be metal and semiconductor, or they could be two distinct semiconductors (or the same semiconductor with different types of conduction). The materials used to make the diodes and transistors of solid-state electronics and microelectronics (see also electronics: Optoelectronics) are essentially the same as those used to build the various layers of solar cells. Microelectronic devices and solar cells utilize identical fundamental technologies. However, since the power generated in solar cell fabrication is proportionate to the illuminated area, one aims to build a large-area device. Naturally, the aim of microelectronics is to build electronic components with ever-tinier dimensions so that they can operate faster and denser inside integrated circuits, or semiconductor chips. There are some parallels between the photovoltaic process and photosynthesis, which is how light energy is transformed into chemical energy in plants. In many applications, solar cells store some of the energy they develop in the presence of light for use in situations where light is not readily available, as they are obviously unable to produce electricity in the dark. Electrochemical storage batteries are a popular way to store this electrical energy. The process of photosynthesis and this sequence of transformations from light energy into excited electron energy and then into chemical energy stored in molecules are remarkably similar.

9.3.3. Solar Panel Design

The majority of solar cells have an area of only a few square centimeters and are covered in a thin layer of clear plastic or glass to keep them safe from the elements. Due to the fact that a standard 10 cm × 10 cm (4 inch × 4 inch) solar cell only produces approximately two watts of electricity (15 to 20% of the light incident on their surface), cells are typically combined in parallel to increase the current or in series to boost the voltage. The standard configuration of a solar, or photovoltaic (PV), module is 36 interconnected cells laminated to glass inside an aluminum frame. To create a solar panel, one or more of these modules may be connected and assembled. Due to unavoidable inactive areas in the assembly and performance differences between individual cells, solar panels are marginally less efficient at converting energy per surface area than individual cells. Every solar panel has standardized sockets on the back so that its output can be joined with those of other solar panels to create a solar array. Numerous solar panels, a power system for handling various electrical loads, an external circuit, and storage batteries can make up a full photovoltaic system. In general, photovoltaic systems can be divided into two categories: standalone and grid-connected.

A solar array and a battery bank that are directly connected to an application or load circuit make up a stand-alone system. The absence of any electrical output from the cells at night or in cloudy conditions necessitates the use of a battery system, which significantly raises the overall cost. Although load requirements may vary, each battery stores direct current (DC) electricity at a fixed voltage specified by the panel specifications. DC-to-AC inverters provide power to alternating current (AC) loads, while DC-to-DC converters supply the voltage levels required by DC loads. Stand-alone systems are the best option for remote installations where it would be too costly to link to a central power plant. Pumping water for feedstock and supplying electricity to lighthouses, telecom repeater stations, and mountain lodges are a few examples. There are two ways in which grid-connected systems combine solar arrays with public utility power grids. Utilities supplement power grids during midday peak usage by using one-way systems. Businesses and individuals use bidirectional systems to meet part or all of their power needs, feeding any extra energy back into the utility power grid. The fact that grid-connected systems don't require storage batteries is a big benefit. The system's increased complexity, however, outweighs the corresponding decrease in capital and maintenance costs. The low-voltage DC output of the solar array must be interfaced with a high-voltage AC power grid using inverters and additional protective equipment. Rate structures for reverse metering are also required when energy from commercial and residential solar systems is fed back into the utility grid.

The most straightforward way to install solar panels is on a fixed mount, which is a slanted support frame or rack. A fixed mount should have a tilt angle from horizontal of roughly 15 degrees less than the local latitude in summer and 25 degrees more than the local latitude in winter in order to operate as efficiently as possible. It should face either north or south in the hemisphere. In more complex deployments, the panels are continuously reoriented by motor-driven tracking systems to track the Sun's daily and seasonal movements. Only large-scale utility generation employing highly efficient concentrator solar cells with lenses or parabolic mirrors that can magnify solar radiation a hundred times or more justifies the use of such systems. Even though sunlight is free, when designing a solar system, the cost of materials and available space must be taken into account. For example, less-efficient solar panels require more space and more panels to produce the same amount of electricity. In space-based solar systems, material cost and efficiency trade-offs are especially noticeable. Satellite panels need to be exceptionally durable, dependable, and immune to radiation damage found in Earth's upper atmosphere. Furthermore, it is more important to minimize the liftoff weight of these panels than it is to reduce fabrication costs. The capacity to manufacture cells in "thin-film" form on a range of substrates, including glass, ceramic, and plastic, for more flexible deployment, is another consideration in the design of solar panels. From this point of view, amorphous silicon is highly appealing.

Specifically, other photovoltaic materials and amorphous silicon-coated roof tiles have been incorporated into architectural designs as well as recreational vehicles, boats, and autos.

9.3.4. Development of Solar Cells

The experiments conducted in 1839 by the French physicist Antoine-César Becquerel led to the development of solar cell technology. In his experiments with a solid electrode in an electrolyte solution, Becquerel discovered the photovoltaic effect when he noticed that voltage formed when light fell upon the electrode. After about fifty years, Charles Fritts created the first real solar cells by covering the semiconductor selenium with an incredibly thin layer of gold that was almost transparent. Less than 1% of absorbed light energy was converted into electrical energy by Fritts's extremely inefficient energy converters. Even though they were inefficient by modern standards, these early solar cells gave rise to the idea of abundant, clean power in the minds of some.

In 1891 R. Appleyard wrote of the blessed vision of the Sun, no longer pouring his energies unrequited into space, but by means of photo-electric cells..., these powers gathered into electrical storehouses to the total extinction of steam engines, and the utter repression of smoke.

By 1927, a new type of solar cell was demonstrated, using copper and copper oxide as the materials. In the 1930s, both selenium and copper oxide cells were used in light-sensitive devices like photometers for photography, but had low energy-conversion efficiencies. The breakthrough came in 1941 with the development of the silicon solar cell by Russell Ohl. Thirteen years later, American researchers showed a silicon solar cell with 6% efficiency. By the late 1980s, silicon and gallium arsenide cells with over 20% efficiency were produced. A concentrator solar cell in 1989 achieved 37% efficiency by focusing sunlight onto the cell. Connecting cells of different semiconductors in series can further improve efficiency, but at a higher cost and complexity. Today, there are a wide range of solar cells with varying efficiencies and costs available.

9.3.5. Organic Solar Cells

Third-generation solar cells made of organic polymer material as the light-absorbing layer are known as organic solar cells (OSCs), one of the newest photovoltaic (PV) technologies. Additionally, a summary of the primary organic components found in the photovoltaic devices' active layer is provided. The main benefit in the creation of OSCs is the organic material's solubility. The production of OSCs at a low cost and in large quantities is made possible by solution processing using roll-to-roll (R2R) techniques in ambient conditions. The goal of organic photovoltaic (OPV) solar cells is to offer a low-cost, low-energy-production photovoltaic (PV) solution using resources found on Earth. In theory, this technology could also be less expensive to use for electricity production than first- and second-generation solar technologies. The transparent or colored OPV devices can be produced using different absorbers, which makes this technology especially attractive to the building-integrated PV market. Although organic photovoltaics have reached efficiencies close to 11%, long-term reliability issues and efficiency restrictions continue to be major obstacles.

OPV cells employ molecular or polymeric absorbers, which produce a localized exciton, in contrast to the majority of inorganic solar cells. An electron acceptor, like a fullerene, which has molecular orbital energy states that promote electron transfer, is used in conjunction with the absorber. The exciton that results from absorbing a photon moves to the interface between the material that absorbs photons and the material that accepts electrons. Enough driving force is provided at the interface by the energetic mismatch of the molecular orbitals to split the exciton and produce free charge carriers (an electron and a hole).

9.3.5.1. Operating Principle

Thin layers of organic materials with a thickness in the range of 100 nm are used to create organic solar cells. The goal of employing organic dyes is to apply straightforward production methods instead of the costly silicon used in traditional photovoltaics. Furthermore, organic solar cells are suitable for flexible and portable systems since they can be prepared on plastic foil. The first electrode, electron transport layer, photoactive layer, hole transport layer, and second electrode are the main layers that make up organic solar cells. Generally speaking, a solar cell collects light, splits the produced electrons and holes apart, and then outputs electricity at the contacts. The direct generation of free charge carriers in inorganic solar cells is the primary distinction between the operating principles of organic and inorganic solar cells. After light is absorbed in organic materials, excitons are created, which typically have binding energies between 0 and 5 eV. Excitons are typically separated at the interface between two distinct organic layers (heterojunction), as an organic solar cell lacks the requisite electric field ($> 10^6$ V cm^{-1}) to overcome this binding energy. In order to effectively separate the excitons while preventing any energy loss during the process, the energy alignment of these two materials must be optimized.

Broadly absorbing molecules from the visible and near-infrared regions of the electromagnetic spectrum are used in small-molecule OPV cells. For the electron-donating system, highly conjugated systems like squaraines, polyacenes, and phthalocyanines are usually employed. The electron-accepting systems that are frequently used are fullerenes and perylene dyes. The most popular method for producing these devices is vacuum deposition, which yields tandem and bilayer architectures. Small-molecule systems that are processed in a solution have recently been created. Long-chained molecular systems are used by polymer-based OPV cells as the electron-donating material (e.g., P3HT, MDMO-PPV), as well as the electron-accepting system consisting of derivatized fullerenes (e.g., PC60BM, PC70BM). Small exciton diffusion lengths characterize these systems, as do small-molecule OPV cells. Nevertheless, a large interface surface area in the active device gets around this restriction. Utilizing small-molecule absorber dyes, dye-sensitized solar cells represent a hybrid organic-inorganic technology. To regenerate the dye, these dyes adsorb onto an appropriate material that accepts electrons, like zinc oxide or titanium dioxide, in combination with an electrolyte.

9.3.6. Polymer Solar Cells

A minimum of three layers make up the polymer solar cell: an active layer, which contains the actual semiconducting polymer material, a back electrode printed onto a plastic substrate, and a transparent front electrode. For the past 20 years, researchers have studied polymer solar cells, also known as plastic solar cells, which use conjugated polymers as light absorbers, electron donors, acceptors, and/or hole transporters. Right now, slightly above 10% is the highest energy conversion efficiency ever recorded. New polymers with different molecular structures and their applications in photovoltaic devices are being thoroughly investigated in order to further improve performance. Polymer solar cells have a structure that is initially similar to traditional silicon-based solar cells with a planar junction. People assume that this device, which has a simple coating of p-type and n-type organic semiconductor materials, operates as a P–N junction solar cell. However, the light absorber needs to be thick enough to capture all of the incident light. Bulk heterojunction, where donor (p-type) and acceptor (n-type) form interpenetrated phases, has been successfully developed to achieve high performances in order to counterbalance this problem. In 1995, the first polymer solar cell with a high power conversion efficiency of 2.9% under 20 mW cm^{-2} illumination was achieved by blending poly (2-methoxy, 5-(2′-ethylhexyloxy)-1,4-phenylene vinylene) (MEH-PPV) with C60 and its derivatives.

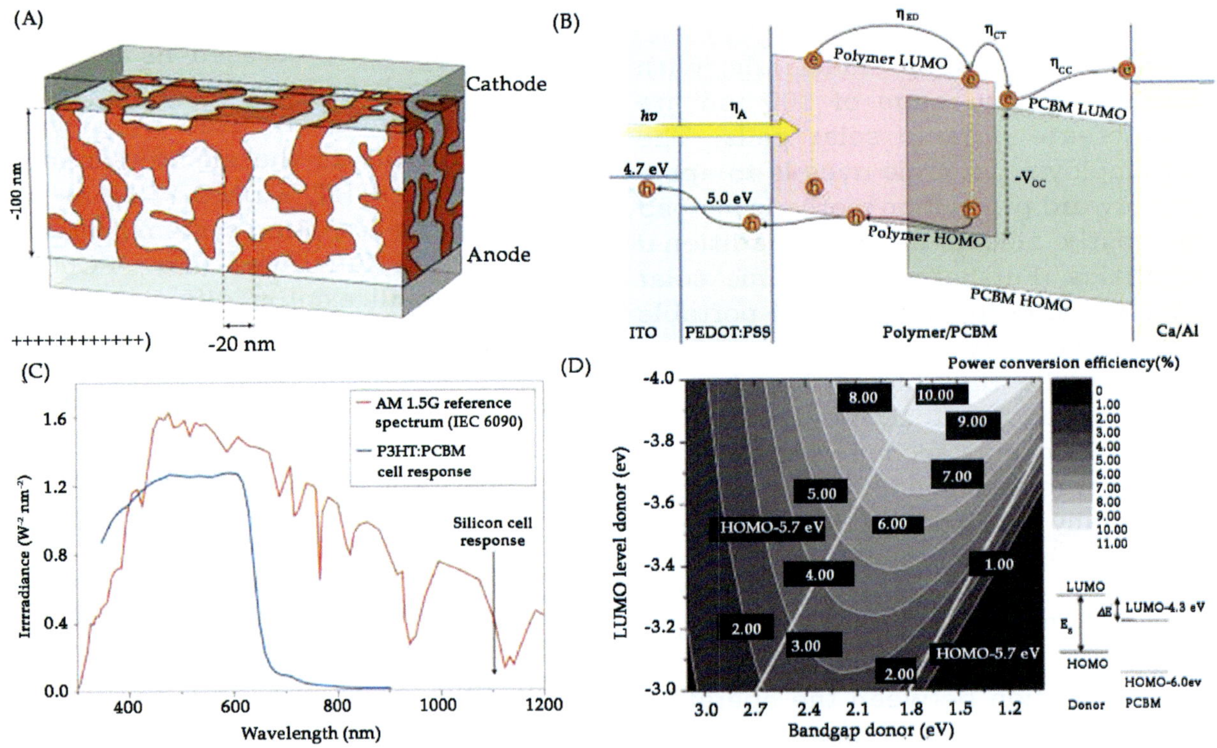

Figure 9.7. (A) Typical structure of polymer solar cell. (B) Energy diagram level and charge transport in a polymer solar cell. (C) AM1.5 reference spectrum and typical absorbing curve in polymer solar cell. (D) Contour plot showing the theoretic power conversion efficiency (contour lines and colors) versus the bandgap and the LUMO level of the donor polymer (PCBM as acceptor).

Source: https://ars.els-cdn.com/content/image/3-s2.0-B9780128110911000057-f05-01-9780128110911.jpg.

The structure and basic operation of the modern polymer solar cell are shown in Figure 9.7. Anode and cathode are positioned above and below the electron and hole extraction layer in a typical polymer solar cell, which is sandwiched between a donor/

acceptor bulk heterojunction light-harvesting layer. In Figure 9.7B, the energy diagram is shown. Producing free carriers from incident photons and excitons is a common feature of a photovoltaic process. Conjugated polymers first absorb light and produce bonded electron-hole pairs known as excitons. Excited hole-electron pairs, or excitons, in polymer solar cells will separate into free carriers at the donor/acceptor interface before migrating—driven by the integrated electric field—into charge extraction and material transportation. These free charge carriers, holes and electrons, will then be gathered in the cathode and anode, respectively. In order to complete the conversion from light to electricity, these free charge carriers will finally move to an external circuit. In this case, the polymer serves as a photoactive layer for the generation and transport of charges as well as light absorption.

The photovoltaic performances of solar cells are assessed using power conversion efficiency, which is calculated by dividing the maximum output power by the incident power. Open-circuit voltage (V_{OC}) and short-circuit current density (JSC), two metrics that measure the performance of polymer solar cells, are derived from the inherent characteristics of the photoactive polymer.

9.3.6.1. Effects of Material Type and Composition in Blend Thin Films

PSCs, or polymer solar cells, have photoactive layers made of bulk-heterojunction (BHJ) blends of n-type acceptor materials and p-type polymer donors (i.e., polymer acceptors or small molecules) are potential sources of power for flexible electronics. Since the type and composition of the material have a significant impact on the mechanical properties of the BHJ blend thin films, representative BHJ blend systems in the PSC field should be studied. Since PCBM is not stable against mechanical, thermal, or photo stresses, it has been known to cause malfunctions and deterioration of device performance in PSCs. Nevertheless, PCBM has been widely used as the n-type acceptor in PSCs. All-polymer solar cells, which substitute polymeric acceptors for PCBM acceptors, are one way to get around this restriction. Using the DCB and pseudo-freestanding tensile testing, the mechanical properties of all-PSCs and PCBM-PSCs are compared with respect to the fracture energy, elastic modulus, and crack onset strain, based on the same and representative polymer donor PTB7-Th (Figure. 9.8). P(NDI2HD-T) and $PC_{71}BM$, two distinct acceptor materials, were utilized for PCBM-PSCs and all-PSCs, respectively. Regardless of the acceptor contents, much higher Gc values of ≈ 2.6 J m^{-2} were observed for the all-PSCs compared to those of the PCBM-PSCs. On the other hand, when the PCBM content increased, the G_c values of PCBM-PSCs significantly decreased, indicating a significant decline in fracture resistance even in optimal device conditions. In a similar vein, it was discovered that all-PSCs had better crack onset strain values than PCBM-PSCs. The ductile polymer acceptor, which can dissipate significant mechanical energy through plastic deformation, is the primary source of all-PSCs' excellent fracture resistance and ductility. In contrast, the addition of PCBM embrittles the BHJ films due to the rigid and large aggregates of PCBM, which provide an easy crack pathway. These findings demonstrated how crucial the acceptor material type is to the realization of mechanically strong and flexible PSCs.

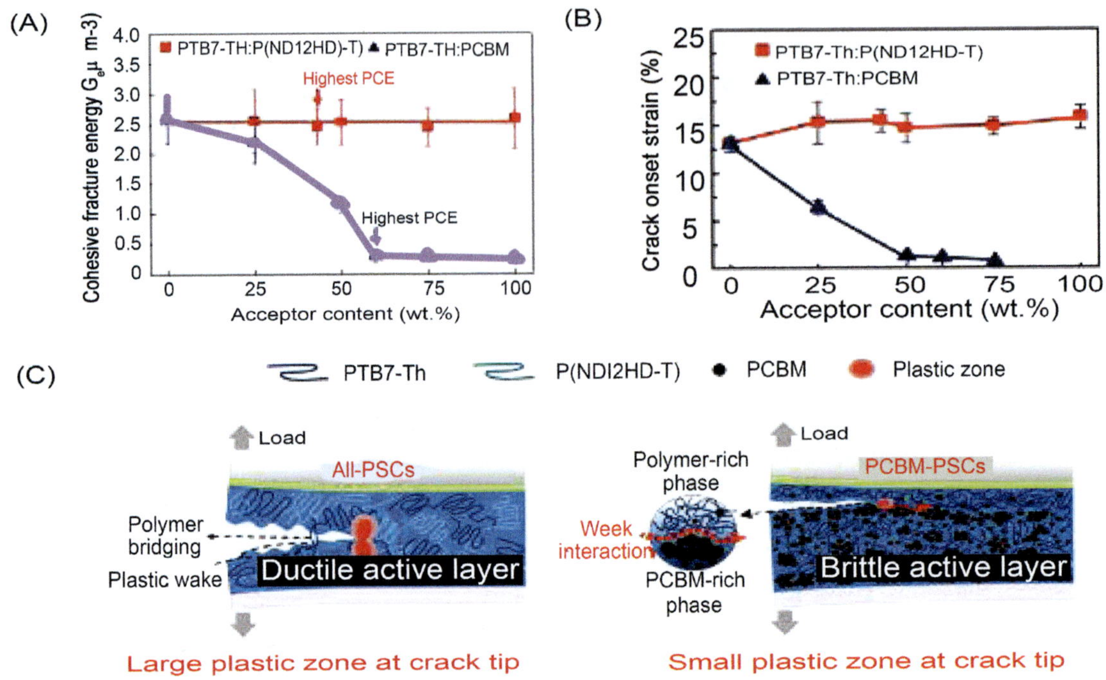

Figure 9.8. (A) Cohesive fracture energy measured by the double cantilever beam test and (B) crack onset strain measured by the pseudo-freestanding tensile testing for all-polymer and small molecule-polymer blend thin films. (C) Schematics of fracture mechanisms in all-polymer solar cells (PSCs) and phenyl-C_{71}-butyric acid methyl ester-PSCs blend thin films.

Source: https://ars.els-cdn.com/content/image/3-s2.0-B9780128188903000072-f07–14–9780128188903.jpg.

9.3.6.2. Prominence of Conjugated Polymers

In order to achieve low cost and high power, more attention is being paid to polymer solar cells. Transparent indium tin oxide is a material for electrodes that is frequently used. ITO exhibits both transparency across a broad range of the solar spectrum and fine electrical conductivity. The low mechanical strength of ITO is a drawback, though. ITO's expensive price and restricted availability present additional issues. Therefore, research on polymer solar cells has focused on finding long-lasting, reasonably priced electrode materials with favorable optical and electrical characteristics. As a result, as substitute electrode materials, researchers have concentrated on using conjugated polymers, nanocarbons (graphene/carbon nanotube), and metal nanostructures. Metals like Au and Ag have been successfully employed as cathode materials. Nevertheless, there are issues with durability, weight, and cost when using metal electrodes. Consequently, conducting polymer electrodes for solar cells have become the focus of research efforts. Both the neat blend and the combination with a-PANINs had their solar cell efficiencies assessed (Table 9.1). The solar cells' power conversion efficiency (PCE), fill factor (FF), and short circuit current (Jsc) were examined.

Table 9.1. Characteristics of a Photovoltaic Cell Measured Under 100 Mw Cm^{-2} Solar Illumination

Solar Cell	Short Circuit Current (Jsc) (Ma Cm^{-2})	Fill Factor (FF) (A.U.)	Power Conversion Efficiency (PCE) (%)
ITO/PEDOT-PSS/P3HT: PCBM/Al	6.33	0.44	1.67
ITO/PEDOT-PSS/a-PANINs/ P3HT: PCBM/Al	7.09	0.45	1.91

Source: https://www.sciencedirect.com/topics/materials-science/polymer-solar-cell

Higher values of Jsc, FF, and PCE have been observed at 7.09 mA cm^{-2}, 0.45, in devices based on a-PANINs.In order to facilitate the movement of electrons and charges within the matrix, PANI was able to create an interpenetrating grid. The molecules' interdiffusion through the devices was the reason behind the improved performance of a solar cell derived from PANI. The heterojunction's increased crystallinity improved photovoltaic performance. The conducting polymers have demonstrated strong electrocatalytic activity and charge transfer. Under 100 mW cm^{-2}, the DSSC demonstrated a solar cell conversion efficiency of 6.58%. For solar cells, the FTO-based electrodes were effectively used. The PCE of the DSSC using a PANI electrode is 11.6%. A polymer solar cell with conducting polymer electrodes is low cost, has a big surface area, and can transport ions and electrons. High-performance conducting polymer-based solar cells provide an affordable and sustainable energy source in the future. It is probable that the maximum efficiency of these solar cells will fall between 7 and 10%.

9.3.6.3. Transparent Conductive Electrodes

Graphene and its derivatives are increasingly being used in polymer solar cells as transport conductive electrodes in place of conventional electrodes based on ITO. At the moment, research on polymer solar cells using PET is at its peak. The substrate in question is PET. To achieve this, rGO is thermally annealed to be deposited on the PET substrate. The spin coating of this plasma-treated PET/rGO is used to provide a hydrophilic surface for the fabrication of photovoltaic cells. The GO films have a thickness of 16 nm and a transmittance of 65%. They also have a J_{SC} of 4.39 mA/cm^2, a PCE of 0.78%, and a V_{oc} of 0.56 V. Consequently, the manufactured device shows consistent performance and can withstand up to 1200 bending cycles.

The elevated temperature, lack of mechanical stability, and high cost of raw materials strongly prohibit the practical applications of ITO. For these reasons, numerous studies have been conducted to investigate the use of nanomaterials such as graphene and carbon nanotubes as an alternative to ITO-based applications in transparent conducting electrodes. Among these, graphene-based electrodes have been investigated using both macro-scale graphene created by CVD and liquid suspension of graphene. In terms of cost, the graphene coatings based on liquid suspensions are highly profitable. They offer reasonable prices for roll-to-roll processing, spin coating, and printing. Additionally, the CVD-deposited graphene exhibits the best physical properties with film resistance and transmittance of up to 30 Ω/Sq and 30%, respectively. Because there is less sheet resistance, the transparency of the graphene-based transport of conductive electrons is more challenging than the transparency level of the transport electrons based on metals. On the other hand, graphene-based TCEs are inexpensive, highly processable, and stable. A hybrid structure based on metallic and graphene morphology has been developed for energy harvesting applications in recent studies to overcome all of these drawbacks. It has been reported that hybridizing graphene with inorganic and organic materials, such as ITO and poly (3,4-ethylene dioxythiophene) polystyrene sulfonate (PEDOT: PSS), improves the performance of TCEs.

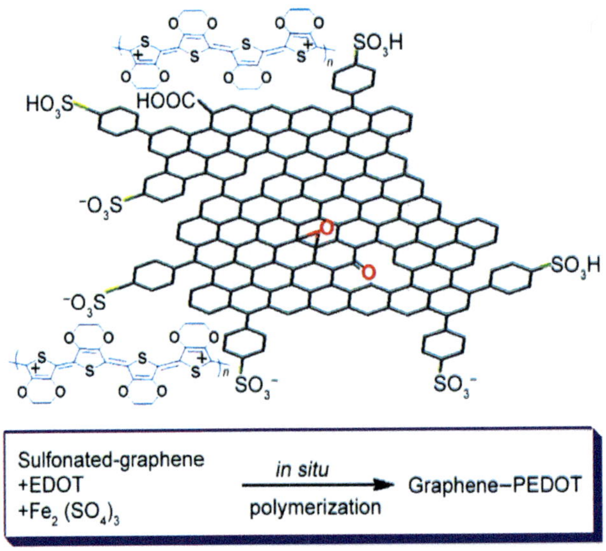

Figure 9.9. Nano-composite of Sulfonated graphene (SG) / poly (3,4-ethylenedioxythiophene): poly (styrenesulfonate) (PEDOT) and the conditions of it synthesis-reaction conditions.

Source: https://www.mdpi.com/polymers/polymers-10–00217/article_deploy/html/images/polymers-10–00217-g002–550.jpg.

Using in situ polymerization, transport conductive electrons based on sulfonated graphene (SG)/PEDOT composites were created (Figure. 9.9). The best transparency, conductivity, thermal stability, and processability were demonstrated by the produced composite. At a wavelength range of 400–1800 nm, the composite demonstrated conductivity and transmittance of 0.2 S/cm and 80%, respectively, for a very thin film thickness. This time, the measured conductivity was significantly higher than that of the commercially used PSS (10^{-6}–10^{-5} S/cm) in PEDOT. Furthermore, the resultant composite exhibits outstanding conductive stability under inward bending conditions after being treated with PMMA.

Graphene doped with CVD with four layers is combined with the commercial PEDOT: PSS and added in polymer solar cells as a cathode. The active layer was composed of poly[(5,6-difluoro-2,1,3-benzothiadiazole-4,7-diyl)-alt-(3,3000-di (2-octyl dodecyl)2,20;50,200;500,2000-quaterthiophen-5,5000-diyl)] (PffBT4T-2OD) as donor and phenyl-C71-butyric acid methyl ester (PC70BM) as acceptor. The obtained assembly showed the J_{sc} of 10.5 mA/cm^2, V_{oc} of 0.72 V, a PCE of 2.8%, and FF of 0.37, respectively, which is not comparable to the cells based on ITO. In some of the researches, the spray coating technique is used with the assistance of vibrations to fabricate the composite of graphene doped PEDOT: PSS. Comparing the obtained results to the solar cells based on ITO, the transparency level was remarkably high. Moreover, they exhibit the best conductivity up to 298 S/cm, or roughly 10 orders of magnitude better than un-doped PEDOT: PSS. Due to strong π–π interactions, PSS in the best charge mobility channels demonstrated the high charge mobility of graphene sheets between PEDOT layers, as demonstrated by this amazing modification. Furthermore, these π–π forces eliminate the flaws and inadequacies in PEDOT: PSS. Furthermore, doping graphene offers a broad range of adjustable properties, including increased hardness, wear resistance, strength, and stability.

9.4. PHOTODETECTORS

All that is in a photo is light. All that a detector is an apparatus for detecting things It transforms the energy of light pulses, also known as radiation, into electrical signals, such as voltage and current. It is also referred to as photosensors occasionally. It absorbs incident light in order to detect incident photons or radiations, as the name suggests. This process operates on a concept known as photodetections. Photodetectors, which are also known as photosensors, are devices that measure electromagnetic radiation, such as light. Numerous types of photodetectors exist, and they can be categorized based on different performance metrics like spectral response or by the mechanism of detection, like photoelectric or photochemical effects. A p–n junction is commonly used in semiconductor-based photodetectors to transform photons into charge. In the depletion region, the absorbed photons form electron–hole pairs. A few types of photodetectors are photodiodes and phototransistors. A portion of the light energy absorbed is transformed into electrical energy by solar cells.

9.4.1. Classification

Photodetectors can be classified based on their mechanism of operation and device structure. Here are the common classifications:

9.4.1.1. Based on mechanism of operation

Photodetectors may be classified by their mechanism for detection:

- Photoconductive effect: These detectors work by changing their electrical conductivity when exposed to light. The incident light generates electron-hole pairs in the material, altering its conductivity. Photoconductive detectors are typically made of semiconductors.
- Photoemission or photoelectric effect: Photons cause electrons to transition from the conduction band of a material to free electrons in a vacuum or gas.
- Thermal: Photons cause electrons to transition to mid-gap states then decay back to lower bands, inducing phonon generation and thus heat.
- Polarization: Photons induce changes in polarization states of suitable materials, which may lead to a change in the index of refraction or other polarization effects.
- Photochemical: Photons induce a chemical change in a material.
- Weak interaction effects: Photons induce secondary effects such as in photon drag detectors or gas pressure changes in Golay cells.

Different configurations can be used with photodetectors. Overall light levels may be detected by a single sensor. It is possible to measure the distribution of light along a line using a 1-D array of photodetectors, such as those found in spectrophotometers and line scanners. Using the pattern of light in front of it, a 2-D array of photodetectors can be utilized as an image sensor to create images. Usually, an illumination window covers a photodetector or array; occasionally, it has an anti-reflective coating.

9.4.1.2. Based on Device Structure

Based on device structure, photodetectors can be classified into the following categories:

MSM Photodetector: A semiconductor layer is positioned between two metal electrodes to form a metal-semiconductor-metal (MSM) photodetector. A sequence of alternating fingers or grids is formed by the interdigitation of the metal electrodes. Materials like silicon (Si), gallium arsenide (GaAs), indium phosphide (InP), or antimony selenide (Sb2Se3) are commonly used to create the semiconductor layer. To enhance its properties, multiple techniques are combined, including plasmonics, etching, substrate switching, and vertical structure manipulation. Antimony Selenide photodetectors demonstrate the highest possible efficiency.

Photodiodes: The most prevalent kind of photodetectors are photodiodes. These are PN junction semiconductor devices. In the junction's depletion region, incident light creates electron-hole pairs, which results in a photocurrent. Additional categories for photodiodes are as follows: a. PIN Photodiodes: These photodiodes function better because they have an extra intrinsic (I) region that extends the depletion region between the P and N regions. b. Schottky Photodiodes: A metal-semiconductor junction is utilized in place of a PN junction in Schottky photodiodes. They are frequently employed in high-frequency applications and provide fast response times.

Avalanche Photodiodes (APDs): APDs are customized photodiodes with avalanche multiplication built in. Near the PN junction, they have a high electric field region that results in impact ionization and the creation of extra electron-hole pairs. The detection sensitivity is increased by this internal amplification. APDs are extensively utilized in high-sensitivity applications like long-distance optical communication and low-light imaging.

Phototransistors: Transistors with a base region sensitive to light are called phototransistors. The base current of the transistor, which regulates the collector current, changes in response to incident light. Applications requiring both detection and signal amplification can benefit from the amplification provided by phototransistors.

Charge-Coupled Devices (CCDs): CCDs are imaging sensors composed of an array of tiny capacitors. Incident light generates charge in the capacitors, which is sequentially read and processed to form an image. CCDs are commonly used in digital cameras and scientific imaging applications.

CMOS Image Sensors (CIS): CMOS image sensors are based on complementary metal-oxide-semiconductor (CMOS) technology. They integrate photodetectors and signal processing circuitry on a single chip. CMOS image sensors have gained popularity due to their low power consumption, high integration, and compatibility with standard CMOS fabrication processes.

Photomultiplier Tubes (PMTs): PMTs are photodetectors based on vacuum tubes.

They are made up of a photocathode that, in the presence of light, emits electrons, and then a number of dynodes that, by secondary emission, multiply the electron current. PMTs are used in low-light detection applications like scintillation detectors and particle physics experiments because of their high sensitivity.

These are some of the common photodetectors based on device structure. Each type has its own characteristics, advantages, and applications in various fields, including imaging, communication, sensing, and scientific research.

9.4.2. Properties

In contemporary technology, photodetectors are essential components that play a critical role in the conversion of light signals into electrical signals. These gadgets are necessary in many different fields, such as scientific research, imaging, sensing, and telecommunications. The ability to utilize the photoelectric effect, in which incident photons release electrons from a material and produce a detectable electrical current or voltage, is the basic idea behind photodetection. The sensitivity of photodetectors—their capacity to recognize even very dim light—is one of their most important characteristics. The lowest detectable signal or the device's responsivity, which expresses the electrical output generated per unit of incident light power, are frequently used to quantify this sensitivity. In situations where picking up on weak signals is critical, like in astronomical observations or medical imaging, sensitivity is essential. The spectral response, which indicates how sensitive the photodetector is to various light wavelengths, is another crucial characteristic. Depending on the needs of the application, photodetectors can be built to function in a variety of electromagnetic spectrum regions, from ultraviolet to infrared. The materials used in the construction of the photodetector determine its spectral response, which can be optimized for particular wavelengths using material engineering and device design. Additionally, in applications where quick detection and signal processing are required, response time is crucial. With response times ranging from picoseconds to nanoseconds, photodetectors make it possible to use them in time-resolved spectroscopy, laser rangefinders, and high-speed communication systems. A few examples of these variables are the intrinsic carrier lifetime of the device, the transit time, and the electronics design that go along with it.

The range of light intensities over which a photodetector can reliably detect and measure signals is also known as its dynamic range. In applications like imaging and remote sensing, where the incident light levels vary greatly, a wide dynamic range is preferred. In order to ensure accurate signal representation across the entire range of light intensities, achieving a high dynamic range frequently requires striking a balance between sensitivity and linearity. Additionally, a variety of noise characteristics, such as thermal noise, shot noise, and dark current noise, can be displayed by photodetectors. Reducing noise is crucial to raising the signal-to-noise ratio and boosting the low-light performance of the photodetector. Noise reduction strategies can help reduce noise impacts and enhance overall device performance. These strategies include cooling the photodetector, refining electronic circuitry, and using signal processing algorithms. Apart from these characteristics, photodetectors could also have spatial resolution, polarization sensitivity, and quantum efficiency—a measure of the effectiveness of converting photons into electrons. Photodetector performance is constantly being improved by developments in materials science, nanotechnology, and device engineering, which results in increased sensitivity, speed, and adaptability for a variety of uses. Photodetectors will continue to be crucial parts of the technological advancement of fields like sensing, imaging, and telecommunications, opening up new avenues for scientific and technological breakthroughs.

There are a number of performance metrics, also called figures of merit, by which photodetectors are characterized and compared:

- Quantum efficiency: The number of carriers (electrons or holes) generated per photon.
- Responsivity: The output current divided by the total light power falling upon the photodetector.
- Noise-equivalent power: The amount of light power needed to generate a signal comparable in size to the noise of the device.
- Detectivity: The square root of the detector area divided by the noise equivalent power.
- Gain: The output current of a photodetector divided by the current directly produced by the photon's incident on the detectors, i.e., the built-in current gain.
- Dark current: The current flowing through a photodetector even in the absence of light.
- Response time: The time needed for a photodetector to go from 10% to 90% of the final output.
- Noise spectrum: The intrinsic noise voltage or current as a function of frequency. This can be represented in the form of a noise spectral density.
- Nonlinearity: The RF output is limited by the nonlinearity of the photodetector.
- Spectral response: The response of a photodetector as a function of photon frequency.

9.4.3. Subtypes

Grouped by mechanism, photodetectors include the following devices:

9.4.3.1. Photoemission or Photoelectric

Photoemission is the process by which electrons can be released from a material when it interacts with electromagnetic radiation. Therefore, photoemission can be defined as a phenomenon in which energy is supplied to release electrons from specific solid materials, such as metal. As a result, this helps provide a basic understanding of the photoemission process. As we proceed, we will concentrate on a number of important subjects, including the photoemission effect and threshold frequency. One can form a basic understanding of the concept of photoemission based on the description provided above. The phenomenon where electrons are expelled from the metallic surface when light is incident is known as the photoelectric effect. It should be mentioned that the term "photoelectrons" refers to electrons that are released. Now, one can also learn about the threshold frequency with the aid of the aforementioned phenomenon. The light frequency that has sufficient energy to release an electron from an atom is known as the threshold frequency. Consequently, the lowest frequency of incident radiation below which electron emission may be rendered impossible is known as the threshold frequency.

To understand the threshold frequency in terms of an equation, one can focus on the following points –

For describing threshold frequency, one can write the photoelectric effect's equation as

$$K_{max} = hv - W$$

Here, W = Metal's work function

$$W = hv_o$$

Now, as observed, vo can be used for representing the electromagnetic radiation's photoelectric threshold frequency.

There are several ways to apply the threshold frequency of the photoelectric effect, including photoelectron spectroscopy, night vision devices, and image sensors. We'll talk about the photoemission effect once more right now. Photoemission is the term for the process by which electrons are expelled from a solid surface through electromagnetic radiation; this process occurs less frequently from a liquid surface. The photoemission effect can also be referred to as the external photoelectric effect. Certain regions, like the visible and ultraviolet regions, can be heavily involved in electromagnetic radiation. In addition, it can be argued that x-ray and infrared regions may also hold some significance.

Here are certain notable features regarding the phenomenon of photoemission –

1. Photon absorption as well as the photoelectron generation method can be instant in nature. Thus, it can be inferred that one cannot find an identifiable time lag between the two.
2. At a particular frequency, the measure of photoelectrons that are ejected every second is related to the incident radiation's intensity.
3. The photoelectron kinetic energy is dependent on the surface's work function and the incident photon frequency; however, it is independent of the incident intensity.

9.4.3.2. Characteristics

- Gaseous ionization detectors are used in experimental particle physics to detect photons and particles with sufficient energy to ionize gas atoms or molecules. Electrons and ions generated by ionization cause a current flow which can be measured.
- Photomultiplier tubes contain a photocathode that emits electrons when illuminated. The electrons are then amplified by a chain of dynodes.
- Phototubes contain a photocathode that emits electrons when illuminated, causing the tube to conduct a current proportional to the light intensity.
- Microchannel plate detectors use a porous glass substrate as a mechanism for multiplying electrons. They can be used in combination with a photocathode like the photomultiplier described above, with the porous glass substrate acting as a dynode stage.

9.4.3.3. Semiconductor

Semiconductors are a class of materials distinguished by their intermediate conductivity properties between insulators and conductors, which play a critical role in modern electronics. These components serve as the foundation for electronic devices, allowing for precise manipulation and control of electrical currents. Because of the covalent bonds that each atom makes with its neighbors to form a crystalline lattice, semiconductors have special electrical properties. Although electrons in this lattice are bonded to their corresponding atoms, they are also capable of being excited to higher energy states, which in some circumstances permits the flow of electric current. The ability of semiconductors to conduct electricity more effectively than insulators but less efficiently than conductors is one of their distinguishing features. The energy band structure of semiconductors has a tiny energy gap between the conduction band, where electrons are free to move and conduct electricity; and the valence band, where electrons are firmly bound to atoms, is the source of this intermediate conductivity. The electrical characteristics of a semiconductor are determined by its band gap, which can be adjusted through material engineering and doping to customize conductivity for particular uses.

Doping, the process of intentionally introducing impurities into a semiconductor crystal, is a key technique for modulating its electrical behavior. By incorporating impurity atoms with either extra electrons (n-type doping) or missing electrons, or "holes" (p-type doping), the conductivity and other electrical properties of the semiconductor can be altered. Doping allows for the creation of semiconductor devices such as diodes, transistors, and integrated circuits, which form the foundation of modern electronics.

Additionally, semiconductors have fascinating optical characteristics such as light emission, transmission, and absorption that make them essential components of optoelectronic devices like solar cells, photodiodes, and light-emitting diodes (LEDs). A semiconductor's optical characteristics are largely determined by its band gap, which controls the energy needed to move electrons from the valence band into the conduction band. Semiconductor materials can be tuned for particular light wavelengths by manipulating the band gap, which makes it possible to convert electrical and optical signals effectively. In the field of microelectronics, where miniaturization and integration are important goals, semiconductors are also essential. The semiconductor industry works hard to advance manufacturing processes like chemical vapor deposition and photolithography so that devices can be made faster, smaller, and more energy-efficient. These developments have sparked a revolution in a variety of industries, including computing, healthcare, and transportation. Electronic devices with greater processing power, smaller size, and better functionality are now widely available.

- Images are captured by active-pixel sensors, or APSs. A common component of webcams, cell phone cameras, and some DSLRs, APSs are typically manufactured using the complementary metal-oxide-semiconductor (CMOS) process.
- Radiation detectors made of cadmium zinc telluride can function in the direct-conversion (or photoconductive) mode at room temperature, while those made of some other materials—especially germanium—need to be cooled with liquid nitrogen. Because Cd and Te have high atomic numbers, they have higher sensitivity to x-rays and gamma-rays and better energy resolution than scintillator detectors. These are their relative advantages.

- Charge-coupled devices (CCD) are image sensors which are used to record images in astronomy, digital photography, and digital cinematography. Before the 1990s, photographic plates were most common in astronomy. The next generation of astronomical instruments, such as the Astro-E2, include cryogenic detectors.
- HgCdTe infrared detectors. Detection occurs when an infrared photon of sufficient energy kicks an electron from the valence band to the conduction band. Such an electron is collected by a suitable external readout integrated circuits (ROIC) and transformed into an electric signal.
- LEDs which are reverse-biased to act as photodiodes. See LEDs as photodiode light sensors.
- Photoresistors or Light Dependent Resistors (LDR) which change resistance according to light intensity. Normally the resistance of LDRs decreases with increasing intensity of light falling on it.
- Photodiodes can operate in photovoltaic mode or photoconductive mode. They are often combined with low-noise analog electronics to convert the photocurrent into a voltage that can be digitized.
- Phototransistors act like amplifying photodiodes.
- Pinned photodiodes are a photodetector structure with low lag, low noise, high quantum efficiency, and low dark current. They are widely used in most CCD and CMOS image sensors.
- Quantum dot photoconductors or photodiodes can handle wavelengths in the visible and infrared spectral regions.
- Semiconductor detectors are employed in gamma and X-ray spectrometry and as particle detectors.
- Silicon drift detectors (SDDs) are X-ray radiation detectors used in X-ray spectrometry (EDS) and electron microscopy (EDX).

9.4.3.4. Photovoltaic

The process of turning light into electricity through semiconducting materials that display the photovoltaic effect is known as photovoltaics (PV), and it is a topic covered in photochemistry, electrochemistry, and physics. Commercial applications of the photovoltaic effect include photosensors and power generation. Solar modules, which are made up of several solar cells each, are used in photovoltaic systems to produce electricity. There are four types of PV installations: floating, wall-mounted, rooftop-mounted, and ground-mounted. The mount can be fixed or it can track the sun's path across the sky with a solar tracker. Since photovoltaic technology produces significantly less carbon dioxide than fossil fuels, it aids in reducing the effects of climate change. Solar photovoltaics (PV) offer distinct benefits as an energy source. Firstly, once installed, it emits no pollution or greenhouse gases; secondly, it can be scaled up or down based on power requirements; and thirdly, silicon is abundant in the Earth's crust. However, other materials needed to manufacture PV systems, like silver, may limit further advancements in the technology. One of the other main obstacles found is land use competition. In addition to requiring energy storage devices or high-voltage direct current power lines for worldwide distribution, using photovoltaics (PV) as a primary source comes with a number of other unique drawbacks, like variable power generation that must be balanced. While not as much as when using fossil fuels, production and installation do produce some pollution and greenhouse gas emissions.

9.4.3.5. Polarization

A characteristic of transverse waves known as polarization indicates the geometrical orientation of the oscillations. The oscillation's direction in a transverse wave is perpendicular to the wave's direction of motion. Vibrations along a taut string, such as those in a guitar string, are a basic illustration of a polarized transverse wave (see image). The vibrations can be vertical, horizontal, or at any angle perpendicular to the string, depending on how the string is plucked. On the other hand, longitudinal waves, like sound waves in a liquid or gas, do not show polarization because the displacement of the oscillating particles is always in the direction of propagation. Gravitational waves, transverse sound waves (shear waves) in solids, and electromagnetic waves like light and radio waves are examples of transverse waves that show polarization.

Light is comprised of an electromagnetic wave with oscillating electric and magnetic fields that are perpendicular to one another. The direction of the electric field determines the polarization of the wave. In linear polarization, the fields oscillate in one direction, while in circular or elliptical polarization, they rotate at a constant rate in a plane as the wave travels. Sources like the sun, flames, and incandescent lamps emit unpolarized light, consisting of equal mixtures of polarizations. Polarized light can be achieved by passing unpolarized light through a polarizer, allowing waves of only one polarization to pass through. Some materials, such as those exhibiting birefringence, dichroism, or optical activity, can affect light differently based on its polarization. These materials are used to create polarising filters. Light also becomes partially polarized when it reflects at an angle from a surface.

9.4.3.6. Graphene/Silicon Photodetectors

It has been shown that a graphene/n-type silicon heterojunction has strong rectifying behavior and high photoresponsivity. A hybrid photodetector is created by combining graphene with silicon quantum dots (Si QDs) on top of bulk Si. Si QDs decrease the photodetector's optical reflection while raising the built-in potential of the graphene/Si Schottky junction. Si QDs' optical and electrical contributions allow the photodetector to function better. By utilizing the special qualities of silicon and graphene, it is a promising development in the field of optoelectronics, enabling high-performance light detection capabilities. Graphene is a two-dimensional honeycomb lattice made up of a single layer of carbon atoms with remarkable optical, mechanical, and electrical properties. Conversely, silicon is a well-known semiconductor material that is frequently utilized in electronic devices because of its accessibility, dependability, and suitability for current fabrication techniques. Graphene and silicon are combined in photodetectors to take advantage of their complementary properties. Because of its extreme thinness and high carrier mobility, graphene is a highly effective material for light absorption and photo-induced carrier generation. Furthermore, graphene can absorb light in a broad spectrum of wavelengths, from ultraviolet to infrared, thanks to its broadband optical absorption, which makes it useful for a variety of applications.

In the meantime, graphene can be integrated into photodetector devices using silicon as a substrate and platform. Because of silicon's well-established semiconductor industry infrastructure, integrating graphene into existing electronic systems and enabling scalable production is made easier. In addition, the bandgap characteristics of silicon can be tailored to enhance the absorption properties of graphene, thereby enhancing the photodetector's performance in particular spectral ranges. Compared to conventional photodetection technologies, graphene/silicon photodetectors have a number of benefits. First of all, because of the effective carrier generation and transport properties of graphene, they show high sensitivity and quick response times. Because of this, graphene/silicon photodetectors can detect light signals quickly and precisely, which makes them appropriate for applications requiring high-speed communication and real-time data acquisition. Moreover, due to graphene's high optical absorption and low noise properties, graphene/silicon photodetectors perform exceptionally well in low light. This makes them perfect for uses where it's crucial to detect weak signals, like biomedical imaging, remote sensing, and astronomical observations.

This creates possibilities for the incorporation of photodetection features into flexible displays, wearable electronics, and other cutting-edge technologies. Even with all of their benefits, there are still issues with improving the scalability and performance of graphene/silicon photodetectors. These include cutting fabrication costs, raising the effectiveness of light absorption and carrier collection, and improving the uniformity and quality of graphene films. The goal of current research is to overcome these obstacles and enhance the potential of graphene/silicon photodetectors for a variety of uses.

CASE STUDY

A CASE STUDY OF SOLAR POWERED CELLULAR BASE STATIONS

A cellular base station is a wireless system that makes use of microwave radio communication technology. The base station acts as an interface between two mobile phones or between a mobile phone and a fixed phone. Generally, these base stations are made up of several antennas mounted on a metallic tower and a house of electronics at the base of the tower. The antennas are connected to the base with cables.

The base station antennas are used for transmitting as well as receiving radio signals to and from mobile phones. Generally, these antennas are mounted on the top of the tower so that obstacles such as trees, high-rise buildings, hills, etc., do not obstruct the radio signals. Usually, three antennas are mounted on the top of the tower to cover the specified region. Among the three antennas, two are used for receiving and one is for transmitting. Although these antennas operate at different frequencies, they are well separated from each other to avoid interference from emitted power.

At the bottom of the tower, there is a small house of electronic circuits comprising power amplifiers used to generate strong signals. These power amplifiers are connected to the mounted antennas with long cables. The base station also has supporting components such as base station controllers using computers and AC/DC rectifiers to convert AC power to DC power. Many base stations have a DC power backup system in the form of batteries connected either in series or parallel. These batteries supply power to the base station during a blackout or power failure.

The area covered by base station signals is called a cell. Based on the amount of area covered, the base stations can be classified as Macro cell base station, Micro cell base station, Pico cell station, and Femto cell base station. Femto cells cover the smallest area and they are deployed in a room. Pico cells are deployed in offices or shopping malls and they cover more area than femto cells. Micro cells can cover blocks of buildings in an urban locality and cover more area than pico cells. Macro cells cover the largest area among all the cells and generally they are deployed in rural areas or on highways.

Power Requirements of Cellular Base Stations

The power needed for the radio base station depends on the number of calls at that time. For different regions, traffic is different. During nighttime, signal traffic is low when compared to daytime. So, it is difficult to find the actual power needed for the base station. The power needed for the radio base station (RBS2202 in one cabinet) varies between 800W to 3200W. The average power for the whole year is 1400W for a Macro base station. The equivalent average power in MWh per year for a macro base station can be computed as 12.3 MWh. If the site container contains more than one cabinet, then the power needed may be more. As in RBS (radio base station), there are air conditioner and heat exchanger, electronics also present. So, the power is used by all the components. The highest amount of power needed for the macro-RBS is 3200W. Micro base station needs 110 watts. If minilink is there, the power is 150 watts. RBS and electrical exchanges are connected by minilink.

CLASS ACTIVITY

SOLAR CELL EFFICIENCY CHALLENGE

Objective: To investigate and compare the efficiency of different solar cell designs through experimentation and analysis.

Materials Needed: Solar cell prototypes, light source, multimeter, data recording sheets, and safety goggles.

Procedure:

- Divide students into small groups and provide each group with a different solar cell prototype or model.
- Instruct students to set up a testing area where they can expose their solar cell prototypes to sunlight or an equivalent light source.
- Have students measure the voltage and current output of their solar cell prototypes using the multimeter or other measuring devices. Encourage them to take multiple readings to ensure accuracy.

SUMMARY

- Optical devices encompass a broad range of instruments and tools that manipulate light for various purposes, spanning from basic lenses to complex optical systems. At their core, these devices rely on the principles of optics, the study of light and its behavior.
- Holography creates three-dimensional images by recording interference patterns of light waves, while photonic integrated circuits manipulate light signals for information processing and communication.
- Optical absorption is a fundamental process which is exploited when optical energy is converted into electrical energy. Solar cells and photodetectors are the best examples of converting electrical energy into optical energy.
- The electron-hole pair generation rate is a critical parameter in the field of semiconductor physics and optoelectronics, representing the rate at which electron-hole pairs are created within a material due to the absorption of photons.
- A solar cell is a key device that converts light energy into electrical energy in photovoltaic energy conversion. In most cases, a semiconductor is used as the solar cell material.
- Solar cells made from multi- or monocrystalline silicon wafers are large-area semiconductor p–n junctions. Technically, solar cells have a relatively simple structure, and the theory of p–n junctions was already established decades ago.
- Grid-connected systems integrate solar arrays with public utility power grids in two ways. One-way systems are used by utilities to supplement power grids during midday peak usage.
- Organic solar cells (OSCs) are one of the emerging photovoltaic (PV) technologies and are classified as third-generation solar cells with organic polymer material as the light-absorbing layer. The main organic materials in the active layer of the photovoltaic devices are summarized as well.
- The polymer solar cell is a layered structure consisting of, at a minimum, a transparent front electrode, an active layer – which is the actual semiconducting polymer material – and a back electrode printed onto a plastic substrate.
- Power conversion efficiency is used to evaluate the photovoltaic performances of solar cells, which is defined by dividing the maximum output power by incident power.
- A polymer solar cell is a technology that is being increasingly focused on to attain low cost and high power. A commonly used electrode material is transparent indium tin oxide. ITO has fine electrical conductivity in addition to transparency in a wide range of the solar spectrum.
- A photo is nothing but light. A detector is nothing but a device that detects something. Therefore, we can directly conclude that a photodetector can be defined as a device that is used to detect light radiations by absorption.

- Photodetectors are indispensable devices in modern technology, serving a crucial role in converting light signals into electrical signals. These devices are essential across various fields including telecommunications, imaging, sensing, and scientific research.

REVIEW QUESTIONS

1. What is the photon absorption coefficient?
2. How is the electron-hole pair generation rate defined?
3. What are the basics of solar photovoltaic cells?
4. How does the structure and operation of solar cells work?
5. What are graphene/silicon photodetectors and how do they function?

REFERENCES

1. Drießen, M., Amiri, D., Milenkovic, N., Steinhauser, B., Lindekugel, S., Benick, J., Reber, S., & Janz, S. (2016). Solar cells with 20% efficiency and lifetime evaluation of epitaxial wafers. *Energy Procedia, 92*, 785–790. https://doi.org/10.1016/j.egypro.2016.07.169.
2. Gong, J., Darling, S. B., & You, F. (2015). Perovskite photovoltaics: Life-cycle assessment of energy and environmental impacts. *Energy & Environmental Science, 8*(7), 1953–1968. https://doi.org/10.1039/c5ee00615e.
3. Hu, Y. (2014). Modeling sources of nonlinearity in a simple pin photodetector. *Journal of Lightwave Technology, 32*(20), 3710–3720. https://doi.org/10.1109/JLT.2014.2348141.
4. Oku, T., Kumada, K., Suzuki, A., & Kikuchi, K. (2012). Effects of germanium addition to copper phthalocyanine/fullerene-based solar cells. *Central European Journal of Engineering, 2*(2), 248–252. https://doi.org/10.2478/s13531-012-0024-1.
5. Pearce, J. M., Podraza, N., Collins, R. W., Al-Jassim, M. M., Jones, K. M., Deng, J., & Wronski, C. R. (2007). Optimization of open circuit voltage in amorphous silicon solar cells with mixed-phase (amorphous+nanocrystalline) p-type contacts of low nanocrystalline content. *Journal of Applied Physics, 101*(11), 114301–114301-7. https://doi.org/10.1063/1.2734507.
6. Pearsall, T. (2010). *Photonics Essentials* (2nd ed.). McGraw-Hill. ISBN 978-0-07-162935-5.
7. Richter, A., Hermle, M., & Glunz, S. W. (2013). Reassessment of the limiting efficiency for crystalline silicon solar cells. *IEEE Journal of Photovoltaics, 3*(4), 1184–1191. https://doi.org/10.1109/JPHOTOV.2013.2270354.
8. Wong, L. H., Zakutayev, A., Major, J. D., Hao, X., Walsh, A., Todorov, T. K., & Saucedo, E. (2019). Emerging inorganic solar cell efficiency tables (Version 1). *Journal of Physics: Energy*. Advance online publication. https://doi.org/10.1088/2515-7655/ab2338.

CHAPTER

10 Semiconductor Microwave and Power Devices

LEARNING OBJECTIVES

After studying this chapter, you will be able to:

- Give an overview of semiconductor microwave
- Explain the power semiconductor device
- Discuss the DC-DC converters

KEY TERMS FROM THIS CHAPTER

Cellular networks
Gallium nitride
Oscillators
Silicon carbide
Transmission
Electromagnetic compatibility
Microwave
Renewable energy
Thyristors

10.1. INTRODUCTION

Semiconductor microwave and power devices play a crucial role in modern electronics by facilitating high-frequency communication and efficient power management. Essential materials like silicon (Si), gallium arsenide (GaAs), and gallium nitride (GaN) are at the heart of these technologies, each with unique benefits. Silicon-based devices are widely used for their long-standing availability and cost efficiency, making them the primary choice in the power device market. These devices are essential components in power supplies, inverters, and motor drives, benefiting from established infrastructure and manufacturing techniques.

Because of its exceptional electron mobility and direct bandgap, GaAs is particularly well-suited for use in microwave and millimeter-wave applications. This makes it a perfect choice for high-frequency and high-speed communication systems, such as satellite communications, radar systems, and cellular networks. GaAs devices offer lower noise figures and higher gain when compared to silicon devices, making them essential for applications that require signal amplification and minimal loss.

Gallium nitride (GaN) has become a revolutionary technology, particularly in power electronics and RF applications. Due to its wide bandgap, GaN enables devices to function at higher power densities and temperatures by allowing for higher breakdown voltages. GaN-based transistors, like high electron mobility transistors (HEMTs), provide outstanding performance in RF amplifiers, radar systems, and power converters. These transistors are highly sought after in 5G technology and electric vehicle powertrains, where efficiency and thermal regulation are crucial.

The limits of these devices are still being pushed by developments in semiconductor packaging and fabrication technologies. New developments such as GaN-on-Si and silicon carbide (SiC) technologies are increasing application domains, decreasing costs, and improving performance metrics. Semiconductor microwave and power devices will continue to be at the forefront of technological advancement, driving improvements in the consumer electronics, telecommunications, defense, and renewable energy sectors as the need for greater efficiency and faster communication grows.

10.2. OVERVIEW OF SEMICONDUCTOR MICROWAVE

Semiconductor microwave devices are essential components of high-frequency technology and are used in radar, sensing, and communication applications. These devices are vital in systems that demand quick data transmission and accurate signal processing. They function in the microwave frequency range, which is normally between 300 MHz and 300 GHz.

GaAs and GaN, two important semiconductor materials, are used in semiconductor microwave devices. GaAs is well known for having a direct bandgap and high electron mobility, which make it perfect for high-frequency and high-speed applications. GaAs-based electronics are widely used in radar systems, cellular base stations, and satellite communications. Examples of these devices are metal-semiconductor field-effect transistors (MESFETs) and high electron mobility transistors (HEMTs). These devices perform exceptionally well in high-gain and low-noise amplification, which are essential for preserving signal integrity and efficiency in microwave applications.

GaN is becoming more and more popular due to its wide bandgap and high thermal conductivity, which allow it to withstand higher temperatures and power densities. GaN-based electronics perform better in terms of bandwidth and power efficiency, especially GaN HEMTs. They play a key role in the creation of electronic warfare systems, 5G communication networks, and sophisticated radar systems. Because GaN has a high breakdown voltage and can function well in harsh environments, it is a material of choice for both military and commercial applications.

The capabilities of semiconductor microwave devices have been further improved by advancements in fabrication technologies, such as monolithic microwave integrated circuits (MMICs) and hybrid microwave integrated circuits (HMICs). Compact and effective microwave systems depend on MMICs because they combine several microwave components into a single chip, minimizing size and enhancing performance. Technological progress in telecommunications, defense, space exploration, and other industries where high-frequency operation and dependability are critical is fueled by the ongoing evolution of semiconductor microwave devices.

10.2.1. What is Microwave?

A microwave is a type of electromagnetic radiation that has wavelengths that are longer than infrared waves but shorter than other radio waves, which were the original source of the radiation. Its wavelength spans, roughly speaking, one meter to one millimeter, or frequencies between 300 MHz and 300 GHz. The range between 1 and 100 GHz (wavelengths between 30 cm and 3 mm) or between 1 and 3000 GHz (30 cm and 0.1 mm) is a more widely used definition in radio-frequency engineering. The prefix micro- in microwave refers to the small size (shorter wavelength) of microwaves in comparison to radio waves used in earlier radio technology, not to a wavelength in the micrometer range.

The distinctions between microwaves, terahertz radiation, far infrared radiation, and ultra-high frequency (UHF) are somewhat arbitrary and are applied differently in different scientific domains. Microwaves cover the whole super high frequency (SHF) band, which is at least 3 to 30 GHz (10 to 1 cm). UHF and extremely high frequency (EHF; millimeter wave; 30 to 300 GHz) bands are also included in a more comprehensive definition. The bands of radio frequencies in the electromagnetic spectrum between 30 and 300 gigahertz (GHz) are known as extremely high frequencies by the International Telecommunication Union. Frequencies in the microwave range are often referred to by their IEEE radar band designations: S, C, X, K_u, K, or K_a band, or by similar NATO or EU designations.

Since microwaves travel in a straight line and do not follow the earth's surface like ground waves or diffract around hills, they are only able to travel a maximum of 40 miles (64 km) due to the visual horizon. This is in contrast to lower frequency radio waves, which travel in a wave-like pattern. The atmosphere's gases absorb them at the high end of the band, limiting useful communication ranges to about a kilometer.

Microwaves are widely used in modern technology, including industrial heating, collision avoidance systems, garage door openers, keyless entry systems, radar, point-to-point communication links, wireless networks, microwave radio relay networks, medical diathermy and cancer treatment, remote sensing, radio astronomy, particle accelerators, spectroscopy, and microwave ovens.

10.2.2. Microwave Semiconductor Devices

Electronic components intended to function at microwave frequencies—generally between 1 and 300 GHz—are known as microwave semiconductor devices. These gadgets are essential for many uses in satellite communication, radar systems, telecommunications, and other fields. They are able to produce, switch, and amplify microwave signals.

They make it possible for high-frequency signals to be processed and transmitted effectively, which is crucial for contemporary communication systems and radar technology. To meet the growing demand for better wireless communication systems and higher data rates, they remain a focus of research and development in the field of electronics.

Table 10.1. Types of Microwave Semiconductor Devices

Microwave Integrated Circuits	Custom-Designed Circuits Incorporating Microwave Semiconductors and Passive Components on a Single Chip; Used in High-Frequency Signal Processing.	
Microwave Transistors	Bipolar junction transistor (BJT)	Allows a small current injected at one of its terminals to control a much larger current flowing between the terminals, enabling amplification or switching.
	Field-effect transistor (FET)	Uses an electric field to control the flow of current in a semiconductor.
Microwave Power Amplifiers	Components that amplify microwave signals to high power levels; used in radar systems, satellite communication, and wireless communication.	
Microwave Switches	Switches used in signal routing and phase shifting for communication systems.	
Microwave Mixers	Mixers employed to downconvert or upconvert microwave frequencies for modulation, demodulation, and frequency conversion for communication systems.	
Microwave Diodes	PIN diodes	Used as RF switches and attenuators
	Schottky diodes	Known for their fast-switching characteristics; used mainly in balanced modulators as well as in mixers.
	Gunn diodes	Solid-state devices that generate microwave signals through the Gunn effect, which results in oscillations at microwave frequencies; used in microwave signal generators.
	Point contact diodes	Used in mixers and detectors for low-signal applications.
	Impact Avalanche Transit-Time (IMPATT) diodes	Generates microwave power through impact ionization and transit-time effects; used in high-power microwave amplifiers and oscillators.
	Varactor diodes	Capacitance of the varactor diode depends on reverse bias applied to it; manufactured with gallium arsenide.
	Step recovery diodes	Operate up to frequency range of about 10 GHz and power rating up to 50 Watts; manufactured with gallium arsenide or silicon microwave semiconductor materials.
	Tunnel diodes	Used to produce low-power oscillators. When tunnel diodes are forward biased, they produce negative resistance.

10.2.3. Microwave Semiconductor Devices and PCB Design

Designing a printed circuit board (PCB) for microwave semiconductor devices involves a set of critical considerations to ensure optimal performance, signal integrity, and minimal losses. Some key PCB design considerations for microwave semiconductor devices include:

- Impedance Matching: Ensure all transmission lines, interconnects, and components are correctly impedance matched to the transmission lines used (typically 50 ohms for most RF and microwave applications).
- Transmission Line Types: Choose the appropriate transmission line type for your application, such as microstrip or stripline. The choice depends on frequency,

board layer stack-up, and isolation requirements.

- Grounding and Ground Planes: Establish a robust ground plane to provide a low-impedance reference for the microwave signals. Grounding is critical for reducing electromagnetic interference (EMI) and maintaining signal integrity. Consider using solid ground planes or stitching vias to create a continuous ground path.
- Component Placement: Place components carefully to minimize trace lengths, optimize signal paths, and reduce parasitic capacitance and inductance. Components should be located as close as possible to each other to minimize transmission line lengths.
- Thermal Management: Implement thermal management solutions such as heat sinks, vias for heat dissipation, and thermal vias to prevent overheating and maintain device performance.
- Isolation and Crosstalk: Use shielding techniques, such as metal shielding cans or grounded coplanar waveguides, to minimize EMI and isolate sensitive components.
- RF Connectors and Feedlines: Select high-quality RF connectors appropriate for microwave frequencies. Ensure a secure and low-loss connection between the PCB and external equipment. Pay attention to feedline design and ensure controlled impedance along the entire signal path.
- Dielectric Material: Choose a PCB substrate material with low dielectric loss, a high dielectric constant (relative permittivity), and a suitable thermal coefficient for your specific application.
- Via Design: Pay attention to via placement and design. Use plated-through-hole vias for connecting layers while minimizing their impact on signal integrity. Avoid stubs or antipads in high-frequency circuits.
- EMC/EMI Considerations: Implement best practices for electromagnetic compatibility (EMC) and EMI mitigation, which includes proper shielding, EMI filters, and layout techniques to reduce unwanted radiated emissions.
- Simulations: Utilize electromagnetic simulation software to model and optimize your PCB design. Simulations help you predict and correct potential issues before fabrication.
- Testing and Characterization: Perform extensive testing and characterization of your microwave PCB to verify that it meets performance specifications, including network analysis, vector network analyzer measurements, and other RF/microwave test equipment.

Due to the specialized nature of microwave PCB design, it is critical to have engineers with experience in high-frequency electronics and the unique needs of microwave semiconductor devices. Achieving optimal microwave semiconductor device performance requires careful consideration of these factors during the design process, rigorous testing, and proper implementation.

10.3. POWER SEMICONDUCTOR DEVICE

A power semiconductor device is a type of semiconductor that is utilized in power electronics, such as switch-mode power supplies, as a switch or rectifier. When utilized in an integrated circuit, this type of device is also known as a power device or power IC. In commutation mode, a power semiconductor device is typically utilized. Since it is binary in nature, its design is optimized for this type of use; linear operation is typically not recommended. Widely used in radio frequency amplifiers, audio amplifiers, and voltage regulators are linear power circuits. Systems that deliver as little as a few tens of milliwatts for a headphone amplifier or as much as a gigawatt in a high voltage direct current transmission line are known to use power semiconductors.

Electronic devices classified as power semiconductors require an external power source in order to function. Semiconductor materials are not particularly good conductors or insulators. In a circuit, they primarily modify, amplify, switch, or control the flow of electric current or voltage. Power Semiconductor Devices: In order to function, diodes, transistors, thyristors, and sensors need power. A circuit is made up of connecting parts. These parts, known as Power Semiconductor Devices, are able to carry out active tasks like switching, rectification, and amplification.

10.3.1. Power Semiconductor

With a high-power rating, power semiconductors carry out the modified electronic functions of conventional semiconductors. Power semiconductors can withstand high voltage and current with less leakage, voltage drop, and other power losses than simple semiconductor devices.

Basic uses of Power Semiconductors

- Switching to turn ON/OFF electricity
- A component in converters and inverters
- Used in power amplifiers to amplify a signal

The main use of power semiconductors is for switching and converting purposes in power control systems. Most importantly, power semiconductors are part of systems that enable power generation and long-distance transmission and distribution of electricity table 10.2

Table 10.2. Power semiconductors are part of systems that enable power generation and long-distance transmission and distribution of electricity.

Parameter	Semiconductor Device	Power Semiconductor Device
Voltage rating	Low	High
Current rating	Low	High
Power rating	Low	High
Losses	High loss in form of leakage current and voltage drop	Low loss in form of leakage current and voltage drop
Rise and Fall Time	High Slow change from ON-state to OFF-state and OFF-state to ON-state	Low Fast change from ON-state to OFF-state and OFF-state to ON-state

As components in power management subsystems, power semiconductors are typically used as switching devices and rectifiers (to convert electrical signals), as well as to change the voltage or frequency of an electrical current.

10.3.1.1. Si, SiC, and GaN, explained

Even though silicon carbide (SiC) and gallium nitride (GaN) have joined silicon (Si) as semiconductor materials, silicon (Si) is still used in many high-voltage and high-current applications. The bandgaps of the latter two materials are wider, which considerably lowers power loss and boosts efficiency. GaN provides the greatest performance out of all these semiconductor materials.

10.3.2. Types of Power Semiconductor Devices

Power control circuits use power semiconductor devices as on/off switches. In power electronics, a power semiconductor device is a semiconductor that is used as a switch or rectifier, such as in a switch-mode power supply. When utilized in an integrated circuit, this type of device is also known as a power device. While signal devices function at higher switching speeds, power devices function at lower switching speeds. Power electronic circuits make extensive use of power semiconductor devices.

Power semiconductor devices are categorized into three types:

1. Power Diodes
2. Thyristors

3. Power Transistors

Just as the name suggests, uncontrollable power semiconductors cannot be controlled by altering input or any of the terminals. In contrast, fully controllable devices are easily controlled through input voltage or current.

1. Power Diodes

An uncontrollable power semiconductor device that can rectify extremely strong electrical signals is called a power diode. It is capable of handling thousands of kilovolts and hundreds of amperes. A power diode can be created by adding an extra lightly doped intrinsic semiconductor layer to a standard PN junction diode. The drift layer is the thinly doped layer that lies between the diode's P and N layers. To make a power diode capable of withstanding high voltage, add a drift layer. Power diodes are employed in clipper circuits, voltage multipliers, rectifiers, and other circuits.

2. Thyristors

A thyristor is a power semiconductor switch with four layers and three terminals that is semi-controllable. It is made up of an alternating PNPN layer and has anode, cathode, and control terminals. While thyristors can switch ON easily, they struggle to turn OFF. They are commonly utilized for electric power control and as protection circuits in household appliances, electrical tools, and outdoor equipment.

Silicon Controlled Rectifier: An AC signal is changed into a DC signal using a silicon-controlled rectifier, or SCR. This device has three terminals and operates similarly to a two-transistor analogy. An SCR has three terminals: the anode, the cathode, and the gate. The drift layer is lightly doped, the anode and cathode are heavily doped, and the gate is moderately doped. Three PN junctions are formed by the four-layer PNPN arrangement that makes up an SCR. It is a semiconductor device with uncontrollable power that takes longer to turn off than expected because it doesn't stop working until the main current is cut off. When compared to other devices, SCRs have one of the lowest ON-state resistances and the highest conductivity modulation. However, its switching frequency is the lowest. Motor drives and switching circuits both use SCRs. HVDC transmission uses LASCRs, or light-activated silicon-controlled rectifiers, which are a different kind of SCR.

Gate Turn Off: A GTO (Gate Turn Off) Thyristor is a three-terminal and four-layer device that allows the gate to switch off the device. The negative pulse at the gate terminal is used to turn it off. There are three PN junctions formed in the structure of GTO that lead to increased conductivity of the device. GTO has comparatively faster rise and fall times, lower size, and more efficiency than SCR. GTOs are used in AC/DC motor drives, robotics, etc.

MOS-Controlled Thyristor: An MCT (MOS-Controlled Thyristor) is a fully controllable thyristor consisting of two MOSFETs. An n-channel MOSFET and p-channel MOSFET are connected in an MCT Thyristor. Each MOSFET is responsible for turning on and off the state of the device.

Reverse Conducting Thyristor: An RCT (Reverse Conducting Thyristor) is a device that has a fabricated anti-parallel diode on the same IC. It is a semi-controllable unipolar device that cannot block reverse voltage during operation. RCT is used in inverters and high-power choppers.

Integrated Gate Commutated Transistor: An IGCT (Integrated Gate Commutated Transistor) is a thyristor that turns off like a transistor. It is a fully controllable thyristor that acts like a GTO but with a faster turn-off time and negligible conduction losses. IGCT is used in high-power semiconductor devices such as frequency inverters, drivers, and compensators.

Triode for Alternating Current: A TRIAC (Triode for Alternating Current) combines two SCRs in an antiparallel configuration with gate terminals connected. It consists of two main terminals and a single gate terminal that is a combination of both SCRs. There is no meaning of cathode or anode because bidirectional current flows through it. TRIACs are used to control AC signals in a variety of electronics such as fans and lights.

Diode for Alternating Current: A DIAC (Diode for Alternating Current) is a bidirectional thyristor diode that is similar to a TRIAC but with the absence of a gate terminal. Two SCRs are connected in an antiparallel configuration with two main terminals. There is no gate terminal, leading DIAC to operate as an uncontrolled switch. DIACs are mostly used to trigger TRIACs in power electronics.

10.3.2.1. Power Transistors

Unijunction Transistor: A UJT (Unijunction Transistor) is a semiconductor device that has a p-type emitter terminal connected to an n-type bar of two bases B1 and B2. It is sometimes referred to as a Double-base device. For simplification, the UJT circuit is represented by a PN diode connected to two resistances. Just as the name suggests, a UJT device forms a single PN junction. UJT exhibits negative resistance and is used as a relaxation oscillator to trigger SCR Thyristor.

Power BJT: A Power BJT is a four-layer three-terminal semiconductor device that has high current ratings. It has three terminals: base, emitter, and collector. A power BJT serves a variety of applications like power amplifiers, relays, and power control systems.

Power MOSFET: Power metal-oxide-silicon transistors are fully controllable power semiconductor switches designed to handle large amounts of power. Power MOSFET is a three-terminal voltage-controlled majority carrier device that has a vertical channel configuration to increase the current rating. The source and drain are placed on the opposite side of the silicon to increase the power rating of the device. Power MOSFETs offer low gate drive power, rapid switching speed, and wide bandwidth, in addition to being easy to operate and repair. They are the most commonly used type of power transistors that perform well at high frequencies. Power MOSFET dominates 53% of the transistor market, making it one of the most popular power semiconductor devices.

Insulated-Gate Bipolar Transistor: An insulated-gate bipolar transistor (IGBT) is a fully controllable power semiconductor switch used for low-to-medium frequency applications. The IGBT is an integration of a power MOSFET and BJT to offer high efficiency. The device forms a PNPN-configuration semiconductor device that enables voltage control. The gate terminal is insulated with a metal oxide coating, while the emitter and collector form conducting regions inside the device. An additional buffer layer and injection region make an IGBT more efficient than most devices. Unlike a power MOSFET, a bipolar current flows inside an IGBT device. IGBTs have high power ratings, low on-state voltage, and are typically used as discrete devices in power electronics units such as consumer electronics, air conditioners, and electric cars.

10.3.3. Applications of Power Semiconductors

Bipolar junction transistors (BJTs), insulated-gate bipolar transistors (IGBTs), and metal-oxide-silicon transistors (MOSFETs) are the three types of power switches that are either used separately or as components of power integrated circuits (PICs). Power semiconductors are primarily used in power transmission and distribution, automotive and transportation, renewable energy, and consumer electronics industries. Power semiconductors are essential for the efficient and sustainable use of energy because they can transfer energy over large distances with little loss.

10.3.3.1. Reliability and Failure of Power Semiconductors

If the operating voltage or current is too high, power semiconductors may malfunction or sustain damage. An IGBT or power MOSFET's insulating gate oxide layer may be punctured by overvoltage. Experts advise running them 20% below their operating range to guarantee dependable performance. Another common cause of power semiconductor failure is overheating. For example, the inherent heat of electronic components results from ON-state resistance, the resistance in the transistor's operating mode between the drain and source. The power loss and heat increase with the ON-state resistance. In any power electronics system, thermal management must be taken into account. At the moment, research is concentrated on lowering ON-state resistance, carrying out appropriate layer insulation, and preserving high-performance components.

10.3.4. Rectifiers and Inverters

Electronic circuits that can alter the kind of electric current are rectifiers and inverters. Whereas a rectifier changes AC into DC, an inverter changes DC into AC. Typically, they consist of transistors, diodes, switches, or other parts that have the ability to regulate current flow. The designs and specifications of rectifiers and inverters can vary based on the waveform, power, frequency, and input and output voltage.

10.3.4.1. Rectifier

In order to function, many aircraft devices need low voltage DC with high amperage. Vacuum tube rectifiers, dry disk or solid-state rectifiers, motor generator sets, DC engine-driven generators, and vacuum tube rectifiers can all provide this power. A dedicated DC generator is not preferred in aircraft with AC systems, though, as this would require the engine accessory section to power an extra piece of machinery. Rectifiers are thus employed. By restricting or controlling the direction of current flow, a rectifier is a device that converts alternating current into direct current.

Solid-state and dry disk rectifiers are the two main varieties. rectifiers with solid-state and dry disk technology are great sources of high amperage at low voltage. Since dry disk and motor generators are mostly restricted to older models of aircraft, solid-state, or semiconductor, rectifiers have essentially replaced all other types.

10.3.4.2. Half Wave Rectifier

The polarities across the diode and the load resistor will both be positive when an AC signal is swinging positively, as illustrated in illustration A of the input signal. Since the diode in this instance is forward biased, a short circuit can be used in its place, as seen in the illustration. Then, there will be no potential loss across the series diode and the positive portion of the input signal will appear across the load resistor.

The input signal is now shown to be reversed in Illustration B. Keep in mind that there is a polarity inversion across the load resistor and the diode. Since the diode in this instance is reverse biased, an equivalent open circuit can be used in its place. Now, there is no current flowing through the circuit and no voltage drop across the load resistor. The output waveform for a fully sinusoidal input is visible at the far right of Figure 10.1. The input waveform is replicated in the output waveform, but without the wave's negative voltage swing. This kind of rectifier is known as a half-wave rectifier for this reason.

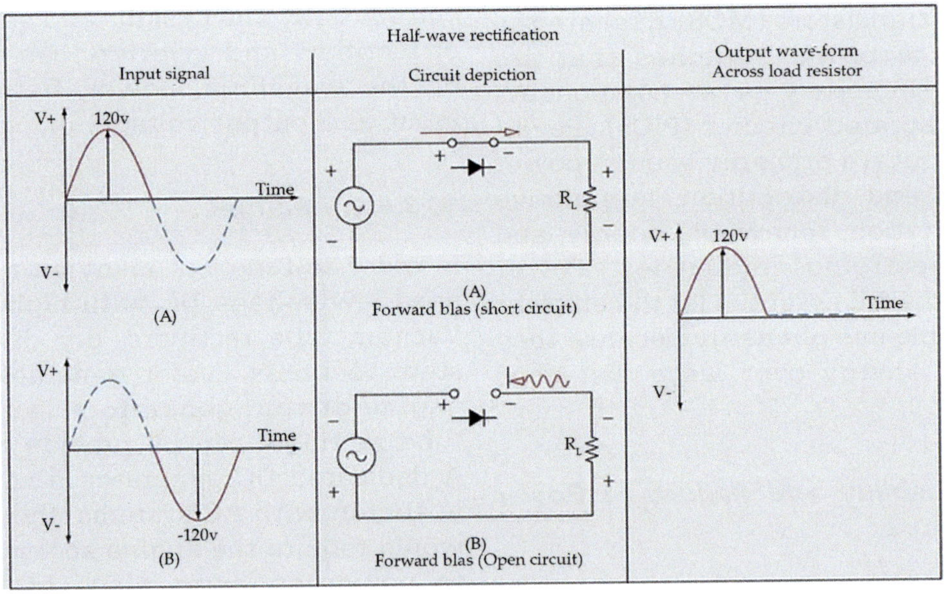

Figure 10.1. Half wave rectifier.

https://static.wixstatic.com/media/ffd442_726e73b854f14528b6441bf70c9853ba~mv2.jpg/v1/fill/w_719,h_445,al_c,q_80,usm_0.66_1.00_0.01,enc_auto/Half%20Wave%20Rectifier.jpg.

10.3.4.3. Full Wave Rectifier

The diode is more frequently used as a rectifier. A full-wave bridge rectifier is the name given to this kind of rectifier. The term "full-wave" describes an output that is not interrupted by gaps as in a half-wave rectifier, but rather is a continuous series of pulses. The initial condition, depicted in Illustration C, involves applying a positive portion of the input signal to the network. Take note of the diodes' respective polarities. Since they are reverse biased, diodes D2 and D4 can be swapped out for an open circuit.

Forward-biased diodes conduct, so they should not be described as an open circuit. It is possible to observe the current path via the diodes, and the resultant waveform develops across the load resistor.

The applied signal's negative phase causes the diodes to reverse their bias and polarity states. As a result, the network depicted in Illustration D. Current now flows through the forward-biased diodes D4 and D2, while the reverse-biased diodes D1 and D3 are effectively open circuits. Keep in mind that the current will flow through the load resistor in the same direction during both of the input waveform's alternations. The waveform's negative swing is flipped up to the positive side of the timeline as a result (Figure 10.2).

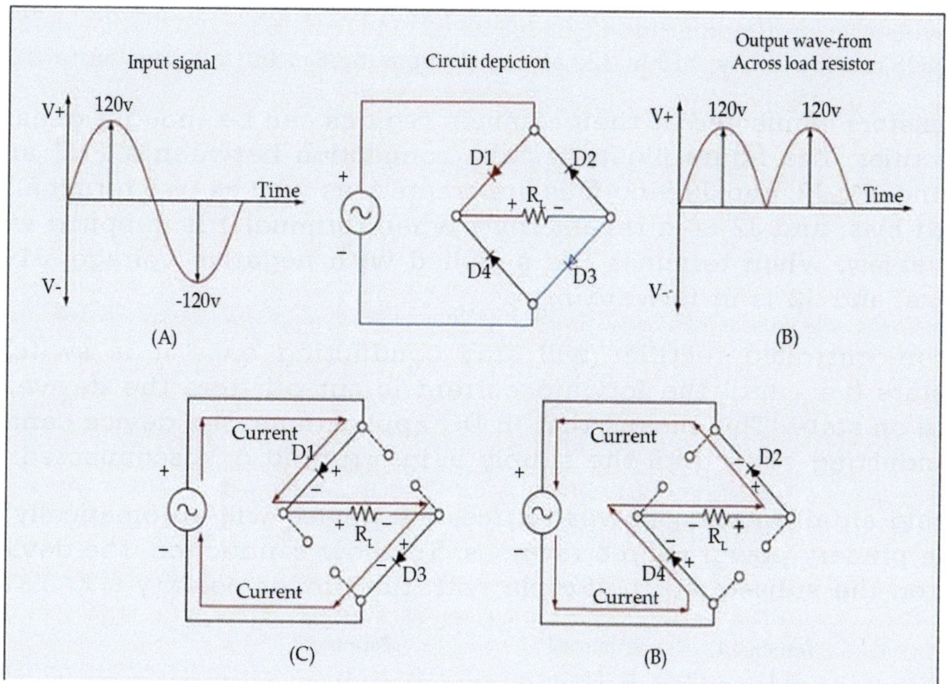

Figure 10.2. Full wave rectifier.

https://static.wixstatic.com/media/ffd442_f20e49aaebc44585baf8c8c13a670f68~mv2.jpg/v1/fill/w_690,h_441,al_c,q_80,usm_0.66_1.00_0.01,enc_auto/Full%20Wave%20Rectifier.jpg.

10.3.4.4. Silicon- Controlled Rectifier / Thyristors

Three terminal components called silicon-controlled rectifiers, also known as thyristors, are used for switching and controlling the amount of AC power. Relays that are controlled by silicon have a very quick transition between conducting and non-conducting states. The silicon-controlled rectifier shows very little resistance when it is on and very little leakage current when it is off. Even when significant power levels are being controlled, this leads to very little power loss within the silicon-controlled rectifier (Figure 10.3).

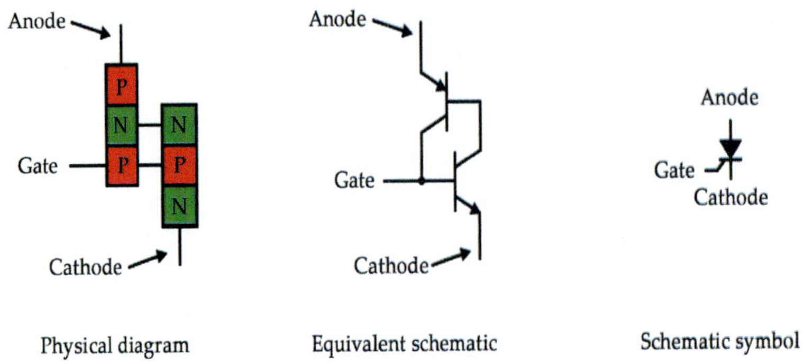

Figure 10.3. Silicon controlled rectifier (scr) or thyristor.

https://static.wixstatic.com/media/ffd442_063c84900317467ca65b6c122070ebf7~mv2.jpg/v1/crop/x_0,y_456,w_3508,h_1586/fill/w_710,h_320,al_c,q_80,usm_0.66_1.00_0.01,enc_auto/SCR.jpg.

Two transistors connected at their common regions can be thought of as the silicon-controlled rectifier. The figure illustrates the connection between a PNP and an NPN. Through joining, J1, J2, and J3 junctions are created, as well as two terminals. J1 and J3 are in forward bias, and J2 is in reverse bias when terminal 1 is supplied with positive voltage; conversely, when terminal 1 is supplied with negative voltage, J1 and J3 are in reverse bias, and J2 is in forward bias.

The silicon-controlled rectifier will stay conducting once it is switched to the conducting state (i.e., until the forward current is cut off from the device, it remains latched in the on state. This means that in DC applications, the device cannot be reset to its non-conducting state until the supply is interrupted or disconnected.

When using an alternating power source, the device will automatically reset itself whenever the primary power source reverses. To allow conduction, the device can then be triggered on the subsequent half-cycle with the proper polarity (Figure 10.4).

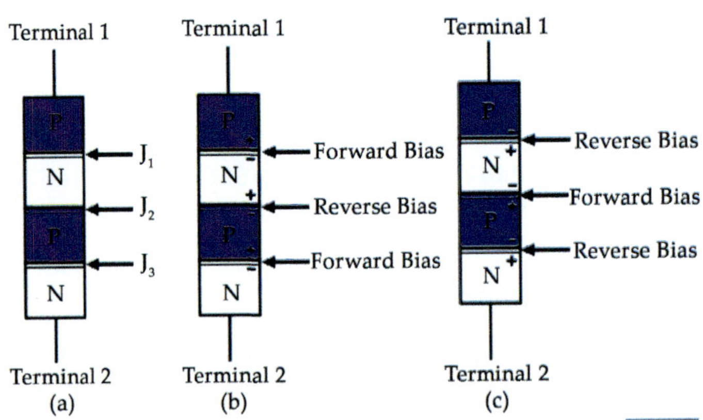

Figure 10.4. SCR construction.

https://static.wixstatic.com/media/ffd442_03a8c365e5104d41809af4f028c9e91e~mv2.jpg/v1/crop/x_24,y_203,w_3466,h_2064/fill/w_707,h_421,al_c,q_80,usm_0.66_1.00_0.01,enc_auto/SCR%20Working.jpg.

When in normal operation, a current pulse applied to the gate terminal triggers a silicon-controlled rectifier into the conducting (on) state. A gate trigger pulse with a quick rise time that originates from a low-resistance source is necessary for the efficient triggering of a silicon-controlled rectifier. When there is not enough gate current available or when the gate current varies slowly, the triggering can become unpredictable (Figure 10.5).

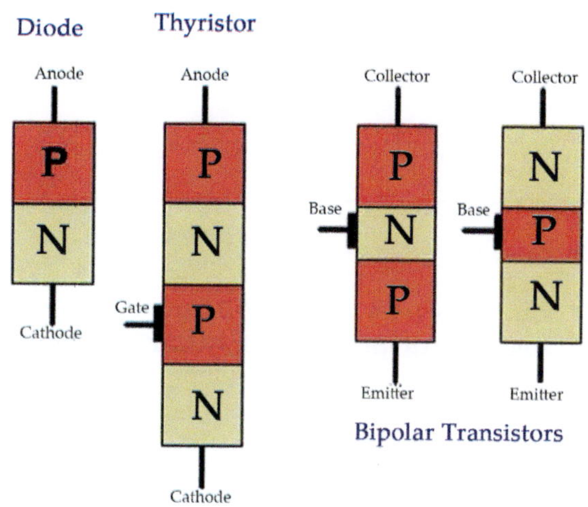

Figure 10.5. SCR vs diode vs transistor.

https://static.wixstatic.com/media/ffd442_951f0d17d81c4b5297b6c62f0021bdc8~mv2.jpg/v1/fill/w_456,h_402,al_c,q_80,usm_0.66_1.00_0.01,enc_auto/Diode%20vs%20Transistor%20vs%20SCR.jpg.

10.3.4.5. Dry Disk Rectifiers

Dry disk rectifiers function by allowing electric current to pass through a junction of two different conducting materials more easily in one direction than in the opposite direction. The resistance to current flow is low in one direction and high in the other. Depending on the materials selected, several amperes may flow through the low-resistance direction, while only a few milliamperes may flow through the high-resistance direction.

Three types of dry disk rectifiers may be found in aircraft:

1. Copper Oxide Rectifier
2. Selenium Rectifier
3. Magnesium Copper-Sulfide Rectifier

10.3.4.6. Copper Oxide Rectifiers

The copper oxide rectifier is made up of a copper disk with a layer of copper oxide formed by heating. It can also be created using a chemical copper oxide preparation applied evenly onto the copper surface. Lead plates are typically used to press against the two opposite faces of the disk to ensure good contact. Current flows from the copper to the copper oxide (Figure 10.6).

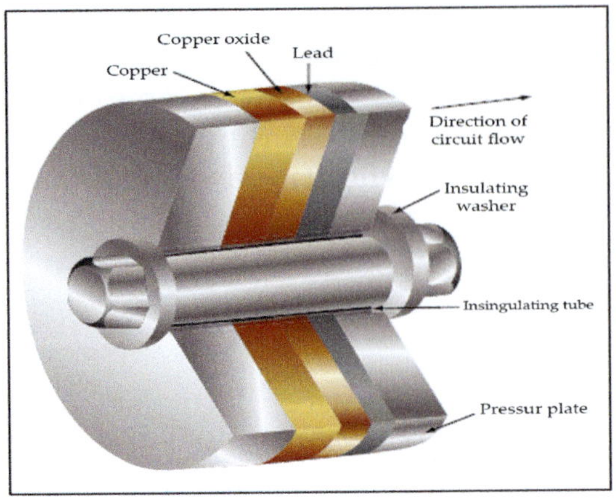

Figure 10.6. Copper oxide dry disk rectifier.

https://lh6.googleusercontent.com/proxy/GycfzlkuKYOJIgwqXdUSxGKYnu3yErC_CZG1aDrEmQ3mGeIOuwD1Yz5RjMDCTABJEObGN3aAiQStI-nwwFk6RE2ITYCym8rJQtKlQO8qD4a0GrQgqXfqYH7Z-jw8lf4I4V_xQL3z03oLQa96sw

10.3.4.7. Selenium Rectifiers

The selenium rectifier is created on an aluminum base sheet, which serves the dual purpose of providing a foundation for the rectifying junction and facilitating heat dissipation. The rectifying junction covers one side of the base sheet, with a narrow strip along the edges and a small area around the fixing hole sprayed with insulating varnish. A layer of low-melting point alloy, known as the counter electrode, is then sprayed over the selenium coating and insulating varnish. Contact with the rectifying junction elements is established through the base on one side and the counter electrode on the other. Mechanical pressure on the rectifying junction helps reduce resistance in the reverse direction, but this is prevented near the mounting studs by the layer of varnish.

In actuality, a rectifier stack is created by connecting several rectifying elements either in series or parallel. A rectifier's ability to handle voltage is increased when its elements are connected in series, and its ampere capacity is increased when they are connected in parallel.

10.3.5. Inverters

An inverter is utilized in certain aircraft systems to transform part of the aircraft's DC power into AC. This AC power is primarily used for instruments, radio, radar, lighting, and other accessories. Most inverters are designed to deliver a current at a frequency of 400 cycles per second, but some are capable of providing multiple voltages, such as 26 volts AC or 115 volts AC. There are two main types of inverters: rotary and static. Both types can be single-phase or multiphase. While a multiphase inverter is lighter for the same power output compared to a single-phase inverter, there are challenges associated with distributing multiphase power and maintaining load balance.

10.3.5.1. Rotary Inverter

Rotating inverters come in a wide variety of sizes, kinds, and configurations. These inverters are essentially integrated DC motors and AC generators. On a common shaft that will rotate inside the housing are the generator field, also known as the armature, and the motor field, also known as the armature.

10.3.5.1.1. Permanent Magnet Rotary Inverter

A DC motor and a permanent magnet AC generator assembly make up a permanent magnet inverter. Each is equipped with a different stator housed in a shared housing. A brush assembly and commutator are used to connect the motor's armature, which is mounted on a rotor, to the DC supply. Mounted on the housing, the motor field windings are linked directly to the DC supply. Without the need for brushes, AC can be extracted from the inverter thanks to a permanent magnet rotor that is installed at the other end of the same shaft as the motor armature and stator windings mounted on the housing.

The six magnetized poles on the generator rotor provide alternating north and south poles around its circumference. The rotor will start to rotate when the armature and motor field are stimulated. The conductors in the AC stator coils will cut the magnetic flux produced by the permanent magnets as the rotor revolves, causing the permanent magnet to rotate within the coils. As each pole passes through the windings, an AC voltage with a shifting polarity will be generated in the windings. To make this kind of inverter multiphase, add more AC stator coils to the housing so that each coil has the appropriate amount of phase shift. The rotary inverter, as its name suggests, has a rotating armature in its AC generator section.

10.3.5.1.2. Inductor Type Rotary Inverter

Inductor-type inverters generate poles equal to the number of stator poles by cutting lateral grooves across the surface of a rotor composed of soft iron laminations. A magnetic field is created by applying DC to the field coils. A low reluctance flux path is created from the field pole via the rotor poles to the AC armature pole and back through the housing to the field pole as the rotor rotates within the field coils and its poles align with the stationary poles. There will be a lot of magnetic flux connecting the AC coils in this situation.

When the rotor poles are positioned between the stationary poles, they create a high reluctance path for flux, primarily through air. As a result, there is a limited amount of magnetic flux linking the AC coils. This fluctuation in flux density within the stator leads to the generation of an alternating current in the AC coils. The frequency of this inverter is determined by both the number of poles and the motor's speed. The voltage is controlled by the DC stator field current (Figure 10.7).

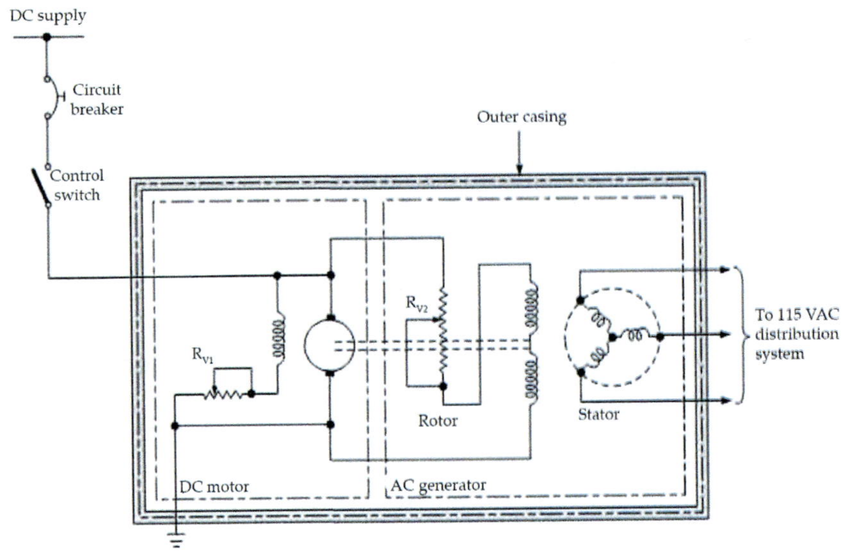

Figure 10.7. Rotary inverter.

https://www.mcico.com/media/wysiwyg/rotary-inverter.png.

10.3.5.2. Static Inverters

Static inverters are increasingly replacing rotary inverters and motor generator sets in applications where continuous DC voltage needs to be converted to alternating voltage. The semiconductor industry's rapid advancements have expanded the capabilities of static inverters, allowing them to be used in voltage and power ranges that were previously not feasible. These applications include power supplies for frequency-sensitive military and commercial AC equipment, aircraft emergency systems, and the conversion of wide frequency range power to precise frequency power.

The use of static inverters in small aircraft has grown significantly in recent years, with the technology now able to meet the same requirements as rotary inverters. For instance, production of 250 VA emergency AC supplies operated from aircraft batteries and 2,500 VA main AC supplies operated from a varying frequency generator supply are now available. These static inverters offer advantages for aircraft applications, such as the absence of moving parts and the ability to be conduction cooled. Additionally, the AC output from these inverters is generated from DC flux lines in the field winding.

Solid-state inverters, also known as static inverters, come in a variety of forms and models that can be categorized based on the power output capacities and the form of the AC output waveform. The output of one of the most widely used static inverters is a regulated sine wave (Figure 10.8).

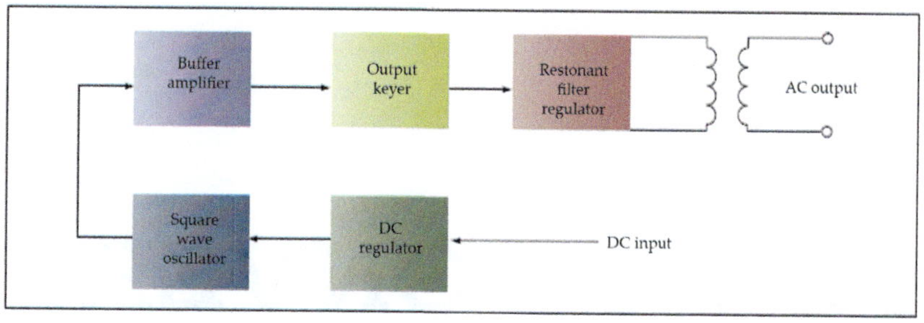

Figure 10.8. Static inverter block diagram.

https://lh6.googleusercontent.com/proxy/eli35b5bPMnp2-Ak1L6qfWYTIRAN3Ct8k8PaQ6IF3XIhb98 2y_utduxeB61cxZm-Q9dKj_YdnMA5MuPgVWFBB50LbJL_XF7tzgs3VOdguM7HAU0Z-A8xH4UO3XO-Z9x2_vf7ejgLQsZNUrpKHA.

10.3.6. Some Applications of Inverters and Rectifiers

Inverters and rectifiers are widely used in various industries such as renewable energy, transportation, communication, computing, and consumer electronics. For example, solar inverters transform the DC output from solar panels into AC power for the grid or local loads. Uninterruptible power supplies (UPS) ensure backup power during outages by utilizing batteries or other DC power sources and inverters to deliver AC power to essential loads. Electric vehicles (EV) rely on inverters to convert DC power from batteries to AC power for motors, while rectifiers convert AC power from the grid or regenerative braking to DC power for battery charging. Additionally, power adapters are essential for converting AC power from wall sockets to the necessary DC power for electronic devices like laptops, phones, and cameras.

10.4. DC-DC CONVERTERS

One level of DC voltage is converted to another level by the DC-to-DC converters. It is vital to specify a voltage for every electronic device because the operating voltage of various components, including MOSFETs and ICs, can vary over a large range. Whereas a Boost Converter provides a higher voltage, a Buck Converter outputs a voltage that is lower than the original voltage.

The circuit's efficiency, ripple, and load-transient response can all be altered by using DC-to-DC converters. The ideal external parts and components typically depend on the input and output specifications as well as other operating conditions. Therefore, standard circuits must be modified or altered during product design in accordance with each product's unique specification requirements. It takes a great deal of knowledge and experience in that field to design a circuit that complies with all specifications. Step-up or step-down DC-to-DC converters are helpful in situations where the output voltage of the regulator is not always equal to the battery voltage. In order to maintain a consistent load voltage across the full battery voltage range during operation, the DC-to-DC converter needs to be able to function as a step-up or step-down voltage supplier.

10.4.1. Working Principle of DC-DC Converter

The DC-to-DC converter operates on a very basic principle. The input current varies unexpectedly due to the inductor in the input resistance. The inductor feeds energy from the input and stores it as magnetic energy if the switch is kept at its highest setting (on).

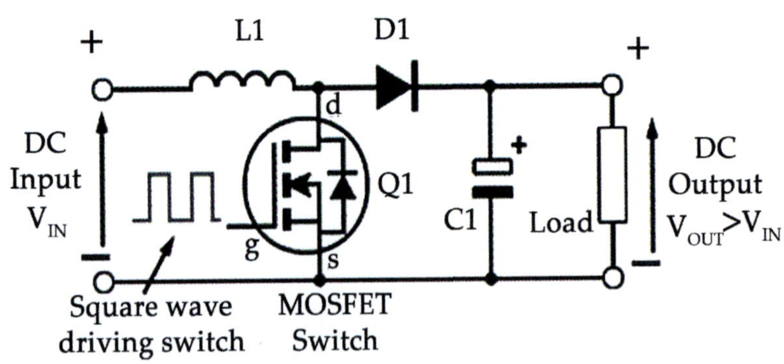

Figure 10.9. DC-to-DC Converters Working Principle

If the switch is in the low (off) position, it will release the energy stored in the capacitor. In this scenario, the capacitor's output is assumed to be high enough for the time constant of an RC circuit on the output side. The long time constant is compared with the switching period to ensure that the steady-state output voltage remains constant. The output voltage should be Vo(t) = Vo(constant) and be available at the load terminal.

10.4.2. Types of DC-to-DC Converters

1. Magnetic Converters

These DC-to-DC converters store and release energy periodically from a magnetic field in an inductor or transformer. The frequency varies from 300 kHz to 10 MHz. By regulating the duty cycle of the charging voltage, it is easier to control the amount of power transferred to a load. Additionally, control can be applied to the input current, output current, or to maintain constant power throughout the circuit. The transformer-based converter provides isolation between input and output effortlessly.

2. Non-Isolated Converters

Non-isolated converters are primarily utilized when there is only a slight voltage change. They connect the input and output terminals to a shared ground, but a drawback is their inability to shield against high electrical voltages and increased noise.

3. Step-down/Buck Converters

In a typical non-isolated step-down or buck converter, the output voltage VOUT depends on the input voltage VIN and the switching duty cycle D of the power switch (Figure 10.10).

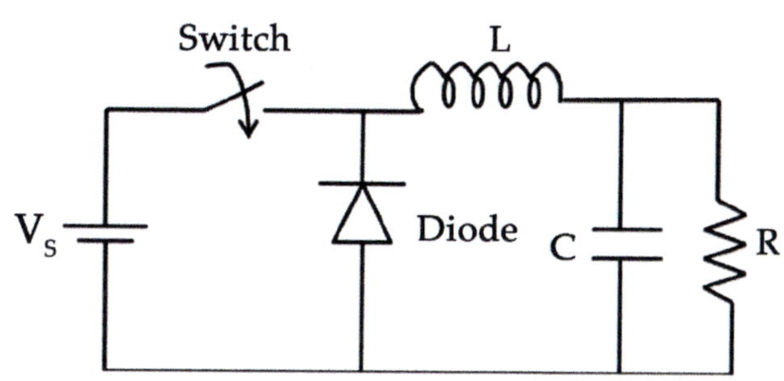

Figure 10.10. Step-down/Buck Converters.

Source: https://www.arrow.com/en/research-and-events/articles/types-of-switching-dc-dc-converters

4. Step-up/Boost Converters

It is used to boost DC to DC converter voltage and it uses the same number of passive components but arranged to step up the input voltage so that the output is higher than that of the input (Figure 10.11).

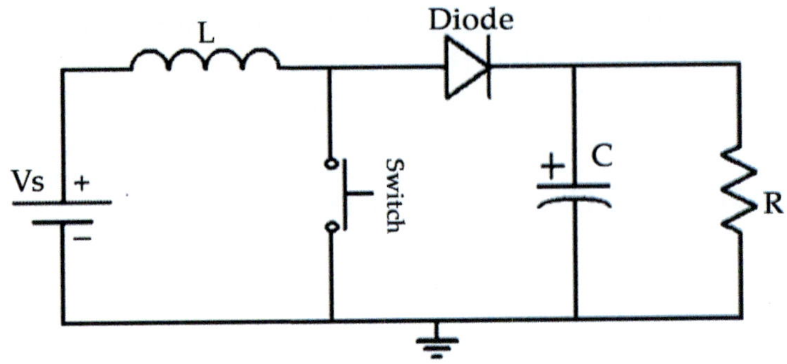

Figure 10.11. Step-up/Boost Converters.

Source: https://www.arrow.com/en/research-and-events/articles/types-of-switching-dc-dc-converters

5. Buck-Boost Converters

This converter allows the input DC voltage to be either stepped-up or stepped-down, depending on the duty cycle (Figure 10.12).

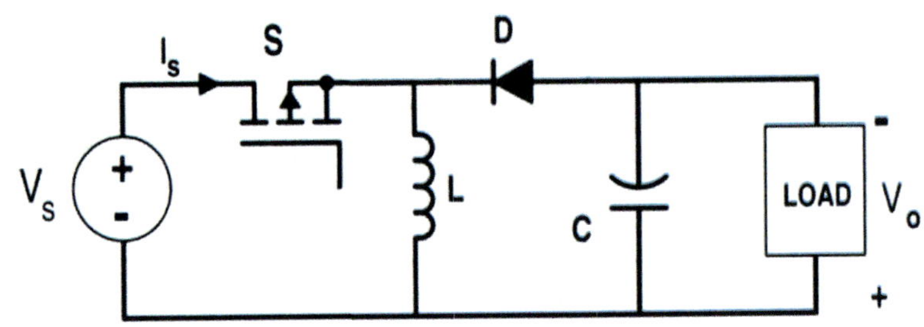

Figure 10.12. Buck-Boost Converters.

Source: https://www.arrow.com/en/research-and-events/articles/types-of-switching-dc-dc-converters

The output voltage is given by the relation as mentioned below:

$$VOUT = -VIN*D/(1-D)$$

From the above expression, we can notice that the output voltage is always reversed in polarity with respect to the input. Therefore, a buck-boost converter is also known as a voltage inverter.

6. Isolated Converters

There is a space between the input and output terminals of the isolated converter. Their isolation voltage properties are high. They are able to filter out interference and noise. They are able to generate the required and cleaner DC output voltage as a result. They are divided into two additional categories.

I. Flyback converters

The working of this converter is similar to the buck-boost converter of the non-isolating category. The only difference is that it uses a transformer to store energy instead of an inductor in the circuit (Figure 10.13).

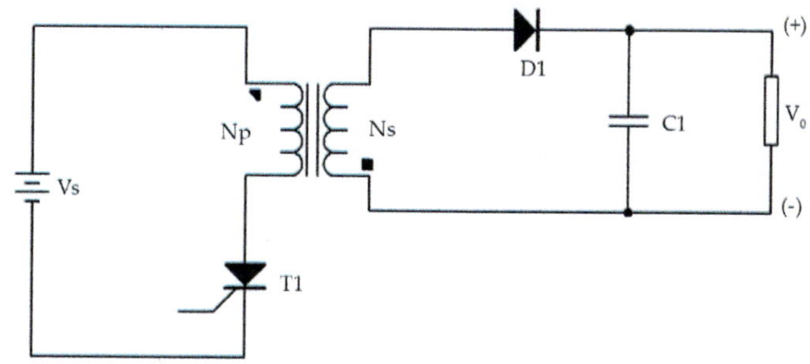

Figure 10.13. Inductor in the circuit.

II. Forward Converters

The working of this converter makes use of the transformer to send the energy, between the input and output in a single step.

Advantages of DC-to-DC Converters

- It simplifies the power supply systems in the circuit.
- It provides isolation in the primary and secondary circuits from each other.
- It provides a technique to extend potential (voltage) as required.
- It is available as a hybrid circuit with all elements in a single chip.
- It is also used in the regulation and control of DC voltage.
- The output is well organized as positive or negative.
- Battery space can be reduced by using a converter.

Disadvantages DC-to-DC Converters

- Switching converters lead to more noise.
- They are expensive as an external circuit is required.
- Choppers are inadequate due to unsteady voltage and current supply.
- More ripple current, More input and output capacitance, higher losses, etc.

10.4.3. AC-DC Converters

In power electronics, AC to DC converters are among the most crucial components. This is due to the fact that these conversions have numerous practical applications. AC-DC converters are electrical circuits that convert alternating current (AC) input to direct current (DC) output. They are employed in power electronic applications where a sine-wave AC voltage of 50 Hz or 60 Hz is used as the power input and a DC output is produced by power conversion.

Rectification is the process of converting an AC current to a DC current. At the load end connection, the rectifier transforms the AC supply into a DC supply. Similarly, transformers are typically used to modify the AC source in order to lower the voltage level and improve the DC supply's operating range.

10.4.4. Concept of Alternating Current (AC) and Direct Current (DC)

10.4.4.1. Alternating Current

The current in an alternating current alternate between flowing forward and backward. An alternating current is a current whose direction alternates on a regular basis (AC). Frequency is not zero for it. It is generated by dynamos, AC generators, etc. (Figure 10.14).

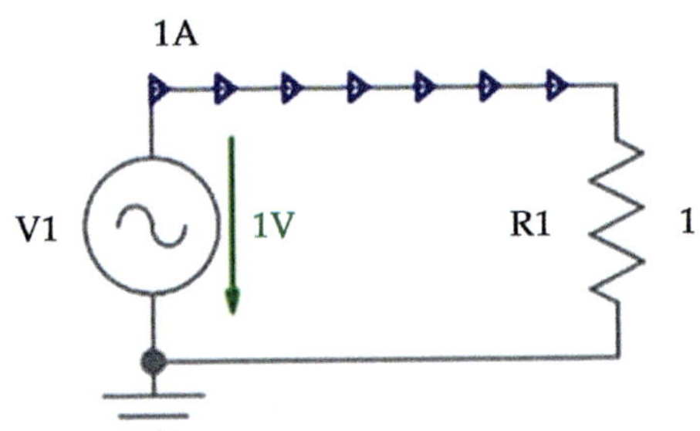

Figure 10.14. Alternating current.

Source: https://how2electronics.com/ac-to-dc-converters-features-design-applications/

10.4.4.2. Direct Current

In direct current, the current doesn't change its magnitude and polarity. If the current always flows in the same direction in a conductor, then it is called direct current. It has zero frequency. It is produced by cells, battery, DC generator, etc. (Figure 10.15).

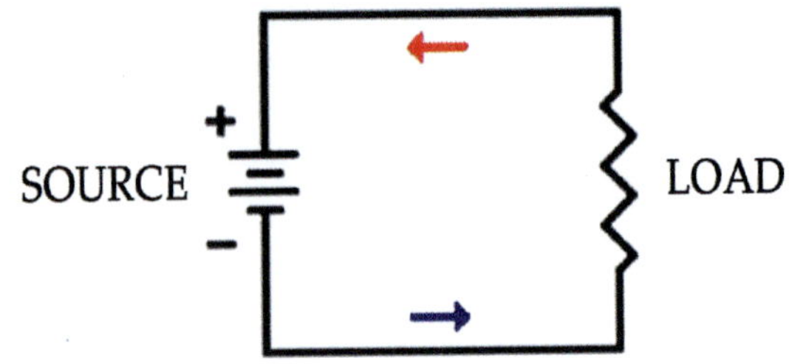

Figure 10.15. Direct current.

Source: https://how2electronics.com/ac-to-dc-converters-features-design-applications/

10.4.5. Simple Steps to Change AC into DC

Now let's discuss about AC to DC converter. Let us consider frequently used converter in the power supply circuit, 230V AC to 5V DC converter.

1. Stepping down the Voltage Levels

In order to transmit power over long distances, voltage may need to be increased. Conversely, equipment that requires lower power may need voltage to be decreased. Step-up transformers are used to raise voltage levels, while step-down transformers are used to lower voltage levels (Figure 10.16).

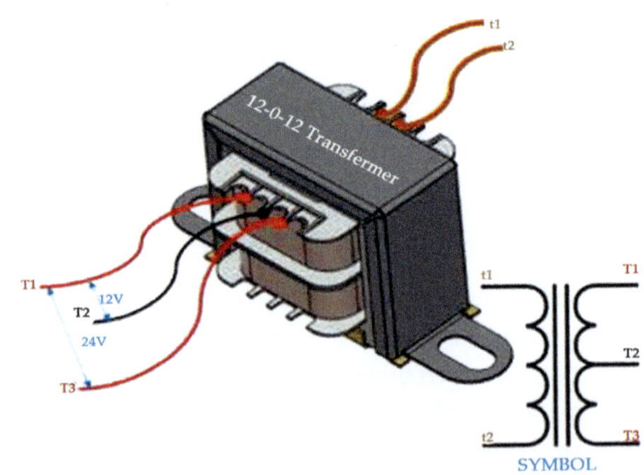

Figure 10.16. Step-down transformers.

Source: https://how2electronics.com/ac-to-dc-converters-features-design-applications/

Consider a transformer with 12V output. The 230V AC power supply is converted into 12V AC by using a step-down transformer. The RMS value and its peak value can be given by the product of the square root of two and the RMS value, which is approximately equal to 17V, the output of the step-down transformer.

2. AC to DC Power Converter Circuit

The rectifier converts the AC supply into the DC supply at the load end connection. There are different types of rectifiers, such as half-wave, full-wave, and bridge rectifiers (Figure 10.17).

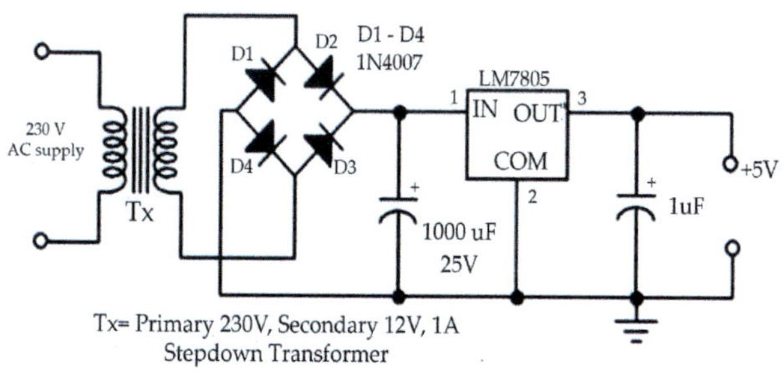

Figure 10.17. AC to DC Converter Circuit.

Source: https://how2electronics.com/ac-to-dc-converters-features-design-applications/

A Full Bridge rectifier is made up of four diodes arranged in a bridge configuration. The diodes only conduct electricity in one direction, known as forward bias, and remain off in the opposite direction, known as reverse bias.

During the positive half cycle of the input AC power, diodes D2 and D4 conduct. During the negative half cycle, diodes D1 and D3 conduct. This process rectifies the input AC power into output DC power. However, the output power is not pure DC and contains pulses.

3. Obtaining Pure DC Waveform

The pulsing DC must be converted to pure DC. The majority of the circuit uses capacitors to achieve that. Energy is stored in the capacitor during the input voltage's rise from zero to its maximum value. When the input voltage drops from its maximum value to zero, the capacitor's energy can be released (Figure 10.18).

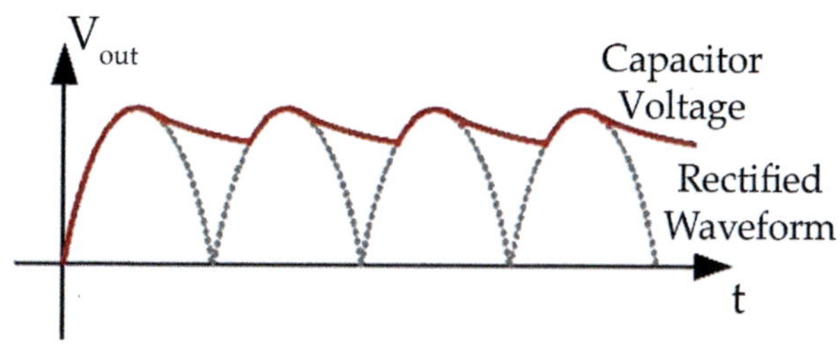

Figure 10.18. Converting the pulsating DC into pure DC using this charging and discharging process of the capacitor.

Source: https://how2electronics.com/wp-content/uploads/2021/06/Waveform.jpg

4. Regulating Fixed DC Voltage

In order to fix the output voltage to the desired value, we typically use a voltage regulator IC. The DC voltage regulator ICs are named 78XX. The last two digits XX represent the output voltage value. For instance, to regulate the output voltage to 5V, we would use the 7805 Voltage Regulator IC. Similarly, to regulate the voltage to 9V, we would use a 7809 Voltage Regulator IC.

10.4.5.1. Applications

AC to DC converters are essential components found in a wide range of electronic and electrical devices. They serve as power supply circuits for various household appliances such as vacuum cleaners, washing machines, refrigerators, and rice cookers. In addition, they are commonly used in everyday items like computers, televisions, and cell phone chargers.

AC to DC converters are crucial as many electronic sensors and modules require

DC supply to operate. These converters are also utilized in medical equipment, factory automation, building automation, process control systems, signage displays, and telecommunication. Furthermore, they have applications in renewable energy management, test and measuring equipment, defense, aerospace, and transportation systems.

CASE STUDY

THE ROLE OF SEMICONDUCTOR MICROWAVES IN MODERN TECHNOLOGY

Introduction

Semiconductor microwaves have revolutionized communication, radar, and various other technologies. Their development has enabled significant advancements in both consumer electronics and military applications. This case study explores the evolution, applications, and impact of semiconductor microwave technology.

Evolution of Semiconductor Microwave Technology

The journey of semiconductor microwave technology began with the invention of the transistor in the mid-20th century. Early microwave devices relied on vacuum tubes, which were bulky and inefficient. The advent of semiconductor materials like gallium arsenide (GaAs) and silicon germanium (SiGe) led to the development of smaller, more efficient microwave devices. The introduction of high electron mobility transistors (HEMTs) and monolithic microwave integrated circuits (MMICs) marked significant milestones, enhancing performance and reducing costs.

Applications in Communication

Semiconductor microwaves are crucial in modern communication systems. They are integral to the functioning of mobile phones, satellite communications, and Wi-Fi networks. The ability to operate at high frequencies and provide high bandwidth makes semiconductor microwaves ideal for these applications. For instance, GaAs-based MMICs are widely used in cellular base stations and satellite transceivers, offering high efficiency and reliability.

Applications in Radar and Defense

In defense, semiconductor microwaves play a vital role in radar systems, electronic warfare, and secure communications. Advanced radar systems use microwave frequencies for high-resolution imaging and target detection. Semiconductor devices like GaN (gallium nitride) transistors are preferred for their high power and efficiency, essential for long-range radar and electronic countermeasure systems. The miniaturization and performance improvements offered by semiconductor technology have enhanced the capabilities of modern defense systems.

Impact on Consumer Electronics

The integration of semiconductor microwaves in consumer electronics has led to the development of compact and powerful devices. Microwave ovens, for example, use semiconductor technology for efficient power generation and control. Moreover, advancements in semiconductor microwaves have facilitated the growth of wireless charging technology, enabling convenient power solutions for a range of electronic devices.

Future Prospects

The future of semiconductor microwave technology looks promising, with ongoing research focused on improving materials and device architectures. Innovations in compound semiconductors, such as indium phosphide (InP) and advanced GaN structures, are expected to further enhance performance. Additionally, the emergence of 5G and 6G networks will drive the demand for high-frequency semiconductor devices, paving the way for faster and more reliable communication systems.

Conclusion

Semiconductor microwave technology has profoundly impacted various fields, from communication and defense to consumer electronics. The continuous evolution and innovation in this domain promise to drive future technological advancements, offering greater efficiency, miniaturization, and functionality. As research progresses, semiconductor microwaves will remain a cornerstone of modern technology, shaping the way we live and interact with the world.

CLASS ACTIVITY

A microwave oven is a common household appliance used for quickly heating and cooking food. It works by emitting microwaves, a form of electromagnetic radiation, which cause water molecules in the food to vibrate and generate heat. This method ensures fast and even cooking, making it a convenient choice for busy individuals. Microwaves are versatile, capable of defrosting frozen items, reheating leftovers, and even cooking meals from scratch. Modern microwaves often come with advanced features like preset cooking programs, grilling options, and sensor cooking, which further enhance their efficiency and usability in the kitchen.

SUMMARY

- Semiconductor microwave and power devices are pivotal in modern electronics, enabling high-frequency communication and efficient power management.
- Semiconductor microwave devices are integral to the high-frequency technology landscape, crucial for communication, radar, and sensing applications.
- Microwave is a form of electromagnetic radiation with wavelengths shorter than other radio waves (as originally discovered) but longer than infrared waves. Its wavelength ranges from about one meter to one millimeter, corresponding to frequencies between 300 MHz and 300 GHz, broadly construed.
- Designing a printed circuit board (PCB) for microwave semiconductor devices involves a set of critical considerations to ensure optimal performance, signal integrity, and minimal losses.
- A power semiconductor device is a semiconductor device used as a switch or rectifier in power electronics (e.g., in a switch-mode power supply). Such a device is also called a power device or, when used in an integrated circuit, a power IC.
- Power semiconductors perform the modified electronic functions of regular semiconductors with a high-power rating.
- Power semiconductor devices are used as on/off switches in power control circuits. A power semiconductor device is a semiconductor device used as a switch or rectifier in power electronics, for example, in a switch-mode power supply.
- An inverter is used in some aircraft systems to convert a portion of the aircraft's DC power to AC. This AC is used mainly for instruments, radio, radar, lighting, and other accessories.
- The process of converting AC current to DC current is known as rectification. The rectifier converts the AC supply into the DC supply at the load end connection.

REFERENCES

1. Davari, P., Blaabjerg, F., Hoene, E., & Zare, F. (2018). Improving 9–150 kHz EMI performance of single-phase pfc rectifier. In *Proceedings of the CIPS 2018, 10th International Conference on Integrated Power Electronics Systems*, Stuttgart, Germany, 20–22 March 2018 (pp. 1–6).
2. Davari, P., Kristensen, O., & Iannuzzo, F. (2018). Investigation of acoustic emission as a non-invasive method for detection of power semiconductor aging. *Microelectronics Reliability, 88–90*, 545–549. https://doi.org/10.1016/j.microrel.2018.06.084.
3. Davari, P., Yang, Y., Zare, F., & Blaabjerg, F. (2016). A multipulse pattern modulation scheme for harmonic mitigation in three-phase multimotor drives. *IEEE Journal of Emerging and Selected Topics in Power Electronics, 4*(1), 174–185. https://doi.org/10.1109/JESTPE.2015.2465711.

4. Davari, P., Yang, Y., Zare, F., & Blaabjerg, F. (2016). Predictive pulse-pattern current modulation scheme for harmonic reduction in three-phase multidrive systems. *IEEE Transactions on Industrial Electronics, 63*(9), 5932–5942. https://doi.org/10.1109/TIE.2016.2560598.
5. Dragicevic, T., & Novak, M. (2019). Weighting factor design in model predictive control of power electronic converters: an artificial neural network approach. *IEEE Transactions on Industrial Electronics, 66(11)*. Advance online publication. https://doi.org/10.1109/TIE.2018.2875660.
6. Dragicevic, T., Wheeler, P., & Blaabjerg, F. (2019). Artificial intelligence aided automated design for reliability of power electronic systems. *IEEE Transactions on Power Electronics, 34(8)*. Advance online publication. https://doi.org/10.1109/TPEL.2018.2883947.
7. Peyghami, S., Davari, P., Mokhtari, H., & Blaabjerg, F. (2019). Decentralized Droop Control in DC Microgrids Based on a Frequency Injection Approach. *IEEE Transactions on Smart Grid, 10(6)*. Advance online publication. https://doi.org/10.1109/TSG.2019.2911213.
8. Taul, M. G., Wang, X., Davari, P., & Blaabjerg, F. (2019). An overview of assessment methods for synchronization stability of grid-connected converters under severe symmetrical grid faults. *IEEE Transactions on Power Electronics, 34(10)*. Advance online publication. https://doi.org/10.1109/TPEL.2019.2892142.
9. Wang, H., Davari, P., Wang, H., Kumar, D., Zare, F., & Blaabjerg, F. (2019). Lifetime estimation of dc-link capacitors in adjustable speed drives under grid voltage unbalances. *IEEE Transactions on Power Electronics, 34*(4), 4064–4078. https://doi.org/10.1109/TPEL.2018.2862692.
10. Zare, F., Soltani, H., Kumar, D., Davari, P., Delpino, H. A. M., & Blaabjerg, F. (2017). Harmonic Emissions of Three-Phase Diode Rectifiers in Distribution Networks. *IEEE Access, 5*, 2819–2833. https://doi.org/10.1109/ACCESS.2016.2562646.

Index

T

U

V

W

X

Printed and bound by CPI Group (UK) Ltd, Croydon, CR0 4YY
18/09/2026
14965063-0005